THE AMERICAN
HORSESHOE CRAB

THE AMERICAN
HORSESHOE CRAB

Edited by

Carl N. Shuster, Jr.

Robert B. Barlow

H. Jane Brockmann

HARVARD UNIVERSITY PRESS

Cambridge, Massachusetts, and London, England

2003

Publication of this book was generously assisted by grants from Associates of Cape Cod, Inc.; Cambrex Bio Science Walkersville, Inc.; Charles River Laboratories, Inc.; and Wako Chemicals USA, Inc.

Library of Congress Cataloging-in-Publication Data

The American horseshoe crab / edited by Carl N. Shuster, Jr., Robert B. Barlow, H. Jane Brockmann.

    p. cm.

  Includes bibliographical references and index.

  ISBN 0-674-01159-7 (alk. paper)

   1. Limulus polyphemus.   I. Shuster, Carl N. (Carl Nathaniel), Jr., 1919–.
II. Barlow, Robert B.   III. Brockmann, H. Jane.

QL447.7.A44 2004   2003056640

595.4′92—dc22

CONTENTS

PREFACE

A decade ago, many of the authors of this book met at the Marine Biological Laboratory (MBL) in Woods Hole, Mass., to discuss the merits
of a book on horseshoe crabs. The decision to write one was easy. The
more debatable question was the scope and theme of the book. Writing
a treatise on *Limulus* was clearly a worthwhile goal; but everyone agreed
that an even more useful pursuit, though perhaps a greater challenge,
would be to write a book that would introduce horseshoe crabs to
a wide audience. It was appropriate that this discussion was held at
Woods Hole, a center of *Limulus* research for more than a century. It
was an ideal setting to address the question, Why a book on *Limulus?*

Three major considerations provided an ample background for de-
liberations over our book:

- Each of the authors had spent much time studying the activities of
  horseshoe crabs in their natural environment—several at the MBL
  or nearby. This attracted the attention of people who live or vaca-
  tion in the areas where the researchers were working. Besieged
  with a host of questions, the authors have been sensitized to the
  depth and breadth of the information others want to know. Some
  of those questions and answers have already surfaced, in fragments,
  in the full range of media from TV specials to news items and
  books for young readers. Yet, with only one notable exception
  (Sekiguchi, 1988), no book has addressed many of the salient features of the life, activities, and significance of horseshoe crabs.
- Horseshoe crabs have an extensive although discontinuous geologic history. They belong to an ancient group of aquatic arthro-

pods that date back to the era of the trilobites but took the other road and survived. In all that time, there have been relatively few species of horseshoe crabs.

- Horseshoe crabs are the most researched of all marine arthropods. Still there is much to be learned about horseshoe crabs and their biological systems. Because one of the objectives of this book is to provide a stepping stone for others who search for a better understanding of horseshoe crabs, care has been taken to identify major previous contributions. Horseshoe crabs are unique. While other arthropod groups have achieved much diversity and specialization in anatomy, physiology, and behavior, horseshoe crabs represent an ancient life-form possibly attained at least some 400 million years ago during the Age of the Trilobites. Not only is the horseshoe crab one of the most distinctive animals and the sole representative of its kind on the American Atlantic Coast, it is also an exciting animal to study.

During the preparation of this book, it became obvious that many volumes would be needed to document more fully what is known about horseshoe crabs and that many more volumes of information are yet to be discovered. In this modest work, we have highlighted topics that we consider significant and those that should answer most of the questions that we have been commonly asked. Oversights are in part related to space constraints, while other topics are outside the range of our individual research expertise and interests.

We hope you will have as much pleasure in perusing the material in this book as the authors have had in writing it.

THE AMERICAN
HORSESHOE CRAB

PROLOGUE

Robert B. Barlow

What animal:

- Has ten eyes?
- Eats through its brain?
- Needs a book to breathe?
- Sees as well at night as during the day?
- Looks and moves like a tank?
- Has a tail that was used as a spear?
- Has a heart that cannot beat on its own?
- Has eyes that know the time of day?
- Chases females who run away?
- Walked with the dinosaurs?
- Refuels migratory birds?
- Tells time with its tail?
- Chews with its legs?
- Can fertilize corn?
- Has blue blood?
- Is called a living fossil?

The answer: a horseshoe crab.

The spring tide is flooding the shoals of North Monomoy Island near Cape Cod. We cut the engine and drift toward shore. Under the boat large dark objects are moving with us. A strange battalion of underwater tanks appears to be storming the beach. Nearer shore we can see that the "tanks" are horseshoe crabs. They're arriving by the hundreds with many joined together in pairs. Some are clasped together in long chains, with a big crab leading the way. Others arrive alone.

As the horseshoe crabs ascend toward shore from the depths, sea birds descend from above. Both groups of animals are ready to get down to serious business: crabs laying eggs and birds eating them. The crabs have spent the entire winter far offshore buried in the ocean floor. The birds spent their winter far from here in South America. They began their spring migration last week, flying more than two thousand miles nonstop and landing here to refuel. They consume enough eggs to reach their spawning grounds in Canada. The crabs also started their shoreward migration last week. The birds are in sync with the crabs.

Witnessing this ancient ritual is like watching *Jurassic Park*. Realizing it has been repeated over and over again for hundreds of millions of years sends a chill down my spine. I expect any minute to see dinosaurs lumbering through the shallows and velocipeds prancing down the beach.

Much has happened since horseshoe crabs first inhabited the coastal waters of major continents. As giant tectonic plates have shifted over the earth's surface, horseshoe crabs have changed little. Just four species now populate the earth, with three closely related ones found along the coasts of Japan, India, and southeastern Asia. Our own *Limulus polyphemus* inhabits the estuaries of the eastern United States and the Yucatán peninsula.

After anchoring in shallow water, we put on wading boots and set out to mingle with our ancient ancestors. We see horseshoe crabs of all sizes scurrying along the bottom. Tiny ones are miniature versions of big ones. All the ones smaller than the size of my foot are alone, but most of the bigger ones are joined together and building nests. The small ones are juveniles. They must wait as many as 8 years to mate if they are lucky enough not to be devoured by sea gulls, turtles, or sharks. It is not known why they mix with these frenzied adults.

We pick up a freely swimming adult very gently by its tail and are immediately struck by the large eyes on the sides of its shell. They look like those of a giant fly. In each eye we see hundreds of tiny black dots. They are the individual light sensors, called ommatidia, with which the animal sees. They function like the rod and cone photoreceptors of our own eyes, but they are much bigger—one hundred times bigger. They are the largest photoreceptors in the animal kingdom. That is why the eye is so valuable. Research has revealed basic mechanisms of vision in the horseshoe crab eye that are common to all animals, including humans. This ground-breaking research was awarded the Nobel Prize. We can also see two more eyes. They are tiny and close together in the middle of the shell, just like a cyclops. This is why the species is named after the mythical one-eyed giant Polyphemus.

Many legs start waving in the air as we turn the animal over. Ten legs are big and two are small. Two big legs in front are shaped like boxing gloves. This is the easiest way to distinguish male from female. The two boxing-glove legs allow the male to clasp onto a female for mating. Two big ones in back, called swimmerets, have vanes that push water back, enabling the crab to swim faster. The two small legs in front, called chelicera, help a scavenging crab pick up food and push it into its mouth—which, surprisingly, surrounds its brain.

We notice that the membranes between segments of the legs are blue. They are blue because their blood is blue—a true blue-blooded crab! Copper gives the blood its blue color. Surprisingly, the oxygen-carrying molecules, hemocyanin, are not in blood cells but float freely in the plasma. The animal has blood cells, called amebocytes, which contain a highly sensitive clotting substance called Limulus amebocyte lysate, or LAL. When ubiquitous bacterial endotoxin invades a cut in the animal's shell, amebocytes lyse and spew out LAL to close the wound. Amebocytes are so sensitive to bacterial endotoxin that the LAL extract is used to test for contamination of surgical instruments and blood, especially if a person is suspected of having a gram-negative bacterial infection.

As we hold the horseshoe crab upside down, it waves thin flaps near the tail. These are the gills for extracting oxygen from the water. There are hundreds of them, and you can flip them just like pages of a book—hence the name *book gill*. By the way, the tail may look like a weapon, but it is not. Although the American Indians used crab tails as spears, the crabs use them to turn themselves right side up. They also use their tails to tell the time of day with little eyes. That's right, the tail can see.

The tide is ebbing and the sun is setting. Horseshoe crabs know this as well as we do, but we must leave before our boat is stranded on the shoals. We find a nest and gently place the male nearby. It joins the intense mating activity as if we never interrupted it.

We didn't exist when horseshoe crabs first appeared on earth. Now we do—can horseshoe crabs survive us? The question is not academic. At the turn of the last century, horseshoe crabs were valued as fertilizer for corn and tomatoes. Harvesting seriously depleted populations until after World War II, when artificial fertilizer became available. Now they are valuable as bait.

Fishermen are harvesting horseshoe crabs in large numbers to bait traps for eels and conch. Restrictions have been put in place for catching haddock, cod, and other fish, but there are no restrictions on catching eels or conch. Consequently, horseshoe crab populations are being depleted. Collecting horseshoe crabs for bait is restricted in many states along the eastern coast of the United States, but not in Massachusetts. Fortunately, the U.S. Federal Wildlife Service does not allow removal of any animals from the Monomoy Wildlife Preserve. The population here on the shoals is safe for now—a consoling thought.

Not consoling is the recent report that the Japanese species, *Tachypleus tridentatus,* is on the verge of extinction. Major parts of its nesting areas in the Inland Sea are polluted or significantly reduced in size by construction. Fortunately, the Japanese government has placed horseshoe crabs on the endangered species list and is cleaning up the Inland Sea.

We weigh anchor and drift off the shoal. Before starting the engine we listen. The gnawing and crunching of males trying to dislodge other males from nesting females is deafening. Just noise to us, but wonderful music to horseshoe crabs. We start the engine and head for Stage Harbor inlet.

A glance back is humbling. Living fossils mating—a yearly event that has replayed over the millennia. The sight evokes a sense of awe. We head out for deep water. The crabs will soon follow. Tomorrow they will return as the tide floods, but in a few days they will stop coming in to nest. Next spring the cycle repeats, and the crabs will return. We will, too.

SYNCHRONIES IN MIGRATION:
SHOREBIRDS, HORSESHOE CRABS,
AND DELAWARE BAY

Mark L. Botton and Brian A. Harrington,
with Nellie Tsipoura and David Mizrahi

A visit to the shores of Delaware Bay in the early spring gives little hint of the spectacle that will soon unfold. The beaches, swept clean by winter storms, are barren, except perhaps for a few remaining carcasses of stranded horseshoe crabs from the previous summer. The shore is silent; only some stray gulls and the occasional swan paddle along the shoreline. As late April arrives, hundreds of thousands of migrant shorebirds begin feeding along the marine coasts and lagoons of South America, preparing themselves for an exhausting flight that will bring them to the Delaware River estuary, thousands of miles away. Simultaneously, horseshoe crabs are massing for their spawning migration that ensures their own survival and also provides the eggs that the shorebirds are anticipating. This is the story of two migrations, intertwined in time and space.

Shorebirds migrate extensively during the course of the year; most species in the Western Hemisphere breed in the far north during the brief Arctic summer and overwinter in wetlands and beaches in Central and South America. Shorebirds make a round-trip migration of over 12,000 km (19,355 mi), although some red knots, one of the longer-distance migrants, may travel 16,000 km each way. Unlike many birds, shorebirds accomplish this migration in a series of long-distance flights that typically exceed a thousand miles nonstop. Fat and protein accumulated at strategic sites just prior to these flights serve as the fuel. In their demanding northbound and southbound migrations, shorebirds must briefly stop and regain energy by feeding at discrete staging areas. Delaware Bay is one of the four most important staging areas in North America, along with the Copper River Delta, Alaska; Gray's Harbor, Washington; and the Bay of Fundy in the Canadian Maritimes (Myers et al., 1987). The timing must be precise: prey abundance at staging areas is cyclical, and there is only a narrow window in the Arctic summer for courtship and reproduction.

How Did Scientists Discover the Horseshoe Crab–Shorebird Relationship?

One of the most remarkable aspects of the horseshoe crab–shorebird relationship in Delaware Bay is that scientists were unaware of its significance until the 1980s. This is by no means a remote area; it is within a few hours' drive of millions of people, and marine laboratories from Rutgers University and the University of Delaware were established many years ago on opposite shores of the Bay. There were some early anecdotal reports: Alexander Wilson remarked that ruddy turnstones in the Bay would feed "almost wholly on the eggs, or spawn, of the great king crab" (Brewer, 1840: 481). E. H. Forbush (1925) observed that the same birds in New England would feed on "the eggs of the great crab commonly called the Horsefoot or Horse-shoe." The famed ornithologist Witmer Stone (1937) described how both turnstones and black-bellied plovers would regularly feed on dead horseshoe crabs along Delaware Bay. Stone also mentions flights of ruddy turnstones across the Cape May Peninsula in the spring, as happens today when they go to roost at night along the Atlantic coastal marshes. Interestingly, no mention of horseshoe crab eggs as food is found in Stone's accounts of any shorebird in the Cape May area, or in the decade-long study by Urner and Storer (1949). Thus, although some early observers noticed that shorebirds ate horseshoe crab eggs, there was no implication that this had special ecological importance, and its magnitude within Delaware Bay was unknown.

During his early studies of horseshoe crabs in 1951, Carl Shuster (1982) observed many birds feeding along Delaware Bay beaches, including ruddy turnstone, red knot, least sandpiper, dunlin, semipalmated sandpiper, sanderling, semipalmated plover, black-bellied plover, and laughing gull. Another 30 years elapsed before scientists began to study the shorebird–horseshoe crab relationship in detail. In fact, the first systematic Bay-wide shorebird count was not made until Wade Wander and Peter Dunne (1981) estimated that there were some

370,000 to 643,000 shorebirds on the New Jersey shore of Delaware Bay at the height of the migration, with perhaps double this number if Delaware beaches were factored in. Bird numbers were low in early May, peaked during the last week of May, and then rapidly declined by early June. Of the twenty species of shorebirds recorded, semipalmated sandpipers, ruddy turnstones, red knots, and sanderlings collectively constituted 95 percent of the total number of birds. The exceptional number of shorebirds placed Delaware Bay among the most important staging areas anywhere in the Western Hemisphere.

The two authors of this chapter came to study the Delaware Bay shorebirds in the late 1970s from rather different perspectives. Botton was a graduate student focusing on the feeding ecology of horseshoe crabs (see Chapter 6). At the suggestion of his Ph.D. committee, he expanded his studies to include the predatory effects of birds on the invertebrate fauna of the tidal flats (Botton, 1984). He used a floating cage that excluded shorebirds and gulls from the tidal flats but at high tide allowed horseshoe crabs and other aquatic predators access to all the flats. Botton detected no significant impacts of bird predation on the tidal flat invertebrates, in either caged or noncaged areas, which suggests that birds were probably passing up clams and worms in favor of the horseshoe crab eggs.

At roughly the same time, Harrington, who had been studying red knots during southward migration in Massachusetts, was sending questionnaires to bird watchers and other researchers asking whether they knew the whereabouts of knots during the northward migration. A contact from the Cape May, N.J., region led him to visit the Bay shore in the spring of 1979, where the assemblage of knots and other shorebirds astounded him. He continued his knot research there and elsewhere through 1990, focusing on identifying the population size and migration routes used for seasonal commutes between North and South America (Figure 1.1).

In the early 1980s, the general public was still largely unaware of the shorebird–horseshoe crab relationship in Delaware Bay. This changed in 1986 with the publication of a popular article by J. P. Myers in *Natural History* that was soon followed by many other feature stories in newspapers and magazines and a film, *Delaware Bay Banquet,* shown on the popular television series *National Geographic Explorer.* Today, nature photographers, writers, filmmakers, television crews, and ecotourists from North America, Europe, and Japan arrive on the shores of Delaware Bay each May—a migration that is as predictable as the animals they seek to observe.

How did scientists fit the Delaware Bay feeding frenzy into the

FIGURE 1.1 Brian Harrington in the process of banding a migratory shorebird. In 1997, at Slaughter Beach, Del., Harrington led a group of International Shorebird Survey cooperators in banding migratory shorebirds. This was one of the functions of the Delaware Bay Shorebird Workshop of that year. The workshop, held at the Bombay Hook National Wildlife Refuge, was presented by the Delaware Coastal Management Program, the Western Hemisphere Shorebird Reserve Network, and the Manomet Observatory for Conservation Sciences.

overall migratory cycle of the shorebirds? Where did the birds come from prior to reaching Delaware Bay; and when they departed in June, where did they go? Early ornithologists tried to deduce shorebird migration patterns mainly by recording the dates of arrival and departure of each species from a network of observers along the coast. But more precise descriptions of migration routes required mark-recapture studies of the vast scale conducted by Myers et al. (1990) for sanderlings, or Harrington (1982b; Harrington et al., 1988) for red knots. In the latter example, we marked over 2,500 red knots, and the presence of banded birds was recorded at their wintering and stopover areas during spring and fall. The majority of Western Hemisphere red knots—some 150,000—winter in southern Argentina; smaller flocks winter in Florida and along the Gulf of Mexico. By late April or early May, the Argentine knots are gathered along the Brazilian coast; from there, they cross the ocean to the southeast coast of the United States, joining up with knots from Florida in Delaware Bay. The breeding areas require yet another long-distance flight, this time to Arctic islands north of Hudson's Bay in northern Canada. Recent advances using molecular markers may further our understanding of population structure in shorebirds (Haig et al., 1997).

What Are the Seasonal Migration Patterns in Delaware Bay?

In the early spring, fishermen working trawls or dredges off the mouth of Delaware Bay catch large numbers of paired male and female horseshoe crabs. Horseshoe crabs first appear along the shoreline in late April. By the first week of May, aggregations of spawning adults lay eggs on the beaches at high tide—setting the table, as it were, for the influx of shorebirds (colorplates 1 and 2).

Weekly aerial surveys, started by Wander and Dunne (1981) and continued from 1986 to the present by Kathy Clark, Larry Niles, and colleagues, have yielded valuable data on the seasonal pattern of shorebird abundance in Delaware Bay (Clark et al., 1993). Comparatively few shorebirds (<50,000) are generally found before the middle of May, and numbers peak by the end of May (Figure 1.2). The majority of the shorebirds depart Delaware Bay by the second week of June. Not all species follow precisely the same temporal pattern. For example, red knots and ruddy turnstones both have a discrete peak migration time, centered on May 27 to May 30. By contrast, sanderlings arrive 1 week earlier, and semipalmated sandpipers are about as numerous on May 29 to May 30 as they are the subsequent week (Clark et al., 1993).

Far fewer shorebirds visit the Delaware Bay area during their

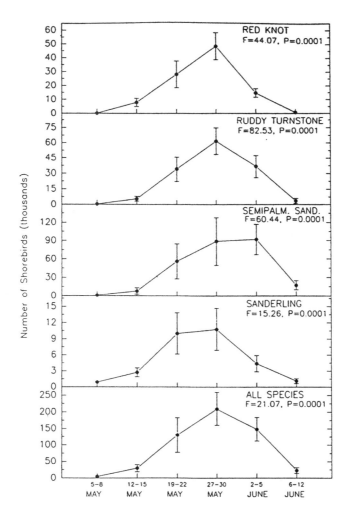

FIGURE 1.2 Peak single-day counts of migrating shorebirds by week, in aerial surveys of Delaware Bay beaches from early May to mid-June, 1986–1992 *(from Clark et al., 1993)*. The data indicate significant temporal variation for each of the four species and a peak in total shorebird abundance on 27–30 May.

southbound journey, though sanderlings are usually seen in respectable numbers near Cape May Point. By late June, horseshoe crab spawning is usually reduced to a trickle: the migration truck stop is closed.

Which Birds Are Most Abundant?

At the peak of the spring migration in late May, there are several hundred thousand shorebirds and tens of thousands of gulls along Delaware Bay, irresistibly attracted by the bountiful horseshoe crab eggs along the beaches. In all, more than twenty-five avian species may be encountered along the Bay at this time of year (Table 1.1). We know that just four species comprise over 95 percent of the shorebirds: red knot,

TABLE 1.1 Species of spring shorebirds in Delaware Bay, and other species of birds that have been observed to feed on horseshoe crab eggs (E) or stranded adults (S). Birds are classified as common (+ + +), occasional (+ +), or rare (+), based on the likelihood of observing the species along the shoreline during May–early June.

Common Name	Scientific Name	Food	Abundance
Great black-backed gull	*Larus marinus*	E,S	+++
Herring gull	*Larus argentatus*	E,S	+++
Laughing gull	*Larus atricilla*	E	+++
Red knot	*Calidris canutus*	E	+++
Ruddy turnstone	*Arenaria interpres*	E	+++
Sanderling	*Calidris alba*	E	+++
Semipalmated sandpiper	*Calidris pusilla*	E	+++
American oystercatcher	*Haematopus palliates*	?	++
Black-bellied plover	*Pluvialis squatarola*	?	++
Boat-tailed grackle	*Quiscalus major*	E	++
Dunlin	*Calidris alpina*	E	++
Glossy ibis	*Plegadis falcinellus*	E	++
Killdeer	*Charadrius vociferous*	?	++
Least sandpiper	*Calidris minutilla*	E	++
Long-billed dowitcher	*Limnodromus scolopaceus*	E	++
Red-winged blackbird	*Agelaius phoeniceus*	E	++
Semipalmated plover	*Charadrius semipalmatus*	?	++
Short-billed dowitcher	*Limnodromus griseus*	E	++
Spotted sandpiper	*Actitis macularia*	E	++
White-rumped sandpiper	*Calidris fuscicollis*	E	++
Willet	*Catoptrophorus semipalmatus*	E	++
Black-necked stilt	*Himantopus mexicanus*	?	+
Greater yellowlegs	*Tringa melanoleuca*	E	+
Hudsonian godwit	*Limosa haemastica*	?	+
Lesser yellowlegs	*Tringa flavipes*	E	+
Marbled godwit	*Limosa fedoa*	?	+
Whimbrel	*Numenius phaeopus*	?	+

Source: Based on Wander and Dunne (1981), Botton and Loveland (1993), Clark et al. (1993), Botton et al. (1994), and unpublished field notes.

ruddy turnstone, sanderling, and semipalmated sandpiper (Wander and Dunne, 1981; Clark et al., 1993; Botton et al., 1994) (colorplate 3). Several others, such as dunlin, least sandpiper, black-bellied plover, and willet, are occasionally seen, but seldom in large numbers; and diligent birders may sight a smaller contingent of rare species. Here, we briefly review some aspects of the ecology of the most abundant migrant shorebirds and gulls in Delaware Bay.

Red Knot *(Calidris canutus)*

In its breeding plumage, the red knot is easily identified by its robin-red breast; it is also larger and plumper than the three other most common shorebirds. Red knots were hunted extensively in the nineteenth century (Bent, 1962), and this no doubt contributed to the declining numbers of "robin snipe," as they were once called locally (Stone, 1937).

The migration system of the red knot has been studied by Harrington and colleagues (Harrington, 1982b, 1996; Harrington et al., 1988) and Morrison (1984). A European wintering population *(Calidris canutus islandica)* breeds primarily on Ellesmere Island in the eastern High Arctic and uses Iceland as the principal staging area during spring migration (Morrison, 1984). The North American race *(Calidris canutus rufa)* visits Delaware Bay during spring migration. The largest overwintering population resides in Argentina, with a smaller number on the Gulf coast of Florida (Harrington et al., 1988). We were able to differentiate populations on the basis of a banding study, although red knots from Argentina, Florida, Delaware Bay, and Massachusetts were not significantly different in wing or bill size (Harrington et al., 1988).

With their comparatively short bills—mean length = 35.6 mm (Harrington et al., 1988)—red knots at Delaware Bay concentrate their feeding on horseshoe crab eggs found in the surface sediments. Elsewhere in their range, red knots frequently specialize on mollusks such as *Littorina* and *Mytilus* in Iceland (Alerstam et al., 1992), *Brachiodontes* in Argentina (Gonzalez et al., 1996), *Macoma* in James Bay, *Donax* (coquina clam) along coasts from Florida to Argentina, and mussel spat in Massachusetts (Harrington, 1996). Bent (1962) also reports that they prey on crustaceans, insect larvae, and other invertebrates.

At times, tightly compacted flocks of many thousands of red knots may roost along Delaware Bay beaches (colorplate 4). The average peak count in Delaware Bay from 1986 to 1992 was 46,513 ± 8,888 red knots, with the maximum number (94,460) in 1989 (Clark et al., 1993). Over this time interval, there was no significant increase or decrease in red knot abundance in Delaware Bay. Based on the data compiled by the International Shorebird Survey (ISS) from 1972 to 1983, red knot abundance decreased by 75 percent; however, the trend was not statistically significant (Howe et al., 1989). Similarly, their abundance has also decreased between 1974 and 1991 during fall migration in eastern Canada (Morrison et al., 1994).

Sanderling *(Calidris alba)*

Sanderlings always seem to be in motion; as Bent (1962) described it, "the surf line attracts them, where they nimbly follow the receding waves to snatch their morsels of food or skillfully dodge the advancing line of foam as it rolls up the beach." In summer plumage, the head, back, and breast feathers are somewhat reddish. During May on Delaware Bay beaches, such birds intermingle with birds that still retain their light gray winter plumage. More than their colors, it is their frenetic activity pattern at the water's edge that most characterizes the sanderling.

Scientific knowledge of the sanderling's migration route is summarized by Myers et al. (1990). Nesting occurs in the Arctic; wintering areas occur along the coastal United States, Mexico, Central America, and South America (Morrison, 1984). The highest wintering counts occur on the Pacific coast of South America and the United States. Most sanderlings wintering along the Pacific coast of South America fly northward through the central United States, with smaller numbers using the Pacific coast. Those birds wintering on the east coast of South America migrate northward along the Atlantic coast, mainly through Delaware Bay. Castro et al. (1992) demonstrated that sanderlings overwintering in coastal New Jersey had a higher body fat content and greater daily energy expenditure than sanderlings that overwintered in warmer climates in Texas, Panama, or Peru.

Horseshoe crab eggs at the tide line are a preferred resource for foraging sanderlings in Delaware Bay. As a wave breaks and its energy is dissipated, suspended eggs momentarily drop onto the sand. Sanderlings quickly pick out as many eggs as they can before the next wave chases them up the beach. Sanderlings are more regularly encountered on beaches in lower Delaware Bay than farther up-bay, and they are also common on the Atlantic coast beaches and on exposed tidal flats (Clark et al., 1993; Botton et al., 1994; Burger et al., 1997). Because few horseshoe crab eggs are found in the latter habitats, sanderlings are probably eating other foods. Sanderlings are known to consume a variety of small surf-zone invertebrates, including amphipods ("beach fleas"), mollusks, and polychaetes (Bent, 1962).

The average peak count in Delaware Bay, from 1986 to 1992, was

14,719 ± 4,355 sanderlings, with the maximum number (33,795) in 1986 (Clark et al., 1993). There has been a statistically significant decline in sanderling abundance in Delaware Bay during this time. Sanderling counts showed a significant decline from 1972 to 1983, based on ISS data (Howe et al., 1989). Sanderlings on fall migration in eastern Canada (1974–1991) showed a decrease as well, with most of the decline occurring from 1974 to 1979 (Morrison et al., 1994).

Ruddy Turnstone *(Arenaria interpres)*

With its orange legs and gaudy breeding plumage, the ruddy turnstone stands out from the other shorebirds along Delaware Bay (colorplate 5). The back is reddish-brown, interspersed with black and white longitudinal bands; the breast and face have an almost clown-like appearance with black and white blotches. Ruddy turnstones are roughly intermediate in size between the sanderling and red knot. They are also the only common shorebirds along Delaware Bay that dig for buried horseshoe crab eggs, often chattering loudly as they defend their plunder against other birds.

The American ruddy turnstone winters from South Carolina and the Gulf of Mexico to southern South America and breeds from northern Alaska through northern Canada to Baffin Island. A Eurasian subspecies is widely distributed on its overwintering grounds, ranging into the South Pacific as well as Europe; breeding occurs on Greenland, northern Scandinavia, Russia, and Siberia (Johnsgard, 1981). Like red knots, many of the ruddy turnstones that breed in eastern Arctic Canada and in Greenland winter in Europe.

Ruddy turnstones have a broad distribution within Delaware Bay; we found that they comprised 20 percent or more of the total number of shorebirds counted on all but one of seven study areas in New Jersey (Botton et al., 1994). They will eat eggs on the surface and also dig for buried eggs. Ruddy turnstones do not choose locations haphazardly; they are readily drawn to fresh horseshoe crab nests, which are shallow, bowl-shaped depressions left in the sand by the female. Even fresh footprints and other kinds of sediment disturbances attract inquisitive ruddy turnstones. Rarely, however, do these birds dig below about 5 cm in search of eggs; although a few fresh egg clutches may be found there, the vast majority are deposited below 10 cm and are therefore out of reach. Ruddy turnstones may also be found in tidal marshes along the Delaware Bay and Atlantic coasts (Burger et al., 1997) where they probably feed on invertebrates associated with mussel beds and mudflats (Stone, 1937). Ruddy turnstones are opportunistic feeders, consuming crabs, clams, small fish, amphipods, barnacles, and other invertebrates (Bent, 1962; Fleischer, 1983). They use a range of foraging techniques in their various habitats, described by Whitfield (1990) as routing (flicking aside pieces of seaweed to expose amphipods and other prey), probing, stone-turning, hammer-probing (used to break open barnacles), digging, and surface pecking.

The peak count of 105,160 ruddy turnstones in Delaware Bay occurred in 1989; the average peak count between 1986 and 1992 was 66,086 ± 9,665 individuals (Clark et al., 1993). Wander and Dunne (1981), however, estimated that there were 106,000 ruddy turnstones on just the New Jersey shore of Delaware Bay. There has been a downward (but not statistically significant) trend in the number of ruddy turnstones in Delaware Bay (Clark et al., 1993). Based on ISS data, Howe et al. (1989) estimated that ruddy turnstone numbers declined 62 percent from 1972 to 1983, but the number of fall migrants in eastern Canada increased slightly from 1974 to 1991 (Morrison et al., 1994).

Semipalmated Sandpiper *(Calidris pusilla)*

Semipalmated sandpipers are the most abundant of Delaware Bay's "peeps," a group of small shorebirds that also includes the least sandpiper and white-rumped sandpiper. The back and head are brownish-black, and the breast is white except for a splash of brown below the head. Thin white stripes on the head, just above each eye, help to differentiate the semipalmated sandpiper from the slightly larger and more robust sanderling.

The migration system of semipalmated sandpipers was reviewed by Morrison (1984). The most important wintering grounds are on the northern coast of South America between northern Brazil and Venezuela, with Surinam and French Guiana being outstanding. Breeding grounds range across the Arctic, with concentrations in Alaska, northern Quebec, Baffin Island, and northern Labrador. Based on morphometric and banding data, it has been suggested that three subpopulations exist. The Alaskan and central Canadian breeders mainly use the interior flyway, with a major stopover in Cheyenne Bottoms, Kansas. The Labrador breeders visit Delaware Bay during spring migration and embayments from the Canadian Maritimes to Cape Cod on their return to South America. The Bay of Fundy is especially important to this group during fall migration.

Wander and Dunne (1981) observed that semipalmated sandpipers occurred on beaches, salt marshes, and mudflats along Delaware Bay. Botton et al. (1994), however, found that these birds were rarely found on open beaches in lower Delaware Bay. Semipalmated sandpipers were more abundant farther up-bay at Moores Beach, particularly on rising and falling tides in the vicinity of tidal creek areas and adjacent sandbars and spits (Botton et al., 1994; Burger et al., 1997). They often occur in mixed flocks with other sandpipers of similar size and plumage. Their short bills—about 19–20 mm (Harrington, 1982a)—limit their feeding to horseshoe crab eggs on the surface. The importance of alternative prey items obtained from mudflats and salt marshes is not known. Elsewhere in its range, the diets of semipalmated sandpipers have been examined in numerous studies (reviewed by Skagen and Oman, 1996); food consisted of insects, crustaceans, polychaetes, mollusks, and many other kinds of invertebrates.

Along the Delaware Bay shoreline, Clark et al. (1993) found that the highest count of semipalmated sandpipers occurred during 1986 (267,348 individuals). There has been a significant decline in their abundance from 1986 to 1992; for the three most recent years, the peak number was only about 50,000 birds. Howe et al. (1989) estimated that semipalmated sandpiper numbers decreased by 53.4 percent from 1972 to 1983 (annual rate of decrease = 6.7 percent), although this trend was not statistically significant. Morrison et al. (1994) found a similar, statistically insignificant decrease for this species in eastern Canada.

Gulls *(Larus sp.)*

Three species of gulls interact with horseshoe crabs and shorebirds in Delaware Bay. The most numerous are the raucous laughing gulls *(Larus atricilla)*, which noisily compete for eggs with each other and with the shorebirds (colorplate 6). At times they seem to spend more of their time vocalizing and fighting than eating eggs. Unlike the shorebirds, which must migrate to northern Canada to breed, laughing gulls nest locally on the nearby salt marshes of the Atlantic coast. Laughing gulls will continue to feed on horseshoe crab eggs for the duration of the spawning season. Later, during June and July, they focus on the trilobite larvae, which swim weakly in the near-shore waters adjacent to the breeding beach. Both eggs and trilobites are regurgitated and fed to their chicks.

Somewhat to our surprise, we have observed that two large gulls, the herring gull *(Larus argentatus)* and great black-backed gull *(L. marinus)*, regularly eat adult horseshoe crabs as well as the crab eggs (Botton and Loveland, 1993). Both gulls have powerful bills that can easily rip apart the book gills and other soft areas on the underside of a stranded horseshoe crab. Once the crab is killed, the gulls consume the eggs and other viscera, leaving the legs and carapace behind as a grue-some reminder of the attack.

How Do Horseshoe Crabs and Shorebirds Select Habitat?

The vast Delaware Estuary is a mosaic of environments. At the mouth of the Bay, near Cape May, N.J., and Cape Henlopen, Del., there are sandy beaches with conditions intermediate between high-energy beaches on the Atlantic Ocean and the beaches farther up-bay. Along the lower portion of the Bay, especially from Villas to Reeds Beach, N.J., and Broadkill Beach to Bowers Beach, Del., there are narrow rib-bons of sandy beach. Farther up-bay, as wave energies decrease, salt marshes dominate; only patches of sand are found. Humans have modi-fied the natural beach geomorphology; in response to beach erosion, some communities have "stabilized" their shorefronts by building bulk-heads and groins made of metal, wood, or stones. Long jetties that maintain navigational inlets (for example, Mispillion River, Del., and Bidwells Ditch, N.J.) further alter the natural movements of sand. In the nineteenth century, vast areas of salt marshes were impounded to create salt hay *(Spartina patens)* farms (Weinstein et al., 1997), and many thou-sands of acres of marsh have been manipulated for mosquito control. Thus, the Delaware Bay shoreline is a mix of beaches and marshes in various states ranging from nearly pristine to extensively altered. Where do horseshoe crabs and shorebirds congregate, and why?

There are unquestionably more horseshoe crabs in Delaware Bay than anywhere else in the world, but they are not uniformly abundant within the system. The distribution of horseshoe crabs was first studied by Shuster and Botton (1985), who counted the number of spawning adults on the full moon in June 1977. Crabs were most abundant along the central Cape May peninsula and from Broadkill Beach to Bowers Beach, Del. (see Figure 14.1 for these and other sites along Delaware Bay). Concentrations of horseshoe crabs farther up-bay were limited to small, isolated sandy beaches, such as Moores Beach, Gandys Beach, and Fortescue.

Botton, working with Bob Loveland and Tim Jacobsen (1988), in-vestigated the importance of beach erosion and geochemical factors to habitat selection by horseshoe crabs. We noted striking differences in spawning intensity on two apparently similar sandy beaches that were less than 1 km apart. But after more detailed sediment analysis, we found that there was a > 20 cm layer of well-oxidized sand at the "pre-ferred" beach, but only a thin veneer of sand overlaying anaerobic peat sediments at the avoided beach. We hypothesized that chemical cues emanating from the beaches might be important in habitat selection by the horseshoe crabs. We tested this hypothesis by collecting interstitial pore water from the preferred and avoided beaches. High levels of hydrogen sulfide released from the peat sediments probably served as a deterrent to mated pairs of horseshoe crabs as they approached the beach on a rising tide. Thus, beaches that were superficially alike could be distinguished by horseshoe crabs on the basis of subtle chemical cues. Only 8.5 km out of the 80.5 km of shoreline surveyed in New Jersey was considered to be optimal for horseshoe crab spawning, based on the scarcity of good sandy beaches and the damaging effects of shoreline construction (Botton et al., 1988).

How do the shorebirds select their habitat? Early in our observations, we learned that certain beaches were more attractive to birds than others. Although their abundance varied with weather conditions, tidal stage, and other factors, certain places, such as Reeds Beach, N.J., always seemed to be hot spots for birds. One obvious hypothesis was that the number of shorebirds might track the abundance of eggs within the Bay. This seemed intuitive; indeed, researchers have found that shorebirds often concentrate in areas of maximal food abundance (for example, Bryant, 1979; Goss–Custard, 1979; Puttick, 1984). But there could be other contributing factors besides food; for example, beach slope, sediment texture, or proximity to roosting sites could also be important variables. Sediment texture is an important parameter when shorebirds forage for infaunal prey such as polychaetes or mollusks (Quammen, 1982; Grant, 1984; Hicklin and Smith, 1984; Kelsey and Hassall, 1989). But sediment texture was not a good predictor of shorebird abundance in Delaware Bay (Botton et al., 1994), probably because feeding on horseshoe crab eggs is not analogous to feeding on benthic invertebrates. The striking green color of the eggs provides strong visual contrast against the pale sand; and eggs, unlike worms or clams, are incapable of escape behavior when poked by a shorebird's bill.

We have found that shorebird abundance in Delaware Bay is very irregular, but it does not correlate directly with horseshoe crab egg abundance (Botton et al., 1994). The broad distribution of foraging shorebirds within Delaware Bay may be a consequence of the superabundance of horseshoe crab eggs throughout the system. Using stomach contents, Tsipoura and Burger (1999) definitively established that *Limulus* eggs are the primary food item eaten by shorebirds in Delaware Bay. We estimated the number of shorebirds that could be sustained along a beach by dividing the number of eggs in the top 5 cm of sand

TABLE 1.2 Estimated number of shorebirds supportable per meter of shoreline in 1990, assuming consumption of 8,300 horseshoe crab eggs per bird per day (based on Castro et al., 1989). The numbers of birds estimated by aerial survey are given for comparison. All beaches were surveyed at 3 m stations from the spring high-water mark to the low-water mark. Unsampled regions on each transect were assumed to have egg abundance comparable to the station above on the tidal gradient.

Beach	Eggs/m Shoreline		Estimated Number of Birds Supportable		Birds / m Based on Clark (1991)	
	24–25 May	1–7 June	24–25 May	1–7 June	1990	5-yr mean
Higbee's Beach	—	0	—	0	—	—
Villas	3,125	62,083	0.4	7.5	6.6	3.4
Norburys Landing	34,167	61,354	4.1	7.4	.3	1
Pierces Point	51,771	26,250	6.2	15.2	0	0
Cooks Beach	49,583	194,271	6.0	23.4	0	1.1
Reeds Beach	499,375	406,165	60.2	48.9	24.8	7.6
Moores Beach	721,354	230,104	86.9	27.7	2.7	4.1

(summed across the intertidal transect) by the daily ration per shorebird (8,300 eggs/day, approximately the number of eggs in two *Limulus* nests), based on the study of sanderlings by Castro et al. (1989). This led to the conclusion that all beaches had enough eggs to support the needs of at least four shorebirds per meter of shoreline, except for Higbee's Beach near Cape May Point, which had almost no eggs (Table 1.2). The distributional pattern of the eggs along the Bay shore reflects both the spawning patterns of the crabs themselves and the subsequent redistribution of these eggs by physical disturbances (further digging activity by horseshoe crabs and wave action) and alongshore currents. Shorebirds often aggregated near shoreline discontinuities such as salt marsh creek inlets and jetties, which act as traps for eggs being transported in near-shore waters.

Increasingly, it is becoming clear that shorebirds make extensive use of all available habitats—beaches, mudflats, and salt marshes—over the course of the tidal cycle (Burger et al., 1997). Shorebirds are also found in impoundments that fringe the Bay shore (Figure 1.3), a behavior that is common in other coastal environments (for example, Velasquez and Hockey, 1992; Weber and Haig, 1996). In fact, some birds such as dunlin and short-billed dowitchers are more commonly observed on back-bay mudflats and salt marsh creeks than on the open Bay beaches (Figure 1.4). Thus, shorebirds do not appear to be selecting habitat in Delaware Bay based solely on the abundance of *Limulus* eggs. A combi-

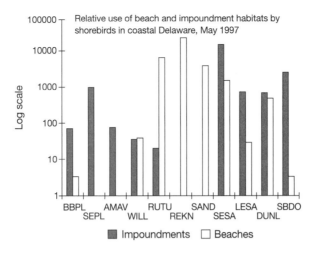

FIGURE 1.3 Relative use of beach and impoundment habitats by shorebirds in coastal Delaware, May 1997. BBPL = black-bellied plover, SEPL = semipalmated plover, AMAV = American avocet, WILL = willet, RUTU = ruddy turnstone, REKN = red knot, SAND = sanderling, SESA = semipalmated sandpiper, LESA = least sandpiper, DUNL = dunlin, SBDO = short-billed dowitcher. *Source: B. Harrington.*

FIGURE 1.4 Back-bay marshes along Delaware Bay are important shorebird habitats. Here, a group of dunlin is feeding along the edge of a salt marsh creek. *Photograph by M. L. Botton.*

nation of habitat types influences shorebirds as they move between various feeding and resting areas during their stopover in the Delaware estuary.

How Much Do Shorebirds Eat?

Considering all aspects of the feeding ecology of shorebirds on their wintering grounds and staging areas is beyond the scope of this chapter. Shorebirds have been extensively studied from the perspective of energetics (Castro et al., 1992), foraging and activity patterns (Puttick,

1984; McNeil et al., 1992), prey preferences (Skagen and Oman, 1996), and the impacts of shorebird predation on benthic infauna (for example, Botton, 1984; Baird et al., 1985; Piersma, 1987; Sewell, 1996). We will concentrate on those aspects of foraging ecology that relate to horseshoe crabs.

Wintering Quarters
Little is known about the activities of those birds that are preparing for their northward flight to Delaware Bay. At a staging area in southern Brazil, red knots feed extensively on small coquina clams *(Donax)*, mussels, and gastropods (Harrington, 1996). Gonzalez et al. (1996) studied the feeding ecology of red knots in a major wintering area, Golfo San Matias in Argentina. Here, the birds forage mainly on rocky substrate covered with dense beds of the mussel *Brachiodontes rodriguezi,* preferring prey of 5–20 mm length. In general, most shorebirds can be classified as opportunists, taking advantage of locally abundant foods in all manner of coastal habitats (Skagen and Oman, 1996). It remains to be learned, however, what some of the specific food resources are for other migrant shorebirds in their staging areas just prior to their departure for Delaware Bay.

Feeding in Delaware Bay
Surprisingly, feeding rates of shorebirds on *Limulus* eggs are unknown, with the exception of a study by Gonzalo Castro and colleagues (1989). They reared captive sanderlings on diets of sand-free horseshoe crab eggs; birds ingested an average of 8,300 eggs per day; but paradoxically, they did not gain weight. Nearly 75 percent of the ingested eggs were passed though the gut, apparently undamaged, which led to a very low metabolic efficiency, 38.6 percent.

Why, then, would it make any sense for a shorebird to concentrate its feeding on something that it can't digest very well? The answer would seem to be the exceptional densities of *Limulus* eggs along Delaware Bay beaches. There may routinely be over 100,000 eggs per square meter in the surface 5 cm of the sand during the peak of the shorebird migration (Botton et al., 1994). Birds save energy by minimizing search and handling time. We believe that the sheer abundance of horseshoe crab eggs compensates for low metabolic efficiency, which allows shorebirds to gain the energy reserves needed to continue their migration to the Arctic (see Research Note 1.1).

Castro and Myers (1993) integrated data on shorebird abundance, energetic requirements, and metabolic efficiency to estimate that shorebirds collectively ate 539 metric tons (mt) of *Limulus* eggs, including 226.1 mt by red knots, 167.0 mt by ruddy turnstones, 99.8 mt by semipalmated sandpipers, and 39.5 mt by sanderlings. This may overestimate egg consumption for several reasons. First, all species were assumed to have a metabolic efficiency of 38.6 percent, based on Castro et al. (1989). In that work, sanderlings were fed on eggs that had been separated from sand; but under field conditions, birds probably ingest sand, which could assist in the mechanical breakup of eggs within the gizzard. If so, actual metabolic efficiency may be higher. In addition, the authors' calculations are based on two important but untested assumptions: shorebirds spend an average of 21 days on Delaware Bay and they feed exclusively on *Limulus* eggs. The authors acknowledged, but did not correct for, the possibility that shorebirds feed on other prey. More recent studies suggest that Delaware Bay shorebirds forage in salt marshes and mudflats, in addition to sandy beaches (Burger et al., 1997).

When horseshoe crab eggs are comparatively scarce and spatially aggregated, competition and aggression among birds may result. Recher and Recher (1969) found that sanderlings established feeding territories around the highest concentrations of eggs. Aggressive interactions took place when nonterritorial sanderlings attempted to cross into others' feeding territory. A similar set of behaviors was found by Mallory and Schneider (1979) in their description of short-billed dowitchers feeding on *Limulus* eggs on a Massachusetts tidal flat. Sullivan (1986) experimentally manipulated the distribution of egg patches and found that ruddy turnstones foraging on evenly distributed crab eggs did not defend them against conspecifics. However, where the distribution of eggs was irregular, about half of the ruddy turnstones defended the patch, with the resident bird prevailing over the intruder in 84 percent of the interactions.

Myers (1986) described how several species of birds interact along Delaware Bay beaches. The focal points for aggressive behaviors were holes dug by ruddy turnstones as they searched for eggs. In intraspecific encounters, duller crowned birds were subordinate to more brightly hued birds. Turnstones successfully maintained their feeding holes against sanderlings and semipalmated sandpipers, over which they have a size advantage, but not against the much larger laughing gulls. But aggression among sanderlings is relatively infrequent, possibly because the abundance of horseshoe crab eggs on Delaware Bay is so great that it is not worth the time for a bird to defend food patches. All shorebirds tended to yield the prime feeding area at the water's edge when laughing gulls were present (Botton et al., 1994). Aggressive interactions or avoidance of interactions among shorebirds and gulls may be an important variable explaining the distribution of birds in the Delaware estuary.

The majority of the shorebirds leave Delaware Bay by early June. During their three-week stay, birds can come close to doubling their mass, and their fat index (fat/total body mass) balloons to 40 percent (Castro and Myers, 1993). More recent research has shown that other factors are also involved in the mass change, including changes in the mass of the gut, liver, and various muscle groups (see Research Note 1.1). But by far, the largest fraction of mass change is due to fat. This energy is needed for the next leg of the birds' migration, which will bring them to their Arctic breeding areas. Recent experiments have shown that birds feed on Delaware Bay until a minimum energetic condition is attained; they then depart en masse when weather conditions are most favorable for the northbound migration (Research Note 1.2).

Breeding and the Southbound Journey

Courtship and breeding behavior in shorebirds have been well documented in Bent (1962), Johnsgard (1981), and others. The various boreal breeding areas have a convergence of shorebirds that have migrated through Delaware Bay, as well as through the central North American flyway or the Pacific coast. The degree to which different migrating groups of shorebirds have evolved genetic differentiation is not well known (Haig et al., 1997). Shorebirds generally arrive on their breeding grounds in late May or early June, about the time when the snow cover recedes.

Certain aspects of shorebird reproduction are remarkably conservative. They typically lay four eggs in simple nests on the ground or in bits of vegetation. The eggs are incubated for about 20 days in semipalmated sandpipers, 22 days in ruddy turnstones and red knots, and 24 to 32 days in sanderlings (Johnsgard, 1981). Both sexes share in the incubation duties. The young are fledged in 16 days for semipalmated sandpipers, 17 days in sanderlings, and 18 days in ruddy turnstones and red knots.

The brief Arctic summer triggers a seasonal burst of productivity. While on the breeding grounds, shorebirds feed extensively on insect larvae, oligochaetes, and mollusks; but they may also consume seeds and other plant material, especially if they arrive before the snowmelt is completed (Johnsgard, 1981; Morrison, 1984). By early July, the southbound fall migration begins.

The Bay of Fundy is the principal East Coast staging area for many of these shorebirds during the fall migration (Hicklin, 1987; Morrison et al., 1994). An estimated 1.2 to 2.2 million shorebirds use the Bay of Fundy (Mawhinney et al., 1993), a mix of returning adults and juveniles making their first southbound flight. Semipalmated sandpipers are the most abundant of thirty-four species of migrant shorebirds, comprising about 95 percent of all individuals counted (Hicklin, 1987). Peak counts of semipalmated sandpipers and red knots occur between mid-July and mid-August. Sanderlings peak from mid-September to early November (Hicklin, 1987). Those arriving later are primarily juveniles. Evidently, few ruddy turnstones use the Bay of Fundy during fall migration.

During their stopover in the Bay of Fundy, shorebirds forage primarily on amphipods, *Corophium volutator,* which live on the extensive mudflats (Gratto et al., 1984; Hicklin and Smith, 1984; Wilson 1989, 1990, 1991). This rich food source enables shorebirds to nearly double their mass before embarking on their return flight to South America.

Spring migrants in Delaware Bay and fall migrants in the Bay of Fundy both exploit seasonally available food items that reach extraordinary numbers; indeed, high densities of preferred prey may be a general characteristic of shorebird staging areas (Botton et al., 1994). Shorebirds can quickly switch their diet to take advantage of locally abundant foods. Within a matter of six to eight weeks, an individual bird might specialize on *Limulus* eggs in Delaware Bay, dipteran larvae in northern Canada, and *Corophium* in the Bay of Fundy.

Conservation Issues: Some Unanswered Questions

Horseshoe crabs are now the basis of an extensive fishery in the middle Atlantic region (see Chapter 15). From several hundred thousand to over a million horseshoe crabs—mainly gravid females—have been taken yearly from the Delaware Bay area since 1992 as bait for eels and whelks. Coincident with this has been a marked decrease in the number of horseshoe crabs counted in annual beach censuses and trawls, and a similar decrease in the number of eggs on Delaware Bay beaches.

Many conservationists are concerned that these trends may devastate migrant shorebirds. We believe that answering several questions would improve predictions that at present are poorly substantiated (but may still be essentially correct).

How Dependent Are Shorebirds on *Limulus* Eggs?

Most authors assume that the energetic needs of the shorebirds can be met by specializing on *Limulus* eggs (Myers, 1986; Castro and Myers,

1993). In actuality, shorebirds forage on beaches, salt marshes, and mud-flats on both the Delaware Bay and Atlantic coasts (Clark et al., 1993; Burger et al., 1997). Delaware Bay attracts far more spring migrant shorebirds than other, equally productive estuaries along the east coast of North America; yet if the main forage base was polychaetes or mollusks, why would shorebirds choose Delaware Bay as opposed to, say, Chesapeake Bay? Certainly, strong circumstantial evidence suggests that horseshoe crab eggs have played a pivotal role in making Delaware Bay a major staging area. However, without additional studies of feeding ecology, it is difficult to evaluate the nutritional importance of horseshoe crab eggs, relative to fauna from marshes and mudflats.

Will It Be Possible to Link Trends in Shorebird Abundance to Conditions in Delaware Bay?

Trying to monitor shorebird and horseshoe crab populations is challenging, and attempts to correlate the two may be an exercise in speculation until more data are available. We can, however, make a number of tentative conclusions and suggest several cautions.

Clark et al. (1993) found that most species of migrant shorebirds in Delaware Bay decreased from 1986 to 1992. This decline took place before the expansion of the horseshoe crab fishery and the decline in egg abundance. Other long-term surveys (Howe et al., 1989; Morrison et al., 1994) also indicate a decrease in shorebird numbers during a time in which horseshoe crab abundance was probably stable (Botton and Ropes, 1987). Does this mean that horseshoe crab numbers are unrelated to shorebird abundance?

Shorebirds may be impacted by environmental stresses on their overwintering, breeding, and staging areas, which make it problematic to associate changes in shorebird abundance with site-specific factors. At the present time, Delaware Bay is the major stopover on the northbound Atlantic coastal migration route. If horseshoe crab egg abundance were to decline below a certain baseline value, this might cause shorebirds to remain longer, which in turn might delay breeding. Another possibility is that birds might leave Delaware Bay on the same timetable as they do now, but with lower fat content. This could affect their ability to complete the northbound migration or reduce their fitness on the breeding grounds. It could take years before a declining horseshoe crab resource in Delaware Bay would be reflected in a numerical response by shorebird populations. All of this suggests that it may be difficult to predict the future status of shorebirds based solely on what happens in Delaware Bay. Over time, could shorebirds find alternative staging areas elsewhere along their route that would provide better food resources?

How "Tight" Is the Shorebird–Horseshoe Crab Linkage?

There is no debating the ancient heritage of horseshoe crabs, and it is quite likely that their intertidal spawning behavior preceded the breakup of the continents and the evolution of long-distance migration in shorebirds. But we simply have no means to determine how long shorebirds have been reliant on horseshoe crab eggs in Delaware Bay. It may be, however, that the phenomenon is much more recent than is commonly assumed.

Knebel and colleagues (1988) reconstructed the paleogeography of Delaware Bay as a function of rising sea level during the past 18,000 years. About 12,000 to 15,000 years ago, the Bay was much narrower than at present, and sandy beaches were absent. Small areas of sandy beach formed near the widening Bay mouth around 10,000 years ago, and beaches began to advance up-bay some 7,000 to 8,000 years ago. We suggest that *Limulus* abundance in Delaware Bay has probably tracked the increasing availability of sandy beach habitat following glacial retreat. Quite possibly, huge numbers of horseshoe crabs were not found in Delaware Bay until a few thousand years ago. If so, it is likely that Delaware Bay was not an important staging area until recently; alternatively, if it was a stopover area, shorebirds were probably feeding on something other than crab eggs.

Even for the past 200 years, we lack solid documentation of the numbers of horseshoe crabs and shorebirds in Delaware Bay. Throughout the 1800s, hunters often killed hundreds, if not thousands, of shorebirds per day along the Atlantic coast of the United States. During the same era, millions of horseshoe crabs were harvested for fertilizer and livestock or poultry feed (see Chapter 14). In the early 1800s, Wilson made note of the feeding of ruddy turnstones and other birds on horseshoe crab eggs (Brewer, 1840). Stone (1937) described ruddy turnstones feeding on stranded horseshoe crabs, but made no mention of shorebirds eating eggs. It is uncertain whether this indicates a relative scarcity of shorebirds and/or horseshoe crabs from Delaware Bay, or a lack of sampling effort. Roger Tory Peterson (personal communication to Harrington) pointed out that Stone was a summer resident who lived on the Atlantic Ocean side; thus, it is possible that he would have missed those migrant shorebirds on the Bay side preceding his arrival in Cape May. Perhaps Stone may have simply had the misfortune to be based in Cape May during the historic nadir of both shorebirds and horseshoe crabs. We know that the horseshoe crab population was limited by the lingering effects of the fertilizer industry, but it is difficult to infer what shorebird numbers were in the 1930s or whether the birds were feeding en masse on horseshoe crab eggs.

In conclusion, the interrelationship between horseshoe crabs and shorebirds in Delaware Bay may only have developed in the recent geological past—a mere few thousand years ago. When viewed against the backdrop of evolutionary and geological time scales, the dependence of migratory shorebirds on horseshoe crab eggs may be an ephemeral phenomenon. However, it would be simplistic and irresponsible to suggest that shorebirds would be immune to a drastic decrease in horseshoe crabs. We cannot presume, because alternative coastal habitats exist, that these would necessarily fulfill the birds' energy needs as well as Delaware Bay does today. As stated by Myers et al. (1987), "the fact that these sites must function in precise sequence both in time and in space means that functional alternatives to current staging areas are unlikely."

PHYSIOLOGICAL ECOLOGY OF SHOREBIRDS DURING MIGRATION THROUGH THE DELAWARE BAY AREA

Nellie Tsipoura

Shorebirds staging on Delaware Bay are very thin when they arrive, but manage to gain enough fuel as fat and muscle protein in a short amount of time to complete the journey to the Canadian Arctic breeding grounds. They offer an ideal system for the study of metabolic adaptations of long-distance migration. Since 1993, we have been investigating several aspects of the physiological ecology of migrating shorebirds. Our research combines a field approach with more technical laboratory work to elucidate how birds achieve this rapid and efficient energy storage.

Our research on Delaware Bay shorebirds includes the following topics:

- A study of shorebird metabolism, focusing on energetic differences between species that undertake migratory journeys of different distance and duration. It has been hypothesized that the high-energetic expenditure of migration leads to higher-than-predicted basal metabolic rates in shorebirds (Kersten and Piersma, 1987). Thus, we would expect that shorebirds undertaking longer migratory flights would have proportionately higher basal metabolic rates relative to predicted values for their body size compared to shorebirds undertaking shorter migratory flights.
- A study of the role of growth hormone and corticosteroids in fat metabolism during migration (Tsipoura et al., 1999). During bird migration, rapid physiological changes take place mediated by hormones. Before migration, birds go through a stage of intensive eating (hyperphagia), which promotes deposition of fat into the body tissues (lipogenesis). We explored how levels of growth hormone relate to physiological status during migration because of its role in breaking down fat and providing the energy for migration. We also studied corticosterone, the "stress hormone," because of its effects on both lipogenesis and feeding behavior. We determined that levels of corticosterone rise under stress, and therefore sensitivity to acute stress is not suppressed during spring and fall migration in these shorebirds, as has been previously seen for breeding birds in extreme environments (Wingfield et al., 1992) and in other migrating birds (Holberton et al., 1996).
- An investigation of the changes in body mass and diet during shorebird stopover on Delaware Bay. Diet information is essential

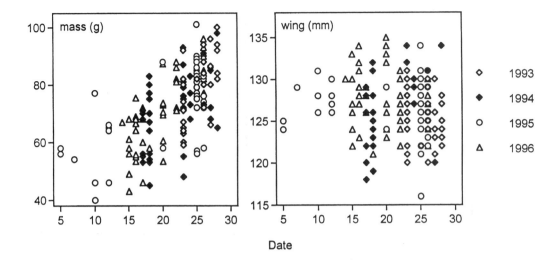

FIGURE 1.5 Body mass (in grams) and wing length (in millimeters) measurements (on vertical scale) and time in days (horizontal scale) for sanderlings in Delaware Bay during the month of May 1993–1996. *By permission of N. Tsipoura.*

for the conservation of sufficient habitats to provide ample food for the migrants. In this research note we provide a brief description of this aspect of the study (Tsipoura and Burger, 1999).

We captured shorebirds along Delaware Bay beaches and marshes and on Atlantic Ocean beaches during spring migration. Capture time and method depended on the species; for instance, semipalmated sandpipers and least sandpipers can be caught in mist nets during the day, whereas sanderlings avoid the nets visually, so they must be caught at night. We also used a net-launching gun and a cannon net to capture sanderlings, red knots, and ruddy turnstones during the day.

We took body measurements, including bill length, flattened wing length, and body mass on all the birds captured, and then banded the birds with U.S. Fish and Wildlife Service metal bands and a combination of color bands before releasing them.

Shorebirds migrating through Delaware Bay increase their body mass by up to 70–80 percent in a short 3-week period, and these increases are not the result of overall changes in body size (Figure 1.5). The timing of these rapid mass gains does not differ significantly from year to year; therefore, it is probably a result of the birds' restrictive breeding schedule, coupled with the timing of the horseshoe crab spawning in Delaware Bay.

We used two different methods to determine shorebird diet: (1) visual analysis and quantification of food items found in the stomach; and (2) laboratory measurements of stable isotope ratios in shorebird blood. Gut samples were collected by inserting a small-diameter Tygon plastic tube through the mouth and into the stomach of the bird and flushing with distilled water, while holding the bird's bill over a jar. This takes about 1 to 3 minutes. Using this method, we collected gut samples without sacrificing the birds. Gut contents from each bird were preserved in alcohol and later viewed on a 1 cc volumetric slide under an inverted microscope at 4 x 10 magnification. We identified individual prey items to the lowest possible taxon, and we estimated the percentage of each item in the gut. We then scanned the complete sample for unusual items by viewing it under a dissecting scope.

We found that horseshoe crab eggs, represented mostly by membranes, made up a significant portion of the shorebird diet in Delaware Bay. Sand was also found in the gut samples, and we believe this is used for breaking and grinding down the outer membrane of the horseshoe crab eggs to facilitate digestion. Polychaetes and ribbon worms were important prey items in sanderlings caught on ocean beaches and in semipalmated and least sandpipers from Delaware Bay marshes. In addition, we found abundant oak leaf scales and pollen in gut samples from both sandpiper species. These items are derived from terrestrial habitats, indicating the importance of marshes as feeding sites.

By measuring ratios of naturally occurring isotopes (^{13}C/^{12}C; ^{15}N/^{14}N, ^{34}S/^{32}S) in shorebird blood, we can attempt to quantify the habitats from which the shorebird prey is derived. Terrestrial, marine, and salt marsh ecosystems differ in stable isotope composition and so does potential shorebird prey: horseshoe crabs and their eggs are likely to have isotope compositions typical of marine environments, whereas marsh invertebrates will have isotope compositions characteristic of the marsh food chain. Stable isotope composition in body tissues of predators can be related to the isotope composition of their diet (Peterson and Fry, 1987); thus, we expect that isotopes in the shorebird blood will reflect those of the prey they are consuming. Based on the gut analysis, we predict that the isotopic signal of the shorebirds will reflect horseshoe crab egg consumption, but will deviate to include a more varied diet that includes prey from marsh habitats.

Our research has increased the understanding of the importance of Delaware Bay as a stopover area for migrating shorebirds as well as of the physiology of shorebird migration. In addition, our work confirms the widely held assumption that Delaware Bay shorebirds feed extensively on horseshoe crab eggs. This point is critical in view of the expanding horseshoe crab fishery and possible recent declines in horseshoe crab populations (see Chapter 15).

Recognizing the dependency of shorebirds on horseshoe crab eggs will ensure better-defined management objectives for the crabs and the protection of other habitats, such as marshes, that are used extensively by the birds.

RESEARCH NOTE 1.2

THE IMPORTANCE OF WEATHER SYSTEMS AND ENERGY RESERVES TO MIGRATING SEMIPALMATED SANDPIPERS

David Mizrahi

At stopovers along the spring migration route, sandpipers and other shorebirds must accumulate enough nutrient reserves to meet the extreme energetic demands of their flight to the breeding grounds. Because invertebrate prey can be scarce when sandpipers arrive on the breeding grounds, and because breeding seasons are short at high latitudes, few opportunities may exist to recoup reserves used during migration. Therefore, sandpipers must arrive on the breeding grounds with sufficient nutrient reserves to compete for and hold territories, in the case of males, and to produce viable eggs, in the case of females. At penultimate stopover sites such as Delaware Bay, shorebirds must accumulate reserves sufficient to complete their flight to the breeding grounds, meet the demands of reproduction, and survive unpredictable environmental conditions once they arrive in the Arctic.

Given the energetic demands of completing migration and initiating courtship and nesting, we might expect migrating sandpipers to store as much fat as physiologically possible before departing from penultimate stopover sites. However, there are costs to carrying excess amounts of fat, such as increased metabolic rates, increased susceptibility to predation, and reduced flight efficiencies as body mass increases. Additionally, the time required to accumulate the necessary energy reserves to complete the migration and begin breeding is constrained by the narrow window available for nesting at high latitudes. How do sandpipers complete migration and arrive in the Arctic with sufficient reserves to initiate breeding while avoiding the costs of carrying excess fat?

Mesoscale or synoptic weather systems (for example, high pressure systems or low pressure systems) producing southerly winds (tailwinds) may provide sandpipers migrating to Arctic breeding grounds with opportunities to reduce migration costs. Flying north with a favorable tailwind can reduce in-flight energy consumption, increase flight distance capabilities, reduce migration time, and reduce the possibility of being blown off course. Conversely, synoptic weather systems that produce northerly winds (headwinds) can have the opposite effect. Tail-wind conditions may represent a kind of "fat equivalent" or "time equivalent" that allows individuals to reach migration destinations carrying surplus nutrient reserves, while avoiding storage of excess reserves during stopovers. Synoptic-scale weather conditions typically persist for several days, so they may represent reliable indicators of conditions that the birds would encounter en route to the breeding grounds.

In my research, I attempt to clarify relationships between the departure of sandpipers from Delaware Bay, energetic condition, synoptic-scale weather systems, and endogenous time programs by answering two questions: (1) do prevailing synoptic-scale weather systems affect the probability of departure of sandpipers from a spring stopover site, and (2) are fat reserves greater on departure days than on other days during the stopover period?

I took an experimental approach to address these questions. During spring migration, I mist-net semipalmated sandpipers along the New Jersey coast of Delaware Bay and test them in Emlen funnels. Emlen funnels are widely used to measure migratory behavior in birds. They consist of a funnel lined with blotter paper set in a circular container. An ink pad is placed at the bottom of the funnel and the entire apparatus is covered with wire or plastic mesh. Once a bird is introduced into the test apparatus, it attempts to escape, leaving ink footprints on the paper lining the walls of the funnel. I use the amount of activity (migratory restlessness, or *Zugunruhe*) exhibited and the orientation of the activity as an index of a bird's migratory readiness on a given day.

Results of my preliminary analyses indicate that synoptic weather conditions producing southerly winds may be important cues for semipalmated sandpipers departing from Delaware Bay in the spring. Nearly 80 percent of the days on which birds depart for the breeding grounds (days when daily test groups show significant northerly orientation) occur when synoptic weather conditions produce southerly winds in the vicinity of the mid-Atlantic coastline. However, these favorable synoptic conditions occur only on 45 percent of all days during my study. These proportions are significantly different from each other, suggesting that semipalmated sandpipers are selective regarding the days on which they migrate.

Fat reserves, although important, may play a secondary role in determining when semipalmated sandpipers depart from Delaware Bay during spring stopovers. When mean percent body fat (fat mass/total mass) of test groups is above the season average, I do not find a significant difference between the energetic condition of test groups on "migration" days and "no migration" days. In some cases, birds departed when their energetic conditions were significantly below the seasonal average, but I noted this most typically when weather conditions were unusually favorable for migration. On the other hand, semipalmated sandpipers migrating on days when weather conditions were suboptimal are usually at or above the season average for mean percent body fat.

I suggest that semipalmated sandpipers remain in Delaware Bay until they attain a certain energetic condition, and then they depart when weather conditions are optimal for migration. This might explain why several investigators have failed to find significant relationships between fat stores and departures of sandpipers from stopover sites. These findings are important when we consider that the next major stop for semipalmated sandpipers after Delaware Bay is probably the breeding grounds in eastern Canada, the closest area being the eastern shores of Hudson Bay, ca. 2,000 km from Delaware Bay. Based on a flight range model proposed by Castro and Myers, semipalmated sandpipers on "migration days," flying under calm wind conditions and at maximum range velocity, have on average about enough fuel to fly ca. 2,100 km, just enough to reach the east coast of Hudson Bay from Delaware Bay. By departing with tailwinds, semipalmated sandpipers may be able to cut their power output or shorten their transport time markedly. A bird using this strategy could arrive in the Arctic with surplus fat without incurring the costs of carrying excess fat reserves en route. Therefore, the evolution of long-range migration strategies in shorebirds may be in part related to their ability to exploit temporal and spatial patterns of synoptic-scale weather systems.

CHAPTER 2

NESTING BEHAVIOR:
A SHORELINE PHENOMENON

H. Jane Brockmann

Each spring, millions of horseshoe crabs lay their eggs on the beaches of the Atlantic and Gulf coasts. There are so many crabs nesting that they often dig up the eggs of those that nested before in the same location. After the tide recedes, birds descend (colorplate 5) to feed on the wrack line of horseshoe crab eggs (colorplate 2). The crabs are a familiar sight to those who walk the beaches of the Delaware Bay, but large numbers of nesting crabs can also be seen on beaches from New Hampshire to the Keys, along the Gulf coast of Florida (colorplate 7), and along the north and west coasts of the Yucatán. Questions abound: Why do horseshoe crabs nest on the high tide? Why do they synchronize their nesting? Why do they use some beaches but not others? How do they find the nesting beaches? How do they lay eggs? (See Exhibit 2.1.)

I began studying horseshoe crabs because I wanted to take my undergraduate Animal Behavior students on an interesting weekend field trip during the spring semester. The horseshoe crabs provided a perfect venue: they came fairly predictably on the new and full moon high tides during March and April and they came in large enough numbers so we could collect sufficient original data to make the field trip a valuable research experience for my students (see colorplate 7). We took advantage of the availability of the University of Florida Marine Laboratory at Seahorse Key, a beautiful island near Cedar Key, Florida, in the northern Gulf of Mexico. I found the horseshoe crabs fascinating; it was as though we were studying Martians. I was intrigued by the idea that this was a scene that had been played out each spring along similar shores for a hundred million years. A literature review revealed that little was known about their behavior, despite the fact that they are so abundant at some times of the year. My students collected terrific amounts of information and eventually these data formed the basis for the grant proposal to the National Science Foundation that launched my research program in earnest in 1990. I now know some parts of the story and answers to some of the questions.

In this chapter I will tell you about the nesting behavior of these animals and our hypotheses about why and how (see Exhibit 2.1) they nest where and when they do. I will discuss what is known from the literature and what my students and I have discovered. I will describe our work in the context of the hypotheses and approaches we have taken and then point out particularly intriguing avenues for future research that remain to be studied with these unusual and fascinating creatures.

Nesting Behavior

A female horseshoe crab arrives at the beach with a male clasping her carapace. The female is ready to lay eggs and she will spawn all of her roughly 80,000 eggs within just a few days (Shuster and Botton, 1985). She buries herself in the sand (Figure 2.1), bending down the front part of her body (prosoma), pressing forward and pushing the sand from underneath with her walking legs and out behind with the last pair of legs specially modified for digging (Vosatka, 1970) (see Chapter 7). It is easy to track the slow movement of a nesting crab through the sand if you place wire flags on either side of the female's hinge (prosomal-opisthosomal hinge). With these flags to indicate the position of nests, it is possible to dig up the eggs from between the flags after the tide (see Figure 2.1). As she digs, the female makes the water-soaked, fine sediments more fluid by paddling with her legs (thixotropy; Shuster, 1993). The eggs are released in a continuous stream into the fluid sand 5 to 20 cm below the surface (Rudloe, 1979; Brockmann, 1990). As the eggs are laid, the male fertilizes them externally with aquatic, free-swimming sperm (see Chapter 3).

Eggs are laid in discrete clusters of 2,000 to 4,000 (highly variable, from just a few to about 8,000 eggs; Shuster and Botton, 1985; Brockmann, 1990). Newly laid eggs are extremely sticky (Brown and Clapper, 1981) and adhere tightly to one another and to sand grains (Rudloe, 1979). After laying one cluster (which takes 3–15 minutes), the female pushes forward 10–20 cm before laying the next batch of eggs. One visit to the beach usually results in a sequence of two to five egg clusters (a range of one to fifteen egg clusters may be laid). Why don't they

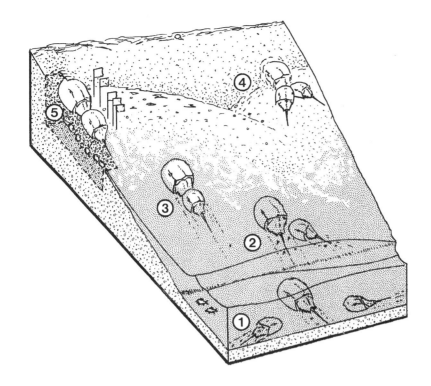

FIGURE 2.1 Horseshoe crab nesting. Male horseshoe crabs search for females offshore (1). An unattached male will attach to a female (2) when he encounters one, and female horseshoe crabs nearly always arrive on the beach with a male attached (3). Other unattached males are attracted to the nesting pair and crowd around them as satellites (4). A nest with four egg clusters 8 cm below the sand surface is marked out with flags that were pushed into the sand alongside the female's hinge while she was nesting. (5) In this way it is possible to return to nests and dig eggs from the sand after the tide has receded *(Figure adapted from Shuster, 1982, and Brockmann, 1990, with permission).*

lay all their eggs at once rather than separating them into these distinct clusters? No one knows for sure, but there are a couple of possible explanations. One is that females are spreading the risk so that if a predator comes on a cluster, it will not take all of her eggs. Another possibility is that egg development is affected by crowding (Barber and Itzkowitz, 1982); eggs that are at the center of a clump or in a large clump develop more slowly than those at the edges or in smaller batches (personal observation), so the female may spread out her eggs to increase the rate of their development.

Females often re-nest several times during one tide, readjusting their location as the tide advances or recedes. The same pairs of males and females may return to the same beach across several tides, remaining buried in the mud near shore between tides. Females usually complete their nesting for the year during one week (one tidal cycle) and do not return again until the following year (Brockmann and Penn, 1992). Males, on the other hand, return repeatedly (either attached to different females or unattached), which partially explains why there are more males at the nesting beach than females.

The eggs the female has left behind develop in the sand for 2 to 4 weeks through four embryonic molts (see Figure 5.2), hatching into "trilobite" larvae (see Figure 5.8; Sekiguchi et al., 1982; Sekiguchi,

1988). The nonfeeding larvae remain in their clusters in the sand for several additional weeks until the next tidal inundation, when they swim into the sea (Rudloe, 1979). Within a few weeks, depending on temperature (Laughlin, 1983), the free-swimming trilobite larvae molt into tiny, spiny juvenile horseshoe crabs (Figure 5.7) that live on the near-shore sand flats (Sekiguchi et al., 1982). This means that when a female chooses a place to nest, she is picking out the site where her off-spring will develop for that important first month and even, perhaps, where they will spend their first year of life. Female nesting decisions (see Exhibit 2.2) are thus crucial to offspring success.

Why Do Crabs Nest in the High Intertidal?

Horseshoe crabs nest near the top of the tide line around the time of the high tide (in many parts of their range) (see colorplate 7). When high tides are higher, the crabs nest higher on the beach; when the tides are lower, they nest a little lower (Figure 2.2; Penn and Brockmann, 1994). For several days around each new and full moon, the high tides are higher than at any other time of the month, and the crabs synchronize their annual spawning cycle to coincide with those tides (Figure 2.3; Lockwood, 1870; Rudloe, 1980; Cohen and Brockmann, 1983; Barlow et al., 1986, 1987). When the two tides in a day are unequal in height, the crabs prefer to nest on the higher one (Barlow et al., 1986, 1987). They do this over several months during the spring when the tides are at their maximum. These nesting patterns mean that, in general, *Limulus* nest about as high on the beach as possible.

High intertidal nesting is really quite puzzling. Very few animals nest so high on the beach and certainly there is no other large invertebrate that crawls from the sea to nest on a beach! Furthermore, it seems likely that such behavior is costly. Exposure to the air may be physiologically stressful (Truchot, 1987; deFur, 1988; see Chapter 9) and sunlight may even damage the animal. When horseshoe crabs come to the top of the tide line during the high tide, they are easily overturned by waves and may be stranded above the waterline after the tide goes out (Shuster and Botton, 1985; Botton and Loveland, 1989; Penn and Brockmann, 1995). Stranding high on the beach is risky because the next tide is less likely to inundate to the same high level and thus rescue the overturned crabs 12 hours later. Gulls attack the overturned crabs, feeding on their gills and removing their legs (Botton, 1993) and the large muscle at the base of the tail (Penn and Brockmann, 1994). Also, the slope of the beach is lower in the high intertidal, and the crabs use beach slope as a cue to orient them back to the sea (Botton and

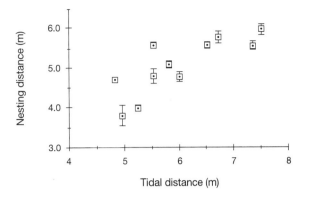

FIGURE 2.2 Location of nesting horseshoe crabs on the beach at different tidal elevations, Seahorse Key, Fla. Each data point refers to one tide when 50–100 crabs nested on the beach. Nesting distance (mean ± SE) is the distance from the center of the nest to the bottom of the beach (the point where the beach flattens). Tidal distance refers to the extent of tidal inundation on the beach for that tide. *By permission from Penn, 1992.*

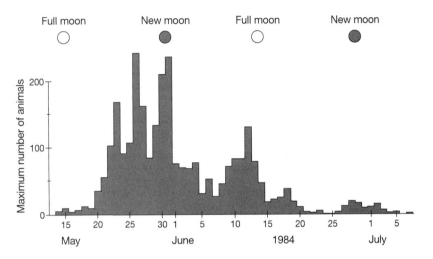

FIGURE 2.3 Seasonal migration of *Limulus* to the beach for nesting usually occurs around the time of the new and full moon high tides at Mashnee Dike, Cape Cod, Mass. Each bar records the number of horseshoe crabs spawning for each day of the mating season. *By permission from Barlow et al., 1986.*

Loveland, 1987, personal observation); without slope, they may become stranded behind the beach. Furthermore, by nesting only on the high tides, the crabs are all nesting at the same time. This results in increased competition with other individuals over mates and nesting sites and may result in crabs digging up the eggs of other females.

Such obvious costs for high intertidal nesting mean that there must be some compensating benefits (see Exhibit 2.3). What might those be? One hypothesis is that when eggs are laid in the high intertidal, they are less likely to be eroded from the sand by waves and wind than those laid lower on the beach. This is the likely explanation for the grunion, a fish that nests in the high intertidal (Thompson, 1919). A second possibility is that by nesting high on the beach, the crabs may be avoiding aquatic nest predators. This is the likely explanation for high intertidal nesting in Atlantic silversides (Tewksbury and Conover, 1987). A third hypothesis is that horseshoe crabs nest where environmental conditions favor rapid development of their eggs (Lockwood, 1870; Badgerow and Sydlik, 1989). What might enhance egg development? The right temperature and moisture conditions are two likely candidates. In sea turtles, which also lay their eggs in the high intertidal, hatching success improves when females dig in well-aerated sand (Ackerman, 1977, 1980; Mrosovsky, 1983; Horrocks and Scott, 1991; Hays and Speakman, 1993), so this is another possible explanation for high intertidal nesting in horseshoe crabs.

My graduate student Dustin Penn and I tested these three hypotheses by measuring the location of nests and by conducting experimental manipulations that determined the effect of beach height on egg development and egg loss by erosion and predation (Penn, 1992; Penn and Brockmann, 1994). We conducted the study at two sites to increase the generality of our conclusions. One was on the northern Gulf coast of Florida where the crabs are active between March and May, and the other was in the Delaware Bay where the crabs are active from late April to late June. The Florida site is on the south shore of Seahorse Key, a small island 4 km west of Cedar Key (29° 06′N, 83° 04′W). This is a low-energy, sandy beach composed of fine, angular quartz sand and some shell fragments (see colorplate 7). The Delaware site is just inside the mouth of the Delaware Bay at Cape Henlopen State Park (38° 47′N, 75° 06′W). It is also a low-energy beach of mixed sand and gravel (see colorplate 2). The size of the sand grains and gravel determines the drainage patterns and oxygen content of beach sediments (Eagle, 1983). The larger grain sizes of the Deleware beach suggest that the sand should be better drained and contain more oxygen than the Florida beach (Maurmeyer, 1978), but we did not know how this would affect egg development.

Egg-Erosion Hypothesis

To test the hypothesis that eggs located lower on the beach are more likely to be washed out of the sand by wave action, we measured the net amount of sand lost (erosion) or deposited at different heights on the beach (in other words, at different distances from the low water line). We also measured how often glass beads (that are comparable in mass and diameter to horseshoe crab eggs) were washed from the sand when buried in different locations on the beach (at 8 cm). We found no evidence that eggs buried lower on the beach were more likely to be washed away by erosion than those buried higher in either Florida or Delaware (Penn and Brockmann, 1994). Erosion was insufficient to uncover horseshoe crab eggs buried at normal depths below the sand surface (Strahler, 1966). One caveat: no storms occurred during this study and we have observed serious beach erosion during storms; so it is possible that the rare, severe storm could be an important factor in the evolution of nest-site selection.

Predation Hypothesis

It is generally assumed that horseshoe crabs nest on beaches to minimize egg loss from aquatic predators such as fish (Rudloe, 1980; Cohen and Brockmann, 1983; Botton and Loveland, 1989; see Chapter 6). Our experiments revealed no loss of eggs or beads from predation at any height on the beach in Florida or Delaware (Penn, 1992). We have observed small schools of striped killifish *(Fundulus majalis)* burrowing into nests that had just been vacated by crabs nesting in shallow water (personal observation; Shuster, 1995) and fish and shorebirds feed on eggs and trilobite larvae that have been dug up by nesting crabs (Shuster, 1960; see Chapter 1). One might predict that horseshoe crab nests located higher on the beach would be more vulnerable to shorebird predation, but this is not the case. Most birds forage along the tide line (Shuster, 1948; Recher, 1966; Recher and Recher, 1969; Burger et al., 1977; Evans, 1988; personal observation), apparently taking advantage of the softer sediments at the water's edge (McLachlan and Turner, 1994). Our observations confirmed that the average location of nests was above where shorebirds normally feed. We conclude that nest predation is not likely to be an important selective pressure on the evolution of nest-site selection in horseshoe crabs. Predation may have played a role in the past, however, when a different complement of predators was present.

Egg-Development Hypothesis

To test the hypothesis that the females nest in the high intertidal because this is where their eggs develop the best, Dustin Penn and I conducted an experiment in which we moved eggs around on the beach (Penn, 1992; Penn and Brockmann, 1994). Clutches laid and fertilized by one pair were dug from the beach, split up, and reburied higher and lower (all at 8 cm depth) than the female had laid them; some eggs were placed as far up on the beach as we had ever observed nesting. To control for our disturbance, we also reburied a batch from each female at the beach height (not the same location) at which she had originally laid her eggs. We measured the rates of development of the eggs (after 10 days) and we took measurements of environmental conditions at different heights on the beach. The results were striking and quite different between Florida and Delaware.

In Florida, eggs buried high on the beach were well developed after 10 days and those laid in the lower parts of the beach were undeveloped. In fact, many of the eggs placed at the lowest extreme of the beach were black and showed no signs of development. It is known that *L. polyphemus* embryos can postpone development under anaerobic conditions (Palumbi and Johnson, 1982), so we transferred these eggs to aerobic conditions in the lab. They did not develop even though most eggs develop well in the lab. All of our measurements of environmental conditions (pH, oxygen concentration, redox potential, temperature, and moisture content) except salinity (which did not vary) were significantly correlated with beach height and with developmental rate, and the variables were also correlated with one another. A stepwise regression analysis found that in Florida a higher oxygen concentration in the sand was the variable that best predicted higher developmental rates for eggs.

In Delaware there was no correlation between where the eggs were buried and egg development except in the extreme upper reaches of the beach. Here many of the eggs were desiccated. At the bottom of the beach, eggs still developed well and our measurements indicated that the sand was well oxygenated. A stepwise regression analysis of our environmental variables and egg developmental rates in Delaware showed that the water content of the sand was the best predictor of developmental rate.

Our different results from the two sites mean that the Delaware crabs do not have the narrow restrictions on nesting that Florida crabs have (Figure 2.4). Horseshoe crabs in both Florida and Delaware nest in locations that maximize egg development; but in Florida 95 percent of the couples nested in a narrow section (40 percent) of the beach,

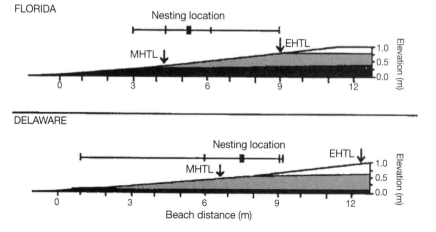

FIGURE 2.4 A comparison of the pattern of nesting of Florida Gulf coast and Delaware Bay horseshoe crabs. "Nesting location" shows the region of the beach over which nesting occurs (mean ± S.D. and range) in Florida and Delaware. MHTL is the mean high tide line and EHTL is the extreme high tide line. Moisture content of the sand is shown as saturated (> 18 percent = black area), moist (5–17 percent = gray area), and drying (< 5 percent = white area). *By permission from Penn and Brockmann, 1994.*

NESTING BEHAVIOR

whereas in Delaware pairs nested over a wide portion (61 percent) of the beach. In Florida, the concentration of oxygen in the sand increased slowly with distance up the beach and did not reach 1 ppm until 3 m, whereas in Delaware 1 ppm occurred everywhere above 1 m (see Figure 2.4). This means that the Delaware crabs can nest lower on the beach without suffering reduced development due to a lack of oxygen. At the top of the beach in Delaware, the moisture content is low, falling to 5 percent at 4 m, whereas in Florida the moisture content remains high (> 5 percent) throughout the area in which the crabs nest.

Conclusions

This study helps to explain, then, why it is that horseshoe crabs nest in the high intertidal, despite its many costs, and why in some areas they nest only on the highest tides of the year (Brady and Schrading, 1997). This hypothesis also provides a possible explanation for why favored nesting sites vary among populations (Shuster, 1982). In Florida (Rudloe, 1980; Cohen and Brockmann, 1983) and Massachusetts (Cavanaugh, 1975; Barlow et al., 1986; Badgerow and Sydlik, 1989), horseshoe crabs nest in a narrow band at the very top of the high tide line, the only area on the beach where their eggs develop well. In the Delaware Bay, however, they nest over most of the beach (Shuster, 1982; Shuster and Botton, 1985; Botton et al., 1992) because a much wider portion of the beach provides suitable conditions for egg development.

Similar arguments can be used to explain why horseshoe crab nesting is spotty, with crabs nesting on some beaches but not others (Shuster, 1982; see also Chapter 8). For example, at Seahorse Key, horseshoe crabs nest in decidedly lower numbers on the western and eastern ends of the island, where the beach is not as wide or steep and thus has reduced oxygen content when compared with the south-facing beaches. At Cape Henlopen, Del., crabs nest in large numbers on offshore sandbars as well as along the shoreline. Careful inspection reveals that these are areas of light, well-aerated sand, even though they have little slope. Our hypothesis is that horseshoe crab females choose beaches with conditions that favor rapid development of their eggs.

Differences in oxygen and moisture in the sand are likely due to differences in beach morphology: sand grain size and wave action determine the drainage patterns of a beach, which greatly affect its oxygen and water content (Gordon, 1960; Brafield, 1964; Eagle, 1983). In Florida the beach sand is fine and therefore has poor drainage, whereas in Delaware the beach sand is coarse (Maurmeyer, 1978) and water flows through it rapidly. Of course, the tides are higher in Delaware than in Florida; but still, the water content of the beach remains lower because of the better drainage. Fine-grained sand like that in Florida also has more surface area for microbial growth. Such growth depletes oxygen and the sand becomes anaerobic (Eagle, 1983; Boaden, 1985). This means that the bottom of the poorly drained beaches in Florida are so high in hydrogen sulfide (swamp gas and gray color due to high microbial activity) that these beaches kill eggs, whereas Delaware's better-drained sands do not support such high levels of microbial growth and therefore the crabs can nest lower on the beach. We hypothesize that differences in crab nest-site selecting behavior are due to differences in beach structure (sand grain size; wave size). Arguing strictly from beach structure, McLachlan and Turner (1994) show that, for animals that depend on high oxygen and moisture conditions, the best area in which to live is a layer extending from just below the surface near the mean tide line to well below the surface at the high tide line. This is exactly where horseshoe crab eggs develop (see Figure 2.4). Below this area, the beach is stagnant; above, the beach is too well drained.

How Do Horseshoe Crabs Find Favorable Beaches?

A study by Botton et al. (1988) demonstrated that crabs are capable of discriminating between beaches of different geochemical regimes. In the Delaware Bay, they avoid beaches that have a layer of peat below the sand. Crabs consistently avoid areas with hydrogen sulfide, indicating low oxygen conditions. Botton et al. (1988) suggest that the crabs might have hydrogen sulfide receptors to detect poor nesting areas. Alternatively, researchers studying respiratory physiology have found external oxygen receptors on the ventral surface, between the legs and on the gills (Waterman and Travis, 1953; Thompson and Page, 1975; see also Chapter 9); carbon dioxide receptors on the flabellum (Waterman and Travis, 1953); and chemosensory structures of unknown function on the flabellum (Hayes, 1971). The legs (particularly the chelae and gnathobases) have exceptionally large numbers of sensory inputs (comparable to antennae) including chemoreceptors (Barber, 1951, 1953; Hayes and Barber, 1982). The animals have a well-developed corpora pedunculata in the brain for processing chemosensory information (Wyse, 1971). Experiments now need to be conducted to determine how crabs locate preferred areas on beaches for nesting and what cues they are really using.

Why Does the Timing of Nesting Differ between Sites?

The timing of nesting behavior differs between sites. In Delaware Bay and Massachusetts waters (Shuster, 1953, 1982; Cavanaugh, 1975; Barlow et al., 1986; personal observation), crabs are known to nest during neap tides (in other words, not on the higher tides of the new and full moon); but only rarely are they seen during neap tides in Florida (Rudloe, 1980; Cohen and Brockmann, 1983; personal observation). There are a number of possible reasons for this, but our results (Penn and Brockmann, 1994) suggest an adaptive explanation (Figure 2.5). In Florida neap tides are lower than in Delaware (tidal amplitude decreases with latitude in the eastern United States), which means that neap high tides rarely reach the aerobic zone of the beach where horseshoe crab eggs can develop. The only way that Florida crabs can nest above the anaerobic zone is to nest on the new or full moon high tides (spring tides). In the better-drained soils of Delaware, even neap high tides allow the crabs to nest in well-aerated sand. Furthermore, wind direction and velocity may have a strong effect on tide height. At Seahorse Key, for example, when the wind is from the north, the tides are always lower than predicted and the crabs do not nest (Brockmann, in preparation).

There is another interesting difference between Florida and Delaware. In Florida the two high tides each day are nearly equal in height and the crabs spawn on both (Rudloe, 1980; Cohen and Brockmann, 1983). However, farther north the two daily tides are disproportionate in size (indeed, the highest neap tides are higher than the lowest spring tides) and northern horseshoe crabs prefer to nest on the higher of the two tides (see Figure 2.5; Barlow et al., 1986; personal observation).

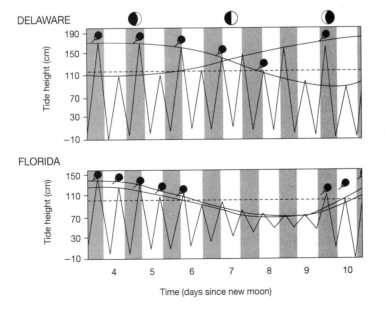

FIGURE 2.5 A model of the timing of spawning in horseshoe crabs. Tidal amplitude decreases during neap tides and then increases during spring tides (based on NOAA data). Tidal inequality is greatest during the spring tides. To deposit eggs in the aerobic regions of the upper beach (above the dotted horizontal line), northern horseshoe crabs can nest on virtually any high tide. Southern populations, however, can nest on only the highest tides, which occur around the time of the new and full moons. *By permission from Penn and Brockmann, 1994.*

During the spring and early summer when the crabs are nesting, the higher of the two daily tides occurs at night. Thus, the northern crabs are much more nocturnal than the southern ones. We hypothesize that the local rhythms of the tides and the location of the aerobic zone on the beach enables us to predict the different timing patterns, as well as the location, of horseshoe crab nesting.

Populations differ in another detail of timing. In Florida the crabs arrive on the beach up to 2 hours before the maximum high tide and stay for about 2 hours after, whereas in Delaware peak numbers of crabs arrive only after the maximum high tide and remain for about 4 hours (Figure 2.6; Penn and Brockmann, 1994). Like the Delaware populations, Massachusetts crabs also spawn largely after the maximum high tide (Howard et al., 1984). There are several possible explanations for nesting during the receding tide:

(1) The incoming tide is called the scour phase because sand tends to be eroded whereas sand deposition occurs on the receding tides (Strahler, 1966). This may mean that nesting on the receding tide would minimize egg loss due to erosion. However, we had little erosion of eggs at any locality so this seems an unlikely explanation.

(2) The mixed sand and gravel sediments in Delaware make digging relatively more difficult, so crabs may reduce the effort needed to dig by nesting during the receding tide when the sediments are more saturated with water and hence softer.

(3) Nesting on the receding tide might also minimize egg loss caused by disturbance from other nesting crabs, which appears to be a major source of egg loss in the Delaware Bay where particularly large

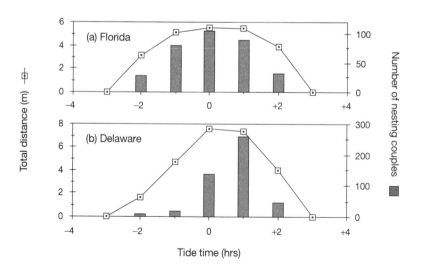

FIGURE 2.6 A comparison of the timing of peak horseshoe crab spawning at two sites, (a) Seahorse Key, Fla., and (b) Cape Henlopen, Del. The bars indicate the mean number of nesting pairs and the points indicate the mean tidal distances recorded during three high tides. Tide time indicates the hours before (−) and after (+) the maximum high tide. Tidal distance refers to the tide height or the excursion of the tide up the beach. *By permission from Penn and Brockmann, 1994.*

NESTING BEHAVIOR

numbers of crabs nest at the same time (Barash, 1993; personal observation). By nesting only on the outgoing tide, the crabs can prevent their nests from being dug up by other females who are nesting on the same tide. Because tidal excursion is so much greater in northern areas, crabs have more time to nest on each tide than they do in Florida. Shuster and Botton (1985) estimate that females lay about twenty-two clusters of eggs to complete their nesting; if they lay one cluster every 10 minutes, this means that they should be able to lay all their eggs in about 4 hours. In Delaware the receding tide often lasts for 4 hours, but not in Florida. In Florida where there are fewer competing crabs, the duration of adequate tidal height is so short that the crabs must begin nesting before the maximum high tide and continue nesting afterward if they are to lay all their eggs.

How Do Horseshoe Crabs Know When and Where to Come Ashore?

Horseshoe crab nesting is strongly associated with lunar periodicity, but that does not mean that the crabs use the phase of the moon as a cue. In St. Joseph Bay, Fla., the highest water is not always associated with the new and full moon (wind direction and velocity can have a profound effect on tide height), yet the horseshoe crabs still arrive for nesting on the highest tides (Rudloe, 1985). This suggests that the crabs are responding to water depth, or perhaps to the velocity of the incoming tide (because tidal currents are faster when tides are higher), rather than following a precise lunar rhythm (Brockmann, in preparation).

Rudloe and Herrnkind (1976, 1980) ran a series of experiments in which they evaluated the response of adult crabs to wave surge, current, and bottom slope, cues that crabs might use to detect tide height as well as the location of a beach. Crabs that were released offshore moved directly into the oncoming surge, suggesting that this was an important cue, although it is possible that they face into the surge to avoid being overturned. This second possibility was rejected after a study conducted in a wave tank. Directional orientation occurred only in the presence of a traveling wave and was lost when the wave was stationary. So horseshoe crabs follow wave surge and current. This study also showed that crabs followed bottom slope toward beaches, and Botton and Loveland (1987) also showed that crabs leaving the beach were disoriented unless they were on a slope of 5–10°. Horseshoe crabs are known to possess mechanoreceptors and proprioreceptors that might be capable of detecting water movements and beach slope (Hayes and Barber, 1967; Eagles, 1973a, b), but no one knows the specific cues they might be using or how the cues are being received. In particular, we do not understand how they might respond to tide height, staying away from the beach on a spring tide (new or full moon high tide that is predicted to be exceptionally high) when the wind prevents the tide from rising.

Do Horseshoe Crabs Return to Their Natal Beaches?

Many think it likely that horseshoe crabs return to breed on the beaches on which they were born (Teale, 1957; Sokoloff, 1978), but there is no direct evidence to support this idea (see Chapter 8). My long-term studies show that the adult crabs return year after year to the same breeding beaches, but a few tagged individuals have been seen on other beaches in the area (1–5 km away). The long-term tagging study of Swan (1996) in the Delaware Bay has found that many crabs move considerable distances, including from one side of the Bay to the other. Probably the best evidence is that horseshoe crabs living on different beaches differ in size, morphology, isozymes, and behavior (Shuster, 1955; Sokoloff, 1978; Riska, 1981; Burton, 1983; Rudloe, 1985; Saunders et al., 1986; Miyazaki et al., 1987; Palumbi, 1992; Botton and Loveland, 1992). Such differences in phenotype suggest that there are genetic differences between geographically separated populations. This can occur only if there are low migration rates between those populations (Pierce et al., 2000). With the new genetic techniques that are now available, it should be possible to study this problem in much greater detail.

There are no studies that tell us anything about how crabs find their natal beaches when far out at sea. Do they use the sun or a magnetic compass like many other species of migrating animals such as sea turtles? Do they use visual cues when in familiar areas? Or do they use odor cues, like salmon (Hasler, 1960)? There is still so much to be learned!

■　　■　　■

We have hypothesized that horseshoe crab females choose nesting sites that will maximize their success by providing adequate oxygen and moisture for their developing offspring. The basis for the hypothesis comes from two sites and it needs to be tested elsewhere. It would be particularly instructive to study nesting in areas without tides (microtidal areas) such as bays and inlets (Rudloe, 1985; Ehlinger and Bush, 2001). When they are nesting in these areas, do the crabs choose times and locations that can be predicted based on the oxygen and moisture regimes of the beaches on which they nest? If the hypothesis is supported, it seems likely that a wide range of traits can be explained. Despite its many costs, high intertidal nesting is favored because this is the area of the beach that provides suitable conditions for offspring; the timing of spawning allows the animals to nest in areas that provide suitable conditions; even the timing of arrival within a single tide might be predicted on the basis of the availability of suitable conditions for egg development. Our hypothesis might also provide an explanation for the distribution of crabs on different beaches. It may even provide an explanation for the biogeography of horseshoe crabs (Shuster, 1979; see also Chapter 8), or for why there are no crabs in most of the Gulf of Mexico. I suspect this may be because the tides in the western Gulf are much smaller than in Florida or the western or northern Yucatán Peninsula. Support for our adaptive hypothesis will also lead to many additional proximate questions. For example, if females nest when the tide or water levels are high, how do they know? What are the sensory systems that they are using to determine tide height?

The nesting behavior of horseshoe crabs illustrates a number of general principles about behavior and how we use the scientific method to study behavior. Throughout we have posed two quite different sorts of hypotheses: proximate hypotheses about how animals behave as they do, and ultimate or adaptive explanations about why animals behave as they do, in other words, the selective advantage for showing particular patterns of behavior (Exhibit 2.1). A complete understanding of behavior (or indeed any trait) requires that we have answers to both proximate and ultimate questions. Proximate and ultimate questions tend to be studied by different types of investigators, such as neuroethologists versus behavioral ecologists; but it is clear that at each step in our study, our understanding would be improved if we were able to integrate the two types of information (see also Chapters 3 and 4 for an integrative approach to the study of horseshoe crab behavior).

EXHIBIT 2.1

QUESTIONS ABOUT BEHAVIOR

The questions we pose about behavior are of two types (Alcock, 2001). Some are about how individuals decide (see Exhibit 2.2) what to do next. For example, How do females synchronize nesting? is a question about the cues that females use to coordinate their behavior (for example, all females nest when the environmental conditions are just right, or females use a chemical, social signal to which other females respond). These are referred to as "proximate" questions; they are questions about the immediate factors influencing the behavior of an individual. Proximate questions may be about external or internal cues (such as the full moon causes the release of a hormone that causes females to come ashore).

But we are also interested in posing another sort of question: Why do females nest synchronously? What are the selective advantages that favor females nesting together? These sorts of questions are referred to as "ultimate" and they are questions about how natural selection favors particular patterns. Although questions are referred to as "proximate" and "ultimate," one kind is not better than the other: a complete understanding of behavior always requires answers to both sorts of questions. So, for example, when we study nesting at the high tide line, we must ask both about the cues females use to come ashore on the high tide and why the decision to nest on the high tide is adaptive (or what makes individuals who lay eggs at the top of the tide line more successful than those who do not). Both questions are equally crucial to our understanding of the behavior.

EXHIBIT 2.2

DECISION MAKING IN ANIMALS

It may sound a little funny to say that an animal makes a decision, particularly an animal like a horseshoe crab. The reason is that you are presuming from the use of this word that the behavior involves the same underlying processes of decision making as in humans, in other words, conscious choice. But the word "decision" is used in a technical sense in the field of animal behavior to mean that the animal has committed itself to a particular course of action for a period of time. The term is meant to be descriptive and not to imply anything in particular about the underlying decision-making mechanisms (such as whether the decision is based on a programmed rule or on past experience).

When a decision is made, it commits the animal to spending time or resources that could have been used in other ways, so the path chosen must be advantageous. Behavioral studies identify decision points, describe the decision-making rules that control what the animal chooses to do, and seek to understand why one behavior is chosen over another. For example, when a female horseshoe crab comes ashore, she could go to the top of the tide line or she could nest lower on the beach, so a decision rule tells her what to do. We can examine the proximate cues she uses to make this decision; for example, the depth of the water may trigger nesting. We can also examine the selective advantage to females for making a particular choice; for example, differences in the height of nesting on the beach may affect reproductive success through differences in rates of egg development. The study of behavior is a study of decision making.

STUDYING BEHAVIORAL ADAPTATIONS

EXHIBIT 2.3

A modern research program in animal behavior operates under the assumption that behavior has evolved. By that we mean that the process of natural selection has been operating on the behavior and favored those decisions that make the animal well suited for its environment. If there is variation in a character, if some of that variation is due to genetic differences between individuals, and if there are differences in the reproductive success of the different variants, natural selection occurs. For example, if the eggs that were laid high on the beach desiccated and died, those females that chose to nest high would leave fewer offspring. This would mean that the proportion of the population that nested high would decrease relative to the proportion that nested lower. It is in this way that favorable traits evolve and that animals become adapted to their environment.

We measure the success of a trait as fitness, the average lifetime reproductive success of individuals possessing the trait. But lifetime reproductive success is difficult to measure for an animal that lives many years and breeds many times. For this reason we usually measure short-term gains that we hope will provide reliable estimates of lifetime reproductive success. When we do this, the language of economics is often used. If a trait is adaptive, then clearly the gains (in number of eggs laid; number of young reared) must outweigh the opportunity costs or what the animal would have gained had it made a different decision.

CHAPTER 3

MALE COMPETITION AND SATELLITE BEHAVIOR

H. Jane Brockmann

One of the most intriguing aspects of the horseshoe crab nesting phenomenon of the Delaware Bay is the large number of males that are found on the beach. Some arrive attached to females, but other males come alone and crowd around the nesting couples so large spawning groups often form (see Exhibit 3.1). The sex ratio on the beach varies markedly, but on some tides there can be twenty males for every nesting female. Why are there so many males on the beach? How do males find females? How does fertilization take place? Why do unmated males congregate around nesting pairs? How do they find them? What is the mating system in this species? (See also Exhibit 2.1.)

These are the questions that have driven my research program for the past 12 years. Although I've found some answers, I still have many questions. This chapter describes the mating behavior and mating system of horseshoe crabs as well as the intense interactions among males.

I was drawn to the study of horseshoe crabs for several reasons. In part I liked the idea of studying a species in which large numbers of animals arrived in a predictable location at a fairly predictable time (see Chapter 2). I was also intrigued by the fact that, despite their abundance, little was known about the behavior of these ancient creatures, which are in many ways quite different from any other group on the planet (see Prologue). Furthermore, at the time that I started this project, sexual selection was gaining importance as an area of interest in the field of animal behavior. Sexual selection is the study of mating systems, mate choice, and differential mating success, usually differences among males in their ability to get mates.

Here was a species with a highly male-biased sex ratio at the breeding site (called the "operational sex ratio") visibly competing with one another—but I wasn't sure about the goal of the competition. I also knew enough about these animals (as a result of undergraduate class field trips) to realize that the males showed two different mating patterns, a phenomenon referred to as "alternative reproductive tactics." Two forms of mating for one sex is a bit mysterious from an evolutionary point of view, because one would expect that one or the other tactic would have higher success and the less successful pattern would evolve out of the population. So in evolution the maintenance of variability always requires a special explanation. This was a problem that I had studied before (Brockmann, 2001), and I was eager to find another system in which I could evaluate hypotheses about alternative reproductive tactics. And finally, I was drawn to the study of horseshoe crabs because I was just intensely curious about the mating behavior of this ancient, large invertebrate that crawls from the sea to spawn.

Mating by Attached Pairs

Male horseshoe crabs attach to the posterior end of the females' carapace (Figure 3.1A) using their first pair of legs (or pedipalps), which are enlarged and modified into bulbous claws for grasping the female (Figure 3.1C; Snodgrass, 1952). Nothing is known about the physiology of grasping; but a lock mechanism in the tibial and tarsal muscles seems likely, similar to what clams use to keep their shell closed with little expenditure of energy (Shuster, 1993). An attached male is difficult to dislodge, will continue to hold onto the female even when overturned in waves, and on occasion will lose a claw rather than release his grip. Pairs may even remain together over the winter, buried in mud banks (Shuster, 1955; Barlow, personal observation). Males that are attached (they are said to be in "amplexus") ride above the female's telson and so it is unlikely that they can feed efficiently. Attached males also cannot bury themselves in mud at low tide in the way that females and lone individuals do, so they probably suffer greater exposure to sunlight and to the rain of aquatic organisms seeking substrates on which to settle. Barnacles, algae, slipper shells, mussels, tunicates, and encrusting bryozoans are all more common on the backs of males than on females (personal observation). So, amplexus has some drawbacks from the male's point of view.

The male grasps a pair of posterior opisthosomal projections (POP) at the back of the female's carapace (Figure 3.1B). Older females show considerable wear as a result of males holding onto them: their POP be-

FIGURE 3.1 The anatomy of horseshoe crab mating. During amplexus in *Limulus polyphemus* (A; the female's telson is removed for clarity), the male grasps the female by a pair of posterior opisthosomal projections (B; indicated by arrows on A) with his first pair of legs, or pedipalps (C), which are modified into a grasping claw. The male grasps the posterior opisthosomal projections (at arrows in A and B) of the female, which may become eroded and broken off (original outlines noted with dotted lines in B). The female's opithosoma shows a mating scar (shaded areas), where it eroded due to the frontal area of the male resting on the female during amplexus. Two aspects of the male's clasping claws (C) are shown: on the right, the left clasper in the "virgin" condition with an attenuated fixed claw that breaks off (ac) during mating. The tibial boss (tb), tibial process (tp), and movable chela (the tarsus) form a three-point grasp, with the boss reducing the amount of roll in an otherwise two-point clasp; on the left, the right clasper after mating. *(A) from Shuster et al. (1961), drawing by Charlotte P. Randall; (B) from J. Brockmann, 1990, with permission; (C) courtesy of C. N. Shuster, Jr., from an unpublished manuscript on the evolution of amplexus.*

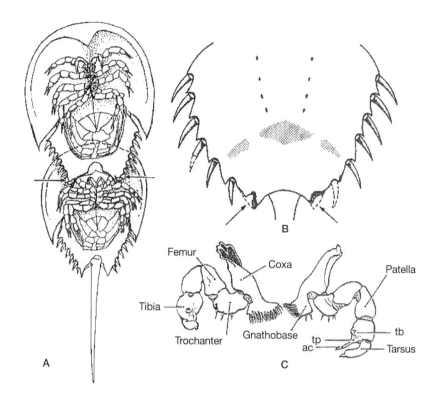

come worn (a characteristic bite is taken out of the outside edge) or broken and a "mating scar," or worn area, develops on the carapace at the center of the opisthosoma where the front of the male's carapace contacts the female (see Figure 3.1C; Shuster, unpublished manuscript). In a few very old females, the opisthosoma is completely worn through (personal observation). Occasionally males hold other parts of the female's anatomy, such as her movable spines, and they often hold her telson with their walking legs. The pair may remain attached for weeks (we have observed up to 51 days in Florida), but our experience is that most stay attached only for a few days during one tidal cycle (Brockmann and Penn, 1992).

As with so many other aspects of horseshoe crab biology, the method of fertilization is unique. No other living arthropod has true external fertilization or the kind of spermatozoa (Fahrenbach, 1973) that horseshoe crabs have. In all other arthropods, the sperm are transferred in packets or deposited in the body cavity of the female; but in horseshoe crabs, the male fertilizes the females' eggs externally while they are being laid. His sperm travel in water exposed to the external environment and are free-swimming just as they are in clams and ma-

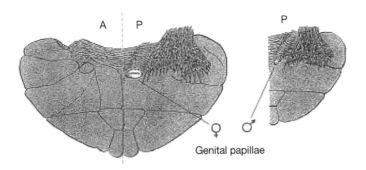

FIGURE 3.2 Eggs are released from paired genital papillae. The posterior surface of the operculum bears the genital pores. The operculum is attached to the ventral part of the prosoma in the region of the hinge. This split diagram shows the anterior face (A) of the female's operculum on the left and the posterior face (P) on the right. The posterior face of the male's operculum is shown, extreme right. *Drawing by C. N. Shuster, Jr.*

rine worms (Jamieson, 1987). This is one of the reasons that I find these animals so fascinating: horseshoe crabs provide an opportunity to understand reproduction in an ancient arthropod lineage before internal fertilization evolved. This form of fertilization may also affect other aspects of the mating system, so it is interesting to compare the behavior of horseshoe crabs with unrelated, externally fertilizing species such as fish and frogs.

Mature eggs are stored in the female's body cavity and released from paired gonopores located under the operculum on the ventral surface (Figure 3.2). Testes are distributed over a large portion of the male's body, and sperm are released from paired genital pores also located under the operculum. Sperm must travel from the attached male to the female to reach her eggs, a distance of at least 20 cm (see Figure 3.1A). Alternatively, sperm could be released over the eggs as the female moves forward in the sand; but data from our paternity analyses, discussed later in this chapter, suggest that this is unlikely. Normal respiratory currents move from front to back in horseshoe crabs (see Figure 7.5; Barthel, 1974), but during spawning these currents reverse (Sekiguchi, 1988; personal observation). Currents, which are probably produced by both the male and the female, carry the sperm forward to the female's eggs, but nothing is known about the details of this process. Nonetheless, it is highly successful: nearly all eggs that females lay are fertilized (Brockmann, 1990). (See Exhibit 3.2 for details on the process of fertilization.)

What Explains the Biased Operational Sex Ratio?

Anyone who has ever watched a mass spawning of horseshoe crabs in the Delaware Bay is struck by the fact that there are so many more males present on the beach than females. In the Delaware Bay, the operational sex ratio is more male biased than elsewhere probably due to human exploitation because the larger females are taken preferentially

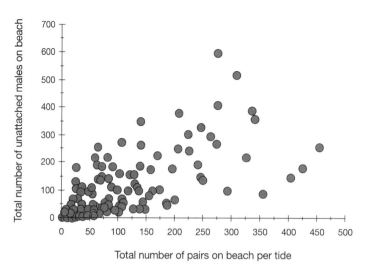

FIGURE 3.3 Operational sex ratio of horseshoe crabs from a nesting beach in Florida during 417 tides. Each data point shows the number of nesting pairs on the beach at Seahorse Key for one tide and the associated number of unattached male horseshoe crabs also present on that tide.

for bait and blood (see Chapters 13 and 14). But even in Florida, where there was no human exploitation of horseshoe crabs, and in other populations, where as far as we know there are equal numbers of males and females in the population as a whole (Koons, 1883; Shuster, 1950), many more males are found on or near the nesting beaches than females (Figure 3.3). In my study of marked animals at a site along the Gulf coast, Seahorse Key, in Levy County, Florida (Brockmann and Penn, 1992), we saw over half of our marked males a second time in the same year, whereas only about a third of the marked females were seen again. Males were not only seen more frequently but were seen over more days (15 days) than females (7 days). This means that, on average, females return repeatedly over only one tidal cycle, whereas males return over two or more cycles (see Figure 3.3). So males outnumber females largely because they come to the beach more often. These biased operational sex ratios set the stage for intense competition among males.

What Are the Unattached, or Satellite, Males Doing?

If you have ever watched horseshoe crabs nesting on the shore, you have probably seen the unattached, or "satellite," males that crowd around the spawning pairs (see Figure 2.1; Shuster, 1953; Shuster and Botton, 1985; Barlow et al., 1987; Sydlik and Turner, 1990). These satellite males take up particular positions around some pairs, strongly favoring a spot over the female's incurrent canal (Figure 3.4; I call this

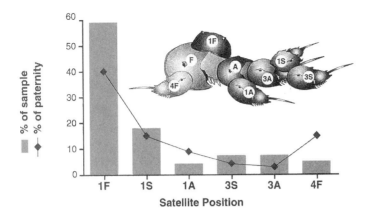

FIGURE 3.4 Mating position and paternity. Unattached (satellite) males (1A, 3A, 1S, 3S, 1F, and 4F) favor particular locations around the nesting pair (F = female, A = attached male, S = satellite). Position 1 is located over the incurrent canal of the female (1F), attached male (1A), or satellite male (1S). The bar graph shows the frequency with which the satellites were found in the different positions. The line shows the paternity associated with the different positions when two to four satellites are present. *By permission from Brockmann et al., 2000.*

"position 1F"). When crabs come ashore, they are at some risk from predation and from being overturned and stranded above the tide line (see stranding, page 65; Shuster, 1982; Botton and Loveland, 1989), so there must be some advantage. I had three hypotheses for what the unattached males were doing: finding unattached females, taking over attached females, and engaging in sperm competition.

Finding Unattached Females Hypothesis. Could unattached males be looking for unattached females? This seems unlikely because there are so few lone females on the beach (Loveland and Botton, 1992). The ones that are there tend to be individuals with worn or broken posterior opisthosomal projections (POP) (Brockmann, 1990), and I suspect they are females who lost their attached males on the way in to shore. These lone females remain on the beach for about half an hour and then leave without laying any eggs, sometimes returning later with an attached male. Interestingly, unattached males pay little attention to these lone females. On the few occasions I have seen males around lone females, they have taken up satellite positions and have not attached, even though no attached male was present.

Takeover Hypothesis. Another possible explanation is that unattached males come ashore to take over females from attached males. Satellite males push and work their way underneath the front margin of attached males (Figure 3.5), but they rarely take over the attached male's position (less than 2 percent of nesting pairs in Florida; Cavanaugh, 1975; Rudloe, 1980; Cohen and Brockmann, 1983). When a takeover occurs, the satellite male works his way underneath the attached male

FIGURE 3.5 Competitive behavior between attached and satellite male horseshoe crabs. The female is the larger crab in the foreground. (A) A single satellite male contacts a nesting couple, bracing himself against the sand with the base of his telson and digging underneath the attached male. (B) The satellite male pushes on the attached male and the attached male responds by tipping his prosoma in the direction of the intruder. (C) The satellite male pushes underneath the attached male and positions himself under the attached male's prosoma over the female's incurrent canal. (D) Occasionally, the satellite pushes entirely underneath the attached male, raising his prosoma and lifting the attached male upward. Sometimes this results in a takeover of the attached male's position by the satellite. *By permission from Brockmann, 1990.*

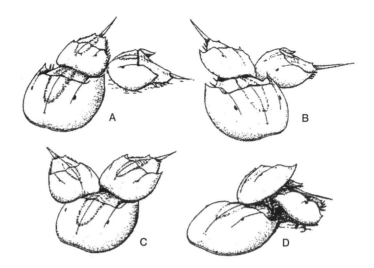

from behind, moving forward toward the female, or he moves from the side toward the female's telson (see Figure 3.5D). This has the effect of raising the attached male, who continues to cling to the female.

In about half the cases in which the satellite gets into this position, the female pulls herself out of the sand and moves out to sea, shedding the satellites as she goes. In a few cases, the satellite, which is not as well anchored as the attached male, will be washed out from under the attached male by a wave. In the few takeovers I have seen, the attached male is held in the air for 5 to 10 minutes before he releases his grip with one claw, the satellite grasps the female's free POP and the attached male releases the other claw, and the satellite grasps the second posterior projection. The now-detached male moves immediately into a satellite position. Once in a while you will see two attached males, one holding onto each POP, but this is very rare (about 1/1,000). So, satellites do take over the positions of attached males on the beach, but the behavior is rare and certainly does not account for the large number of unattached males that come ashore.

Sperm Competition Hypothesis. A third possibility is that unattached males come ashore to compete for fertilizations by engaging in sperm competition with an attached male. Satellite males release sperm (Barlow et al., 1986; Brockmann, 1990), but until recently it was not known whether they fertilized any of the eggs. My colleague Wayne Potts and I used a standard paternity analysis technique by comparing the DNA of all the participants in a nesting group (Brockmann et al., 1994, 2000) (Exhibit 3.3). The results were quite interesting.

When the attached male was the only one present, he fathered all the offspring; but when just one satellite was present, the attached male's success dropped to 51 percent and the satellite fathered on average 40 percent of the progeny (the remainder being unresolved; Brockmann et al., 1994) (see Figure 3.4). When more satellites were present, they did even better: in groups of two to four, on average 74 percent of the young were fertilized by satellite males (Brockmann et al., 2000). The average paternity of each satellite was affected by the presence of others: the success of two satellites spawning with a pair (40 percent each) was higher than the success of three (21 percent each) or four satellites (20 percent each). So when two satellites were present, they reduced the success of the attached male, whereas three or more satellites reduced the success of other satellites.

These numbers surprised us. In other externally fertilizing species with sperm competition in which paternity had been established (only fish), the average success rate of individual satellite males was much lower (5–17 percent; Maekawa and Onozato, 1986; Hutchings and Myers, 1988; Jordan and Youngson, 1992), although in a few species satellites are suspected of fertilizing many eggs (Dominey, 1981). The high success of satellite male horseshoe crabs suggests that it is advantageous for them to come ashore and compete for fertilizations with the attached males and other satellites even when this behavior involves considerable risk.

Do Unattached Males Fertilize Eggs outside Spawning Groups?

External fertilization would seem to provide opportunities for males to fertilize eggs even when they are not present at the time the eggs were laid. Our paternity analyses revealed a few cases. We have several possible explanations for this intriguing result (in each case, the female we observed was the mother). We knew that horseshoe crab sperm could remain viable for at least 96 hours at 4°C (Brown and Knouse, 1973). I ran this study again at a more realistic temperature (28°C) and found that sperm continued to remain viable in seawater for 20 hours after ejaculation (if they do not come in contact with eggs).

Our first explanation for the unexpected fertilizations is that sperm may survive for some time under the concave carapace of a female. However, I never found that a male fertilized eggs after he had left a spawning group, so this seemed an unlikely explanation. The second explanation for the odd fertilizations in our sample was that satellite males released large quantities of sperm so the beach is probably awash in sperm. As the tide recedes or as waves come ashore, seawater percolates through and aerates the sandy intertidal zone of the beach (Riedl and Machan, 1972; McLachlan and Turner, 1994). The tiny sperm contained in this water may be drawn through the sand to a depth that should allow eggs already in the sand to be fertilized. I conducted a study on unfertilized eggs and found that they retained their ability to be fertilized for at least 40 minutes at 28°C, so this is a possible explanation for the odd fertilizations we observed. Because the sperm are so long-lived, it is even possible that sperm trapped in the sand from previous tides may fertilize newly laid eggs. Because the rate of percolation increases with particle size, this may be a more effective means of fertilizing eggs in Delaware, where particle size is larger than in Florida (Penn and Brockmann, 1994) and where there are many more unattached males.

Our third explanation is based on the observation that after a nesting pair has pulled out of the sand, unattached males often move slowly over, turn, circle, and repeatedly cross the depression that was left in the sand by the recently departed pair (Cohen and Brockmann, 1983). The males that do this may be satellites that were involved in the nesting or unattached males that come upon the depression. I suspect that these circling males are releasing sperm that may fertilize eggs in the soft sand below (Cohen and Brockmann, 1983; Barlow, personal communication; personal observation). Clearly, there is much more to be discovered with more detailed paternity analyses.

Do Satellites Compete for Positions That Increase Fertilization Success?

Unattached males have strong preferences for particular positions around the female (see Figure 3.4). Satellites bend, push on, and attempt to dig under the attached male or other satellites (Figure 3.5). The attached male responds by pulling himself closer to the female and dipping his carapace to the side, which prevents the satellite from getting underneath (Figure 3.5A, B). Unattached males also move around the attached couple or they may leave, swim along the shoreline, and join another nesting pair. Once the unattached male gets into a position over the female's incurrent canal (position 1F) (Figure 3.5C), however, he usually remains there throughout the nesting, grasping the attached male or the female's movable spines.

As you might guess, we found that males in this position were much more successful than those taking up other positions around the female: males in position 1F have, on average, 40 percent paternity,

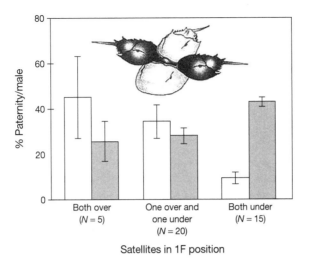

FIGURE 3.6 The effect of over and under positions on paternity. The light bars show the average paternity for the attached male and the dark bars show the average paternity for each satellite when both males are in the 1F position and are either over or under the attached male's carapace. The figure shows the male on the left in the under position and the male on the right in the over position. *By permission from Brockmann et al., 2000.*

overwhelmingly more than in other positions. Some satellites take up positions over the attached male's incurrent canal or over the attached male's telson, but they fertilize only 8 percent of the eggs in a clutch on average; satellites that take up similar positions on other satellites fertilize on average 12 percent. Although males do better when they are in particular positions in the group, the data show that if they change positions while a clutch is being laid, they will be much less likely to father the eggs of that clutch (mean success of animals that moved is 10 percent and of those that did not move, 33 percent; of course, this is only a correlation, so it could be that the animals moved because they were in a less advantageous spot).

When in position 1F, males push and shove on the attached male to get under the front margin of his carapace (see Figure 3.5C). Now we can understand why (Brockmann et al., 2000). Satellites that are over the attached male's carapace have lower paternity (29 percent) than those that are under (42 percent) (Figure 3.6). We can also see why satellites work so hard at getting into the under position: when two satellites are in the over position, the attached male is as successful as when only one satellite is present, but attached male success drops precipitously when one (34 percent) or particularly both (10 percent) satellites are in the under position. This means that when both satellites are under, their average success (43 percent each) does not differ from the average success of a single satellite spawning with the attached male in that position (43 percent; Brockmann et al., 1994). These data also provide insight into why takeovers are so rare: satellites in the right position do as well or better than the attached males!

We found considerable variation in satellite male success, ranging from 0 to 100 percent; attached male paternity also varied from 0 to 100 percent. Strong differences exist between the two males in position 1F: the average paternity for the more successful of the two satellites is 56 percent and the less successful male in position 1F averages 12 percent of the eggs fertilized, which is not significantly different from the success rate of the attached male. What accounts for the variation in success between satellite males and among attached males? The amount of time that the satellite male has been with the couple, conditions on the beach (for example, waves or currents), male condition or size, and the behavior of the female and other males may be involved; but we have little evidence that any of these accounts for differences in paternity. The success of attached males is slightly higher when they nest with larger females, but no other attribute of females or their nests (for example, egg depth, clutch size, or time to complete the clutch) is correlated with male success. We simply don't understand as yet all the variation in paternity.

Do Females Gain from Mating with Multiple Males?

What about the females? Females appear to have no influence over the number of satellites around them: they are usually well buried with all the male-male competition going on overhead. However, the high fertilization success of satellites, the fact that females may be able to control water currents around them, and the fact that females can shed satellites if they choose to leave the beach for a few minutes (personal observation) suggests that females might have some control. Do females obtain some benefit from satellite male fertilizations? There are numerous adaptive explanations in the literature to account for multiple mating by females of other species (Maynard Smith, 1978; Halliday and Arnold, 1987; Ridley, 1988; Westneat et al., 1990; Watson, 1993). While many of these explanations do not apply to horseshoe crabs, a few seem possible. The first can be dismissed immediately: satellites do not improve the number of eggs that are fertilized (Brockmann et al., 1994). Second, females may allow multiple mating because, if they did not, there would be disruptive, time-consuming takeover fights among males, an explanation akin to the sexual harassment hypothesis of Alcock et al. (1977) and Svard and Wiklund (1986).

Third, females may benefit from sperm competition and multiple mating by the increased opportunity for their offspring to be fathered by the highest-quality males (Halliday, 1983; Petrie et al., 1992). When nesting occurs at low densities with few satellites, offspring are fathered by attached males that are younger and in better condition than satellite males (Brockmann and Penn, 1992). Satellite males gain substantial paternity at higher densities when they compete successfully with attached males and other satellites for position 1F (Brockmann et al., 2000). If there is a correlation between male viability and his competitiveness or the quantity or quality of his sperm, then females could also benefit from sperm competition by having better-quality offspring (Knowlton and Greenwell, 1984). Of course, it is possible that multiple-male spawning is strictly the product of male-male competition and females do not gain in fitness from spawning with more than one male. Whatever the explanation, it seems likely that there will be strong selection on males, particularly satellite males, to release large quantities of sperm under such highly competitive conditions (Parker, 1990).

Do Attached and Unattached Males Differ?

Whenever you see two types of behavior within one sex and one population of animals, such as satellite and attached male horseshoe crabs, you have an evolutionary puzzle: how can two different routes to the same functional end (fertilization) be maintained within the same population? We know there are costs and benefits associated with each tactic. It is unlikely, however, that the average success of each tactic would turn out to be exactly equal, so the behavior with the higher average success rate should evolve into the population. One possible explanation for the maintenance of the two types of males would be that males switch readily between attached and unattached behavior. However, most male horseshoe crabs consistently remain either attached or unattached (Figure 3.7; for males that we saw three times or more, 55 percent never changed tactics and a further 20 percent used the same tactic on 75 percent or more of the times that we saw them). So, attached and unattached are fairly discrete categories within a year. Furthermore, the proportions of attached and unattached males differ between populations (Rudloe, 1980; Botton and Loveland, 1992; Brockmann and Penn, 1992). How can such variability be explained? One possibility is that attached and unattached males differ.

In most species in which strong competition exists between males, the larger males mate more often and with larger and more fecund females than do smaller males. This may be because females prefer larger males or because larger males are more successful at male-male competition. However, this is not true of horseshoe crabs: there is no correlation between the size of the female and the size of her mate (Brockmann, 1990; Botton and Loveland, 1988, 1992). Furthermore, size is not

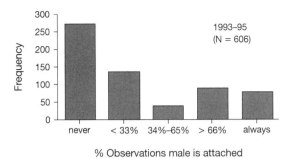

FIGURE 3.7 Frequency distribution of male mating tactics for a population from Florida. The data show the percentage of males that were always observed attached, sometimes observed attached and sometimes unattached, and never observed attached (in other words, always unattached). Data are for 606 marked males (1993 to 1995) that returned at least three times to the beach at Seahorse Key.

associated with whether a male finds a mate or not: attached and unattached males do not differ in any measure of size (carapace width, body weight or density, or claw size) (Table 3.1; Pomerat, 1933; Cohen and Brockmann, 1983; Botton and Loveland, 1988; Brockmann, 1990; Brockmann and Penn, 1992). We also found that when females change mates between tides, the replacement male is not larger than the original male (Brockmann, 1990; Brockmann and Penn, 1992). Similar results led Botton and Loveland (1988, 1992) to propose a "random collision model" in which pairing results from males simply grabbing the first female they encounter and coming ashore with her; if a male does not find a female, he comes ashore unattached. This is a reasonable hypothesis; but after measuring a large number of variables on hundreds of male crabs, my graduate student Dustin Penn pointed out that attached and satellite males look different.

The horseshoe crabs that you see on the beach come in different colors: some are light and some are dark (Wasserman and Cheng, 1996). Many are heavily encrusted with fouling organisms such as barnacles, slipper shells, mussels, other mollusks, and algae (MacKenzie, 1962; Turner et al., 1988; Patil and Anil, 2000), their eyes may be damaged or covered with fouling organisms, and their carapaces may be pitted from algal or bacterial infections (Hock, 1940; Leibovitz, 1986; Leibovitz and Lewbart, 1987; see Chapter 9). Lighter animals are less likely to be damaged or covered by fouling organisms than darker individuals (Table 3.2) and the lighter animals are also more likely to have a slimy surface. The surface of the horseshoe crab shell is covered with pore canals that contain mucous glands (Patten, 1912; Hanstrom, 1926), and some individuals, particularly juveniles, are covered with slime. This slimy mucus may be a mechanical barrier to pathogens (Stagner and Redmond, 1975; Pistole and Graf, 1986; Harrington and Armstrong, 2000), may have antibacterial properties, and certainly retards fouling by barnacles and other settling organisms (Barthel, 1974; Quigley, 1997; Harring-

MALE COMPETITION AND SATELLITE BEHAVIOR

TABLE 3.1 Differences between attached and unattached male horseshoe crabs in various characteristics. (P value given for Mann-Whitney U Test.) Data are from Cape Henlopen, Del. (1991–1993). Sample size is 194 for attached animals and 114 for unattached (except for *Crepidula* data, for which attached n is 74 and unattached n is 53).

| Measure | Mean for Males That Are | | |
	Attached	Unattached	P
Carapace width (cm)	20.8	20.8	0.76
Interocular distance (cm)	12.2	12.3	0.66
Weight (g)	1081	1074	0.73
Telson length (cm)	17.5	17.5	0.39
Telson height (cm)	1.17	1.14	0.04
Size of largest *Crepidula* (cm)	2.68	2.95	0.15

| | Percentage of Sample for Males That Are | | |
	Attached	Unattached	P
with few fouling organisms	55	68	0.06
with light prosoma	67	33	0.0001
with two good eyes	65	54	0.11
with two good POP	88	70	0.0001
with mucus on surface	44	17	0.0001

Source: Modified from Brockmann, 1996.

ton and Armstrong, 1999; Swain, personal communication). Daniel Gleeson (personal communication) has shown that the mucus (microsporum glycine) absorbs UVB (with significant absorption of ultraviolet [UV] light at 310 nm) and therefore is likely to prevent damage to the shell from sunlight, which is known to be harmful to other invertebrates and which penetrates water to considerable depth.

Many horseshoe crabs are also infested with the parasitic flatworm *Bdelloura candida* (colorplate 27; Wheeler, 1894; Chevalier and Steinbach, 1969), which causes damage by laying its eggs on the crabs' book gills (Watson 1980a, b; Groff and Leibovitz, 1982) and reduces respiratory efficiency (Huggins and Waite, 1993). Besides having more fouling organisms and being in worse condition than lighter individuals, darker animals also have many more scars from worm egg cases.

Horseshoe crabs do not molt as adults (Koons, 1883; Patten, 1912)

TABLE 3.2 Correlation between prosomal color and characteristics of horseshoe
crabs showing increased wear with age.

| 1991 | Color of Prosoma | | | N | P |
	Light	Medium	Dark		
Epibiotic growth (no epibionts)	73%	11%	0%	102	0.0001
Eye condition (good condition)	100%	57%	35%	102	0.0001
POP (good condition)	96%	76%	52%	126	0.001
Telson height (mm, worn)	100	99	93	89	0.002
1992					
Epibiotic growth (mean cover)	5%	11%	23%	783	0.0001
Eye condition (good condition)	94%	83%	59%	784	0.0001
POP (good condition)	93%	82%	62%	784	0.0001
Carapace shine (with slime)	93%	55%	1%	662	0.0001

Source: Data from Seahorse Key, Fla.

so the darker males are probably older than the light-colored individuals that have molted into the adult form more recently. The best direct evidence that we have for this hypothesis comes from the slipper-shell mollusk *Crepidula fornicata,* which attaches in its larval stage to the backs of horseshoe crabs, where it lives for many years and continues to grow (Botton and Ropes, 1988). We found that the *Crepidula* are larger, on average, on darker animals (Brockmann, 1996), which means that they had been living there longer and thus that darker individuals are older than lighter ones. It is likely that the darker color of older individuals is due to abrasion that accompanies activity and burying in the mud because the tissue under the outside layer is darker than the surface (Shuster, personal communication). If the mucus secreted from the surface of the crab declines with age or if the secretory apparatus on the surface is damaged by the abrasion associated with feeding, burying in the mud, or UV light, older individuals will be more susceptible to settling by fouling organisms (Williams, 1955; Shuster, 1982; Turner et al., 1988; Sydlik and Turner, 1989) and invading pathogens.

It is also possible that there may be differences in the immune systems of young and old individuals. The immune system of horseshoe crabs has been well studied (Bang, 1979; Pearson and Weary, 1980; Armstrong, 1991; see Chapter 12), but it is not known whether the immunological defenses of horseshoe crabs deteriorate with age, although it is known that individuals differ markedly in their response to bacterial infection (Lindberg et al., 1972). Although differences in age are a likely explanation for the differences in color and condition that we see in horseshoe crabs, it is possible that differences in behavior, such as whether the males have been attached for a long time, could affect male condition.

Unlike size, the age (color, condition) of the male crab was strongly associated with whether he was attached or unattached (see Table 3.1): the light-colored animals were more likely to be attached and the darker, heavily fouled animals were far more likely to be satellites. This is quite remarkable because, in most mating systems, the satellites are usually the younger, smaller males (Howard, 1978; Weatherhead, 1984; Manning, 1985; Alatolo et al., 1986; Kirkpatrick, 1987; Zuk, 1988).

Do Males Differ in How Likely They Are to Be Stranded?

In the Delaware Bay, the beach is littered with thousands of overturned crabs, stranded on the beach after the tide (Teale, 1957). This results from wave action or from being overturned by other crabs. Most crabs can right quickly when overturned; but the ones that cannot are left lying on their backs, vulnerable to death due to desiccation or predation by gulls, which is a significant cause of natural mortality in adult horseshoe crabs (Botton and Loveland, 1989; Botton, 1993). The method of righting did not differ for animals of different ages, but the probability of righting did. Older (darker) crabs or those that had been satellites were significantly less likely to right themselves than younger individuals or those that had been attached (see Figure 3.8; Penn and Brockmann, 1995). This means that the risk associated with coming ashore is greater for older males than for younger ones and unattached males are at greater risk than attached males. Theory tells us (Clark and Mangel, 1984), on the other hand, that older males should be more willing to run these risks because they have a lower expected future reproductive success than younger males.

Nothing is known about the physiology of these differences be-

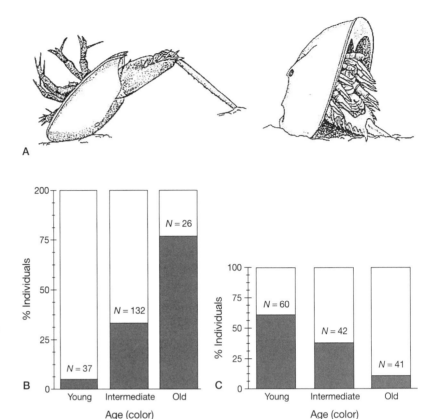

FIGURE 3.8 The association between color and righting behavior. (A) Two different types of righting behavior. (B) Differences in the colors of crabs that were found stranded on the beach (dark bars) after the tide and those that were not stranded (light bars). (C) Differences in the color of individuals that successfully righted themselves (dark bars) and those that did not (white bars) when they were overturned. Both studies were conducted at the Delaware study site. *Used by permission from Penn and Brockmann, 1995.*

tween males. It seems likely, however, that the less active, darker individuals may have reduced energy reserves or oxygen capacity (capacity for activity) when compared with the lighter, younger animals (horseshoe crabs have excellent mechanisms for handling hypoxia but accumulate alanine; Carlsson and Gäde, 1986). Strong individual differences can be found in the oxygen-carrying capacity of the blood of different individuals (Nordlie, personal communication; see Chapter 9), but we do not know as yet whether this is associated with being light or dark, young or old, attached or unattached. Penn (Penn and Brockmann, 1995) made the intriguing suggestion that perhaps the animals that are unable to right themselves are infected with *Microphalus limuli* (colorplate 31), a trematode worm whose larvae live in the tissues of *Limulus* (Stunkard, 1951, 1953). The definitive host of this parasite is gulls, which are known predators on horseshoe crabs that are overturned on the beach (Botton, 1993). Perhaps differences in the righting behavior of the crabs are due to differences in infection with this parasite (which gains by being transmitted to its host when the infected crabs cannot right themselves). Far-fetched perhaps, but such parasite control of host behavior is known to occur in other trematodes (Curtis, 1987; Moore, 2002).

Why Do Males That Differ in Condition Also Differ in Behavior?

Of course, saying that there are age or condition differences between males does not fully explain the behavioral differences we see. Why is it that younger males are more likely to attach to females, whereas older males become satellites on the nesting pairs? We marked males that came to the beach attached and unattached and then recorded what they were doing on the next tide on which we saw them 12 to 48 hours later. The effect was strong: if a male was attached on one tide, he was very likely to be attached on the next tide (and the same held for unattached animals). But of course, this is not really fair. The reason that the attached males were more likely to return with a female on the next tide was that they were still hanging on to the same female from the last tide (81 percent retain the same female across at least one tide). In other words, this study did not demonstrate a real difference in behavior between the two types of males. To get around the problem, we conducted several experimental manipulations in the field (Brockmann and Penn, 1992).

In the first experiment, we leveled the playing field for the two kinds of males to determine how they would behave. For each attached and unattached male we encountered, we detached him from his female, if attached, measured and marked him, and then released him along the shore, so that formerly attached and formerly unattached males went out to sea without a female. Were they equally likely to find and return with a female on the following tide? The answer is no (Table 3.3). There were real behavioral differences between young and old (light and dark) males in how well they found females—or perhaps in how well they remained attached once they had found a female. You cannot tell the difference between these two possibilities without another experiment.

So, in a second manipulation we did the reverse experiment (Brockmann and Penn, 1992). We collected males, detached them, and put them in wading pools that we had placed along the shoreline. We also picked up unattached males and put them in these pools. Females were added to the pools—not with their mates. We counted the number of males attaching after one hour and found clear differences. For-

TABLE 3.3 Results from experiments with male horseshoe crabs. The original attachment status refers to the condition of a male when he was first seen on the beach. We then returned the males to sea in various conditions and observed whether they returned with a female. In the detaching experiment all males were allowed to swim back to sea without a female (Brockmann and Penn, 1992). In the reattaching experiment males were allowed to pair with females in wading pools and were returned to sea with a female (Penn and Brockmann, 1992). In the detaching with bags experiment, males returned without a female and with their claws covered so they could not attach (Brockmann, 2002).

| Experiment | Condition in Which Crabs Were Returned to Sea | Original Attachment Status | Percentage in Each Attachment Status at Resighting | | χ^2 | P |
			Attached	Unattached		
Unmanipulated	Attached: paired	Attached (127)	81	19	39.0	<0.0001
	Unattached: alone	Unattached (80)	38	62		
Detaching	Attached: alone	Attached (129)	89	11	37	<0.0001
	Unattached: alone	Unattached (81)	51	49		
Reattaching	Attached: paired	Attached (89)	82	18	4.3	0.04
	Unattached: paired	Unattached (75)	68	32		

			Percentage Returning to Beach (none could attach)	χ^2	P
Detaching with bags on claws	Attached: alone	Attached (117)	33	6.8	0.01
	Unattached: alone	Unattached (108)	50		

merly paired males reattached far more quickly (average 82 minutes) than formerly unattached males (average 155 minutes). This means that if you were to place one attached and one satellite in a pool together with one female, far more (54 percent) of the formerly attached males would pair than would the formerly satellite males (29 percent).

For the animals that chose to attach in the wading pools, we coaxed them into containers, gently picked them up, and released the newly attached couples at sea. During this release process some males let go of the females, and these were far more likely to be formerly unattached males (19 percent) than formerly attached males (5 percent). (By the way, this study was conducted blind; in other words, we did not know which animals had been attached and which had been unattached when we released them.) What happened on the next tide when the pairs returned to nest on the beach? Those males that were originally unattached were less likely to be with the female they had paired with in the wading pool (40 percent) than those males that had originally arrived on the beach attached (76 percent; see Table 3.3). So, the older, darker males were less likely to find a female and more likely to let go when they did attach compared with the younger, lighter males. It is possible that older males are physically unable to hang on to females or it may be that young and old males use different decision rules about whether to attach.

To differentiate these two hypotheses, we conducted a third experiment (Brockmann, 2002). We gathered up equal numbers of attached (now detached) and satellite males, marked and measured them, and outfitted each with a pair of small plastic bags that fit snugly over their grasping claws to prevent them from attaching to females. We released all at sea and observed their behavior over the next few tides. If attached and satellite males used the same rules about when to return to the beach, then all should return in the same proportions. Because none could attach, one would expect all to return as satellites. This is not what happened. Former satellite males were more likely to return to the beach as satellites and former attached males, even though they were now unable to attach, were more likely to remain at sea (see Table 3.3). These results tell us that young and old males use different decision rules.

Putting all these data together, we can see that mating is not random at all, but rather strongly associated with differences in male age and condition. Figure 3.9 illustrates a model for the maintenance of the alternative mating tactics, attached and unattached, in the population (Brockmann, 2001). We based the model on the condition-dependent sex-allocation theory of Charnov (1982; Brockmann, 1986), but it is similar to models that have been developed for condition-dependent strategies in other species (Barnard and Sibly, 1981; Waltz, 1982; Vickery et al., 1991; Gross, 1984, 1991a, 1991b; Clark, 1994). In this model reproductive success is a function of age or condition: an older male will have higher reproductive success when unattached than when attached; a young male will do better if he is attached (see the lines on Figure 3.9). This means that as the male ages and his condition deteriorates, he reaches a condition at which he switches from one tactic to the other (see the arrow on Figure 3.9). The age (or condition) at which individuals switch is an evolved decision rule for that population and depends on the slopes and intercepts of the two lines. The population will remain at this switch point for the two tactics unless the average reproductive success of one tactic changes relative to the other (such as might happen if the environment changes; Gross, 1991b).

How can this model be tested? Models can be tested in two ways, by examining their assumptions and by testing their predictions. The first assumption is that reproductive success is a function of condition

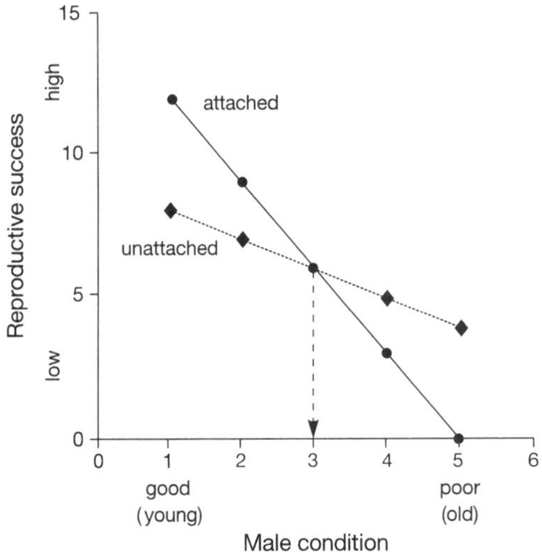

FIGURE 3.9 A model of male mating tactics in horseshoe crabs. The lines show the success associated with tactics for animals that differ in condition. The arrow indicates the condition at which males should switch from the attached to the unattached tactic. *Redrawn from Brockmann, 2001.*

and has the form shown by the lines on Figure 3.9; in other words, younger males have higher reproductive success when attached and older males have higher reproductive success when unattached. This could be evaluated experimentally by measuring the reproductive success of males in good condition that were forced by the experimenter to be satellites (as in the last experiment) and of males in poor condition that were encouraged to be attached (as in the reattaching experiment).

The second assumption is that males switch from specializing in attached to specializing in unattached behavior at a particular age or condition and the point at which they do this is where the two lines cross. This could be evaluated with a long-term study of marked individuals. We could also evaluate predictions of the model by seeing how the behavior changes when conditions change. For example, tides differ in the numbers of pairs on the beach, and therefore in the gain that crabs may expect from coming ashore. Tides also differ in the size of waves and intensity of currents, which are likely to affect the costs for coming ashore. This would mean that the slopes (or shapes) of these curves would change under different tidal and wave conditions and males should adjust their behavior (decision rules) from one tide to the next, depending on their condition and their expected reproductive return for behaving in different ways. One of the factors that may have a strong effect on male behavior is the difference between attached and unattached males in the risk of being overturned when they come ashore.

How Do Satellite Male Groups Form?

Figure 3.10A is adapted from a video of nesting crabs in Delaware Bay (Brockmann, 1996). It shows two females nesting side by side, one with four satellite males and one with none. From our paternity analysis, we know that the male in position 1F on the right-hand female is about as successful as the attached male, but the satellites in other positions on that female would be more successful if they moved over and became a satellite in position 1F on the left-hand female (see Figure 3.4). What explains this seemingly nonadaptive behavior?

To address this question, we first had to determine whether group sizes were random because you can get variable group sizes by random processes alone. If satellites were thrown at random onto the beach and they went to whichever pair was closest, different group sizes would occur by chance. Such a random model is quite unlikely—satellite males

A

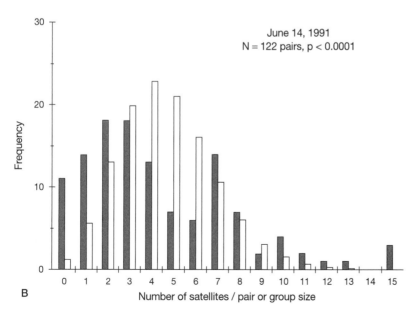

B

FIGURE 3.10 Nesting crabs and satellite males. (A) Two females; the one on the left has no satellite males and the one on the right has four. *Drawn by permission from a video by J. Brockmann, 1996.* (B) Frequency distribution of satellite male group sizes comparing the observed group sizes (satellites/pair, dark bars) for one representative tide (June 14, 1991, N = 122 pairs with 556 satellite males) at Cape Henlopen, Del., with expected group sizes (white bars) based on a random distribution. *By permission from J. Brockmann, 1996.*

are free to move from one pair to another, and they often do—but this random model is just a way for us to begin to understand what is going on. We tested this model by taking a census of group sizes for nesting crabs across thirty-one tides over a 3-year period in Delaware (Figure 3.10B; Brockmann, 1996) and comparing my observed distributions with random distributions. In twenty-five of the thirty-one tides, more pairs than expected had no satellites, more pairs than expected had large group sizes, and fewer pairs than expected had intermediate numbers of satellites. Only when there were few crabs on the beach were the distributions random.

What might explain these nonrandom distributions of satellite males? One simple explanation is that pairs that nest on the beach the longest accumulate the most satellites (Brockmann, 1996). We evaluated this hypothesis by conducting a field manipulation in which we equalized all the couples by removing their satellites. We did this across several nights when large numbers of pairs were nesting on the beach and very large numbers of unattached males were coming ashore. We walked along the beach and, when we came upon a nesting pair, we marked them with wire flags. If satellites were present, we called them "popular," and if no satellites were present, we called them "unpopular." We did not disturb the nesting pairs, but we removed all the satellites from the popular pairs.

About 10 minutes later, a second team of observers arrived and counted the number of satellites that were associated with each marked pair (they did not know how many satellites the pair had attracted previously). If unattached males were arriving at random and associating with whatever couple they encountered, popular and unpopular pairs should be equally likely to gain satellites. This is not what happened. Popular pairs were significantly more likely to regain satellites after they had been removed than unpopular pairs. In one year we uniquely marked all the satellites as we were removing them from forty pairs. Only 2 of the 492 males returned to the same pair, so our results were not due to the same satellites homing in on the same pair or location. Rather, there is something about some pairs or their immediate surroundings that is making them attractive to incoming males. What might that be?

My graduate student Cynthia Hassler thought that the unattached males might be attracted to pairs by chemical cues (Hassler, 1999; Hassler and Brockmann, 2001). She constructed three identical cement models of crabs and placed them near one another along the shore. One was placed over a spot on the beach where a popular female had been removed, one where an unpopular female had been nesting, and one where no females had been (the three models were run simultaneously

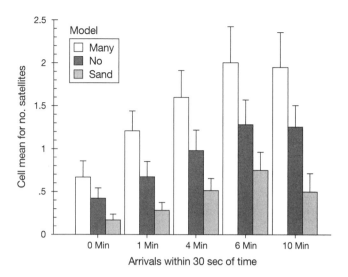

FIGURE 3.11 The number of satellites approaching locations on the beach where "popular" (white bars) and "unpopular" (black bars) females had been nesting. The counts were made for 30 seconds at 1 minute, 4 minutes, 6 minutes, and 10 minutes after the nesting female had been removed and a cement model had been placed over the same location (0 refers to the number of satellites that were originally with the nesting pair). In this experiment, which was conducted at Cape Henlopen, Del., all arriving satellites were removed after they have been counted (N = 16). *By permission from Hassler and Brockmann, 2001.*

in Delaware). The results were striking: satellite males approached the model located over the popular female's spot more frequently than over the other two locations (Figure 3.11). The attractiveness of the spot was lost after about 10 minutes, however.

We also conducted a second experiment in which we placed two cement models on the sand where no crabs had been nesting recently. We then placed a sponge under each of the models. One sponge contained water that had been taken from under a popular female and the other contained seawater from the beach. The results were clear: the model with water taken from under the popular female was more likely to attract unattached males. Taken together, these two experiments provide good evidence that the unattached males are using specific, but as yet unknown, chemical cues to locate pairs on the beach and to distinguish among females. But why would males be more attracted to some females than others?

Why Do Males Form Groups around Some Females?

We know that males form groups around some females and not others, but what advantage do they derive by associating nonrandomly with attached females? We considered several hypotheses: nest location, social satellites, female fecundity, and female size.

Nest Location Hypothesis. Females may pick locations on the beach that are particularly likely to result in good development of their eggs, and the males may detect these locations and accumulate there. If this hypothesis were true, popular females should be nesting in good areas and unpopular females in poor areas. However, Hassler (1999) did not find any differences in the rate of development of eggs laid by popular and unpopular females. The two sorts of females also do not differ in where they nest on the beach, ruling out microhabitat differences.

Social Satellites Hypothesis. Males may be attracted by the presence of satellites, the male equivalent of mate copying (Brooks, 1998). This could occur if grouped males are better protected from predators than single males or if together they are more likely to overcome the defenses of the attached male and thus fertilize more eggs. To test this, we conducted another field experiment in which we walked along the beach marking pairs with two or more satellites. We removed all but one satellite from half of these pairs and all satellites from the other half. My assistants arrived 10 minutes later and recorded the number of satellites that had accumulated around each pair. There was no effect. Newly arriving satellite males are not attracted to the presence of other satellites already around a nesting pair. If they are not attracted by location or the presence of other males, perhaps unattached males are attracted to some females over others.

Female Fecundity Hypothesis. One possible reason that satellites prefer certain females may be that popular females are on their first night at the beach when they are laying more eggs whereas less popular females may be depleted of eggs. We compiled data collected over a number of years, however, and showed that individual females tend to retain their popularity from one night to the next (Hassler, 1999).

Female Size Hypothesis. Next we considered physical differences between females. Maybe the popular females are larger and lay more eggs and males are accumulating around the more fecund females. We measured popular and unpopular pairs that were collected on the same night. Popular females are, on average, slightly larger and heavier than those found nesting without satellites (Agresti, 1996) and their telsons are less worn (Table 3.4; Brockmann, 1996). There are no differences in the sizes or condition of males that are attached to popular and unpopular females.

So why should unattached males prefer to associate with slightly larger females? One logical hypothesis is that these females lay more eggs than smaller females. To determine whether this is true, we observed nesting couples for 30 minutes or until they left the beach, dug up their eggs, separated them from the sand, and measured the volume

TABLE 3.4 A comparison of the characteristics of females found in Delaware nesting with multiple satellites present and nesting with no satellites present. The sample size is 62 for females without satellites and 111 for females with satellites (except for *Crepidula* data, for which the sample size is 43 and 62).

Measure	Mean for Females That Were		
	Alone	with Satellites	P
Carapace width (cm)	25.2	26.8	0.0001
Interocular width (cm)	15.6	16.8	0.0001
Weight (gm)	2191	2604	0.0001
Telson length (cm)	18.9	21.7	0.0001
Telson height (cm)	1.34	1.46	0.0001
Size largest Crepidula (cm)	2.68	2.94	0.2

	Percentage of Females in This Condition		
	Alone	with Satellites	P
with few fouling organisms	73	72	0.7
with light prosoma	47	70	0.008
with two good eyes	42	56	0.2
with two good POP	31	24	0.5
with mucus on surface	24	20	0.0001

Source: Data from Brockmann, 1996.

of eggs laid. We found that large females and females with several satellites do not lay more eggs than smaller or unpopular females, they do not nest longer, nor do they lay at a faster rate. However, Hassler repeated these measurements, making sure that she compared females that were collected on the same night under the same conditions. She found that popular females lay more eggs than unpopular females (Hassler, 1999). Given the paternity data that we collected earlier, she argued that an unattached male that becomes the first or second satellite with a popular female can expect a higher number of fertilized eggs than a male that becomes the first satellite on an unpopular female. However, neither of us has been able to explain group sizes of more than two satellites.

Theory predicts that uneven and nonrandom group sizes should occur when the gains (or costs) from being in some groups are very different from others (Waltz, 1982; Milinski and Parker, 1991); when the best thing to do depends on what other individuals in the population are doing (Pulliam and Caraco, 1984); and when individuals stay in groups only as long as it benefits them (Clark and Mangel, 1984). Under these models, predicted group sizes will change with conditions, and there will be no one optimal group size. Such models would appear to be appropriate for horseshoe crab behavior, but nothing is known, as yet, about the rules that individual males use to join or leave groups. Also, unattached and attached males are not equal competitors (Brockmann and Penn, 1992; Penn and Brockmann, 1995; Brockmann et al., 2000); but it is not clear from the models how this will change their behavior when they are in complex, interacting groups (Milinski, 1981; Parker and Sutherland, 1986). Group behavior in male horseshoe crabs could benefit from some serious dynamic modeling.

What Is the Horseshoe Crab Mating System?

Scientists love to classify things and this includes mating systems. Classifications are useful because they allow us to identify patterns of similarity between species. If two species that are not close taxonomic relatives show similar patterns, we should be able to identify the common selective pressures that have favored the similar patterns. Such comparisons allow us to understand the evolution of those patterns as adaptations. Mating systems are usually classified by the number of mates that individuals have, by the level and type of defense shown, and by the amount of variation in success among males.

In many mating systems in which males are highly competitive, the males are larger than the females. In horseshoe crabs, however, females are almost always substantially larger than males. Sekiguchi et al. (1988) have shown that the proximate explanation for the size dimorphism is that males mature after sixteen molts (9 years), whereas females mature in seventeen molts (10 years). Because males gain little advantage by growing larger (Brockmann and Penn, 1992; Loveland and Botton, 1992) and even the smallest males can clasp females successfully (Botton and Loveland, 1992), selection favors their maturing earlier than females. By maturing later, females may gain a fecundity advantage by being larger in size and able to carry more eggs (Loveland and Botton, 1992; but see Shine, 1988).

Horseshoe crabs are typically presumed to have a chaotic, random, or scramble competition (Alcock, 2001) mating system that shares many features with the "explosive breeding system" of some frogs, toads, and fishes that also fertilize their eggs externally (Brockmann, 1990). First, like *Limulus* (Barlow et al., 1986), male explosively breeding frogs and toads actively search for females (Wells, 1979) rather than sitting and waiting for them to approach, as the territorial species do (Wells, 1977; Davies and Halliday, 1979; Halliday, 1983). Second, also like *Limulus,* male explosively breeding toads have no size advantage, unlike the large male territorial toads, which have a strong advantage when competing for mates (Davies and Halliday, 1979; Wells, 1979; Howard, 1980; Arak, 1983; Höglund and Säterberg, 1989). However, within a species, when the time available for mating increases, male-male competition increases and a size advantage emerges (Höglund and Robertson, 1987; Höglund, 1989). Third, males of the explosively breeding frogs and toads often clasp inappropriate objects such as conspecific males, the fingers of human observers, or sticks (Wells, 1979; Howard, 1980). Such inappropriate attachment is commonly seen in horseshoe crabs: other males, a black Frisbee, a shoe, a branch sticking out in the water, a cinder block or horseshoe crab–sized stone, and dead females are all objects of male interest (Sargent, 1988; personal observation).

Fourth, unlike prolonged breeders, male explosive breeders (frogs and toads) guard their mates by remaining mounted in amplexus for a number of days throughout the short breeding period (Wells, 1977). Horseshoe crabs show similar behavior. Fifth, in some explosive breeders, fighting among males is so extreme that they may even injure the female (Davies and Halliday, 1979; Howard, 1980). We suspect that the broken posterior opisthosomal projections (see Figure 3.1B) on females may result from male takeover attempts, which may explain why females usually interrupt nesting and leave the beach rather than allowing a takeover to occur. Sixth, in frogs and toads and in some explosively breeding fish such as grunion, male alternative reproductive tactics are common, with four or five males surrounding one spawning female (Thompson, 1919). Finally, sperm competition is suspected in a number of explosively breeding toads and frogs (Halliday and Verrell, 1984) and has been confirmed for some fishes (Gross, 1984). As in *Limulus,* unpaired males lurk around the spawning couples, releasing sperm.

The one way in which horseshoe crabs seem to differ from other explosive breeders, however, is that in explosively breeding frogs and toads with scramble competition, no differences have been found between males that are successful and those that are not. However, alternative male mating tactics based on physical condition other than size and age have not been considered.

The profound similarities between anuran explosive breeding systems and horseshoe crabs suggest that two important factors shape the evolution of these breeding systems: the type of reproduction (external versus internal) and the amount of time available for breeding (Emlen and Oring, 1977; Wells, 1977; Halliday, 1983; Halliday and Verrell, 1984; Höglund, 1989). Future studies of horseshoe crabs should examine the effects of changes in the length of the breeding season on the intensity of competition among males and the varying success of males that differ in condition or age. Such a study would be possible given the wide geographic range of this species.

Some Generalizations

The mating system of horseshoe crabs is highly competitive, with male-biased operational sex ratios, large groups around some nesting pairs, strong sperm competition, and males fighting for positions that maximize their fertilization success. We have developed the hypothesis that young males or those in good condition choose to be attached and older males or males in poor condition choose to be unattached, coming ashore as satellites. They do this because young males have higher success when attached and older males have higher success when unattached. This model needs to be tested experimentally and on additional populations. We also need to know more about why males nest in groups. Little is known about female choice or female behavior. We have evidence that some females are more attractive to satellites than others and may produce a chemical that attracts males and promotes sperm competition and male-male rivalry.

The study of behavior requires good observation and description and openness to many possible explanations. Initial descriptions of patterns and correlations allow us to develop proximate and ultimate hypotheses. Next we have to find ways to test the hypotheses, and we do this by asking the animals questions. It is often possible to answer those questions by conducting experimental manipulations in the field. It is literally possible to ask animals about the cues they are using or what affects their reproductive success by conducting an experiment on the beach as the animals are mating (see also Chapters 2 and 4).

At every step in my description of the reproductive behavior of *Limulus,* we have discussed additional questions, alternative hypotheses, or other interpretations of their behavior. Although we have learned a great deal about horseshoe crab behavior, there is still much to be done. It is clear that the answers to our questions about behavior have led not only to more questions about behavior but to more questions about every other aspect of *Limulus* biology, including their structure, phylogenetic history, development, physiology, and immune system.

LANGUAGE AND EXPLANATION IN BEHAVIOR

EXHIBIT 3.1

Everyone loves to use colorful language. We don't want our prose to be cumbersome or too technical so we use lighthearted or evocative phrases. But as soon as we start to use nontechnical language to describe what we see, we find that we are loading our description with words that imply particular explanations. If we say, "the male is shoving the other male and lashing his tail from side to side," we are creating a very different image or hypothesis about what is going on than if we say, "the male sidles alongside and swishes his tail from side to side to maneuver in the current." Sometimes altering just a few words can utterly change our view of the most likely explanation for a behavior. For example, some years ago, animal behaviorists switched from referring to a closely associated male-female pair as "pair-bonding" to describing them as "mate-guarding." This change in language completely altered the adaptive hypothesis that had been used to explain the behavior and, in fact, the new language had a new adaptive hypothesis embedded in the words.

Furthermore, when we describe behavior, we tend to use human terminology and therefore we tend to develop human explanations for behavior. In other words, our explanations tend to be anthropomorphic. If we want to understand horseshoe crab behavior, we need to remain open to a wide variety of explanations and consider many options. This is as true for proximate explanations as it is for ultimate explanations. Horseshoe crabs have very different sensory receptors and live in a different sensory/perceptual world than we do, and we must take that world into account when we propose hypotheses about their behavior.

THE PROCESS OF FERTILIZATION

EXHIBIT 3.2

Newly released sperm do not actively swim; they move passively in water currents until they are within 0.5 cm of the egg (Brown, 1976; Clapper and Brown, 1980). At this distance the sperm suddenly become active (in other words, they are capacitated; Figure 3.12A) when they contact a specific protein released by the eggs (or perhaps when they contact fluids released by the female) (Shoger and Bishop, 1967; Clapper and Epel, 1985). The sperm swim up this chemical gradient (Brown, 1976); and when they contact the egg less than 5 seconds later, they undergo an explosive acrosomal reaction: the spring-like acro-

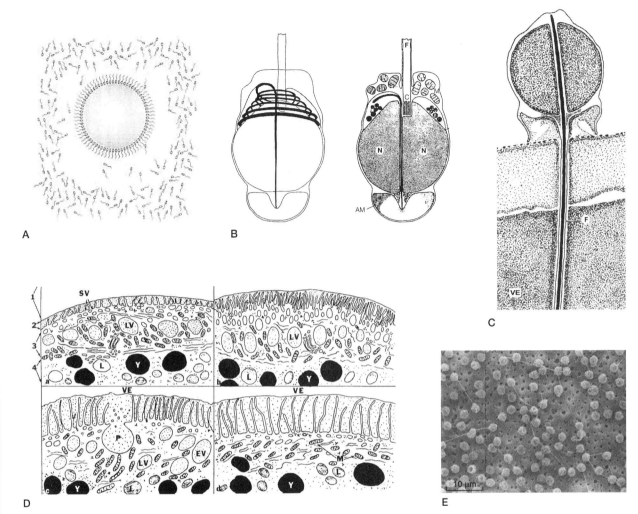

A

B

C

D

E

FIGURE 3.12 *Limulus* sperm, eggs, and fertilization. (A) Near the egg is a region of low density where the sperm have become motile, swum to the egg, and attached. *By permission from Clapper and Brown, 1980.* (B) The structure of the unreacted *Limulus* sperm shows the coiled acrosomal filament rod on the left, on the right the acrosomal membrane (AM) (at the bottom) that will attach to the egg, and at the top the flagellum (F, sperm tail) and associated mitochondria (M), centriole (C), and nucleus (N). *By permission from Shoger and Brown, 1970.* (C) Diagram of the initial sperm-egg interaction: the acrosomal filament has penetrated the egg membrane and the acrosomal membrane has begun to fuse with the egg membrane in an acrosomal reaction. *By permission from Shoger and Brown, 1970.* (D) Diagrammatic representation of the egg cortical reaction in *Limulus.* Region 1 is the outer egg envelope; 2 is the perivitelline space; 3 is the egg cortex; 4 is the lipid-yolk center of the egg. The upper-left frame shows an uninseminated egg; upper right shows the egg 1 to 8 minutes after mixing with sperm, showing that the small vesicles (SV) along the perivitelline space are in the process of fusing and the microvilli have become longer and wider. Lower left shows the inseminated egg after 9 minutes as large (LV) and enlarged vesicles (EV) continue to form; pits (P) also form. The lower right shows the egg 60 minutes after insemination. L, lipid droplets; M, mitochondria; Y, yolk; VE, vitelline envelope. *With permission from Bannon and Brown, 1980.* (E) High magnification of an egg (x3040) showing the spermatozoa that have attached to the egg surface and undergone the acrosomal reaction. The high concentration of attached sperm is obvious. *By permission from Brown, 1976.*

somal filament that is inside the sperm head (Figure 3.12B) suddenly explodes through the cap of the sperm head and penetrates the membranes of the egg (Figure 3.12C; Shoger and Brown, 1970), rotating as it elongates, thus screwing through the egg jelly (Tilney, 1975).

Within 3 minutes of the first acrosomal reaction, the egg undergoes a number of changes that visibly alter its appearance (Figure 3.12D; Bannon and Brown, 1980). Small vesicles near the surface fuse with the overlying layers (Figure 3.12D), which apparently prevents further sperm attachment (Bannon and Brown, 1980; Brown and Clapper, 1980; Brown and Barnum, 1983) but not before up to a million sperm have undergone acrosomal reactions with a single egg (Figure 3.12E; Shoger and Brown, 1970; Brown and Humphreys, 1971; Brown, 1976).

Nothing seems to be known about the process by which the DNA of only one particular sperm gets to the egg nucleus. Brown and Knouse (1973) and Brown (1976) suggest that there may be a selective process by which only some sperm nuclei are allowed to contact the egg cell membrane, a kind of cellular mate choice. Brown and Clapper (1980) found that females differed markedly in the rate of the egg cortical reaction: some are fast (completed in 60 to 90 minutes), whereas other females are slow (125 to 150 minutes). Could this be due to differences in the responses of females to the sperm of different males? Could it be a reaction between particular male and female genotypes? Alternatively, there could be a mechanism by which sperm digest their way through the egg envelope (Brown and Humphreys, 1971). When sperm from multiple males are present, does the female "choose" among them by some physiological process inside the egg or are sperm nuclei from different males competing to reach the egg nucleus? How does the egg prevent multiple sperm nuclei from contacting its nucleus (which would disrupt development)? These would be fascinating questions to pursue!

PATERNITY ANALYSIS IN HORSESHOE CRABS

EXHIBIT 3.3

I observed groups of crabs (a pair with one to four satellites) nesting on the beach and marked the position of their eggs with wire flags (see Figure 2.1). I took a tissue sample from each adult and froze it immediately. I dug the eggs from the sand after the tide, reared the larvae in incubators in the lab until the late trilobite stage (1 to 2 months) (Brown and Clapper, 1981), and froze them. Extracting the DNA from the samples was easy to do with the large adult samples, but it required a special

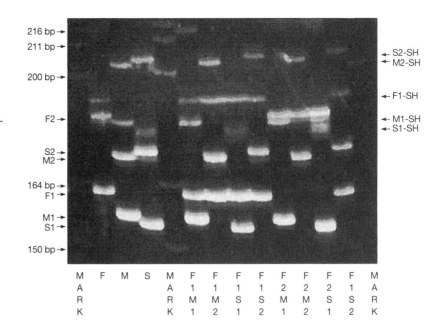

FIGURE 3.13 Acrylamide gel showing the DNA banding pattern for the female (F), attached male (M), and one satellite (S) that nested together. To the right is the banding pattern for eight of their offspring. Since horseshoe crabs are diploid, each parent contains two copies of its DNA, hence F1 and F2, M1 and M2, S1 and S2. Each offspring contains one copy of DNA from its mother (either F1 or F2) and one copy from its father (either M1 or M2 or S1 or S2, depending on whether the attached male or the satellite fertilized the egg). *By permission from Brockmann et al., 1994.*

amplifying technique (PCR) for the tiny larvae (Brockmann et al., 1994).

Initially we tried standard fingerprinting techniques (Jeffreys et al., 1985; Georges et al., 1988), but none worked with *Limulus* DNA. For this reason we used microsatellite analysis (Queller et al., 1993), which means we cloned a piece of *Limulus* DNA, which was then hybridized with the DNA from each participant in a nesting group. These hybrid bits of DNA were then placed on an acrylamide gel and stained. The staining allowed us to see bands of light and dark representing pieces of DNA of different lengths, polarities, and molecular weights (Figure 3.13). We placed together on one gel the DNA from 16 to 24 offspring, along with DNA from each adult participant in that clutch. We first made sure that all offspring belonged to the female involved in the nesting (because contamination between nests is possible), and then we matched up which male belonged with each offspring. This allowed us to calculate the percentage of offspring from each clutch that were fathered by each male participating in the nesting.

SEEING AT NIGHT AND FINDING MATES: THE ROLE OF VISION

CHAPTER 4

Robert B. Barlow and Maureen K. Powers

"**I** have been studying vision in a blind animal for 50 years," commented H. Keffer Hartline in 1978. He did not believe for a moment that the horseshoe crab was blind. He was expressing his frustration for the many years of research on the animal's lateral eyes that had not answered the question, "What does a horseshoe crab see?" This chapter describes what is known about the lateral eyes, their purpose, and why the crabs can see so well at night. Surprisingly, horseshoe crabs do not see with just two lateral eyes. They possess a multitude of visual organs—ten in all—which they use for a variety of purposes.

Hartline Discovers the *Limulus* Lateral Eye

Let us go back to the spring of 1926 and follow H. Keffer Hartline (1903–1983; colorplate 8) as he walked the beaches of Cape Cod in search of an animal with a relatively simple visual system. Strolling along the water's edge, he stumbled across a large female horseshoe crab that was building a nest to lay eggs. He knelt down to look at her eyes and marveled at their size, especially the individual light receptors, the ommatidia. The receptors were so large that he could see them without a magnifying glass. Numbering about 1,000 in each eye, the individual ommatidia (Figure 4.1) are roughly one hundred times the size of rods and cones in the human eye. In awe of their size, Hartline reasoned that if he could see them so easily, he might be able to record their electrical responses to light and understand how these nerve signals provide the animal with vision.

Hartline had gone to the Marine Biological Laboratory (MBL) in

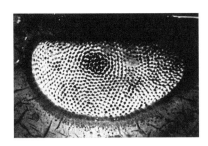

FIGURE 4.1 The compound eye of *Limulus*. A pair of these large eyes, each containing about 1,000 ommatidia (black disks), are perched high on each side of the prosoma and provide *Limulus* with wide fields of vision (Herzog and Barlow, 1996). The crab not only can see to each side but also ahead, behind, and above. The individual ommatidia are the largest known retinal receptors in the animal kingdom. They are roughly 100 times the size of the rods and cones in the human eye.

Woods Hole, Mass., for just this purpose, hoping to identify an animal with which he could begin to understand the neural mechanisms underlying sight. He had first become interested in vision several years before as an undergraduate at Lafayette College, where he discovered "a relationship between certain photochemical laws and the phototropism of animals" during a meticulous behavioral study of light responses of pill bugs. To explore how light-initiated signals in the eye lead to changes in an animal's behavior, he started studying nerve signals in the eye of the frog, but he found them to be too complex and sought a simpler system.

Hartline had successfully recorded a light-evoked electrical signal, termed the electroretinogram (ERG), from the human eye. The amplitude of the ERG provides a good measure of retinal sensitivity. He set about to measure the ERG from the *Limulus* lateral eye to a brief flash of light with a wick electrode placed in contact with the cornea. Because highly sensitive physiological amplifiers did not exist in the late 1920s, Hartline attached the electrode to the most sensitive instrument available, a mechanoelectrical device known as a string galvanometer. Fortunately it was sensitive enough and Hartline succeeded in measuring the response properties of the ERG. He found that its amplitude depended on both the intensity and duration of the light flash and thus obeyed a relationship common to photochemical reactions, the Bunsen-Roscoe Law. He concluded that the visual response of the *Limulus* eye resulted from a chain of unknown cellular events triggered by a photochemical reaction in the retina (Hartline, 1928).

During these early ERG studies, Hartline explored various neural structures of the eye with his electrode. In one experiment he placed it directly on the optic nerve and noticed brief, tiny electrical events that he thought might be individual nerve impulses. The possibility of tapping into the nerve signals the eye sends to the brain excited him, but the electrical events proved to be too small to resolve with his string galvanometer (Barlow, 1986). He was determined to find a more sensitive recording instrument. Within a year he learned about a vacuum tube amplifier built by Gasser and Newcomer that appeared to have sufficient sensitivity to detect small nerve signals. Vacuum tubes were a rare commodity, but he managed to locate one and constructed his own amplifier.

How Were the First Recordings from Optic Nerve Fibers Made?

In 1931, Hartline and his colleague Clarence Graham returned to the MBL armed with a highly sensitive vacuum tube amplifier. They hoped to detect minute nerve signals with the amplifier and observe them by connecting its output to a moving reed oscillograph, the predecessor to the cathode ray oscilloscope. They removed the lateral eye of a juvenile crab, frayed out its optic nerve, and found it easy to record nerve impulses; but even the tiniest bundles gave a mass of impulses. They were frustrated in their attempts to record responses from single axons. Late that summer, the day before they were to leave MBL, they had used all the babies and only two horseshoe crabs remained in their seawater aquarium. They were large adults with crusty shells and dull, scarred eyes—"miserable specimens" compared to the clear-eyed, young ones. They were surprised to find that they could easily dissect single fibers from the optic nerves of the adults, and they succeeded immediately in recording trains of nerve impulses from them. They furiously set about to learn what they could before packing up the lab to leave the following day.

The outpouring of results from subsequent single-fiber experiments was enormous, touching on virtually every aspect of vision and leading to the formulation of basic mechanisms of vision common to all species, including humans (Hartline, 1972). A detailed presentation of these results would readily fill a book twice the size of this one (Ratliff and Hartline, 1974). What follows is a selected review of the most salient work.

The initial recordings by Hartline and Graham from single optic nerve fibers showed that the *Limulus* eye encodes light intensity with the rate of discharge by a single optic nerve fiber such that a simple logarithmic relationship exists between light and the discharge rate (Hartline and Graham, 1932). The discharge from a single fiber followed the Bunsen-Roscoe Law as did the ERG, that is, it exhibited a reciprocal relationship between the intensity and duration of brief light flashes (Hartline, 1934). In essence, the eye can function as an adding machine, summing the influences of individual photons in brief flashes to produce a visual response. Hartline and Graham found that, after the onset of light, the discharge rate of a single fiber was initially high and then decreased to a lower rate, indicating rapid sensory adaptation.

After a hiatus of research during World War II, Hartline and McDonald (1947) found that increasing the duration of the light flash further reduced the eye's sensitivity, while leaving it in darkness increased it. This was the first cellular evidence for the visual adaptation we all experience after entering a dark movie theater on a sunny day. After about 10 minutes, we can see much better. Hartline and McDonald concluded that the horseshoe crab's ability—and ours—to adapt to different light intensities begins in the retina.

Graham and Hartline (1935) also studied the action spectra of single receptors and found that their discharge rate varies with the wavelength of light, peaking in the blue-green region of the spectrum. Ruth Hubbard and George Wald (1960) found that the visual pigment extracted from *Limulus* photoreceptors absorbed light in the same region of the spectrum. This spectral match laid the foundation for understanding the cellular mechanisms of color vision. It was clear to Hartline that these and other visual phenomena originated in the retina of this primitive animal.

First Recordings from Single Photoreceptor Cells

Although the ommatidia of the horseshoe crab's lateral eye are by far the largest in the animal kingdom, they are difficult to surgically isolate. Hartline and Graham (1932) were masters at such delicate surgical manipulations and eventually succeeded in removing an ommatidium from the eye. Then, using a small electrode, they recorded a minute electrical current that they called an "action current." This current coincided with the propagation of impulses in the optic nerve. Hartline and Graham suggested that the action current of an ommatidium might initiate the nerve impulses.

Unfortunately their electrode was not small enough to probe the inner workings of cells within an ommatidium. Recording from a single cell required a new type of electrode, the glass microelectrode, which was developed in the 1940s. This electrode is made by pulling forcibly on both ends of a glass capillary tube as its center is heated. The glass separates, producing two electrodes with tips approaching the wavelength of light in diameter. The tips are so tiny that they can penetrate single cells without damaging them.

By filling these glass microelectrodes with a conducting salt solution and connecting them to an amplifier, Hartline, Wagner, and MacNichol (1952) successfully impaled a single photoreceptor cell, called a retinular cell, and recorded its response to light. They found that a brief flash of light depolarized the transmembrane potential as much as 50 millivolts before returning to its previous level. They believed that this photoreceptor potential was "intimately related to the initiation of nerve impulses." Tsuneo Tomita (1956) and E. F. MacNichol (1956) then showed that this photoreceptor potential results from an increase in cell membrane conductance and is indeed related to the generation of nerve impulses. These germinal studies, continued in Hartline's laboratory and many others throughout the world, led to a detailed understanding of how both invertebrate and vertebrate photoreceptors respond to light.

In 1964 Fuortes and Yeandle discovered that single photons evoked elemental voltage events when they exposed retinular cells to very dim light. These so-called quantum bumps increase in frequency as photon flux increases and sum to form the receptor potential that leads to the generation of optic nerve responses. Several years later Dodge, Knight, and Toyoda (1968) discovered that the amplitude of quantum bumps decreases as their frequency increases, thus adapting the eye to higher light intensities. Their "adapting bump" model is the first comprehensive explanation for retinal light adaptation.

How Was Lateral Inhibition Discovered?

"I turned on the room lights and the optic nerve response decreased," said Hartline, recounting an experiment he performed on the *Limulus* eye in 1949. "Why should the response decrease when I increase the light intensity?" He had seen this phenomenon countless times before, but had not appreciated its significance. Why he was suddenly alerted to the effect of room light is not clear, but he finally grasped its meaning: illuminating ommatidia in the lateral eye can inhibit the responses of neighboring ommatidia. The concept of lateral inhibition was born, a concept that has proven to be a cornerstone of visual system organization in all animals (Hartline et al., 1956).

Over a decade earlier, Hartline replied to George Wald's question about the possible differences between *Limulus* and vertebrate eyes: "In the *Limulus* eye, which lacks anatomical interconnections, Graham has shown that the discharge of impulses from a given sensory cell is not affected by the activity of adjacent elements." Hartline found much to his surprise that this was decidedly not the case. The two scientists would later share the Nobel Prize with Ragnar Granit for research in vision.

Hartline's discovery of lateral inhibition initiated a truly remarkable line of research extending to the present day. Because *Limulus* ommatidia are so large, it was possible to illuminate them individually while recording their responses as well as the responses of neighboring ommatidia (Figure 4.2). With this technique Hartline and his coworker Floyd Ratliff (1957, 1958) found that the steady responses of individual optic nerve fibers could be quantitatively expressed in terms of the algebraic sum of inhibitory influences of neighboring receptors. This achievement stands today as the only complete quantitative analysis of neural integration among an ensemble of sensory receptors. The well-known Hartline-Ratliff formulation has been the starting point for a number of treatments of information processing in more complex

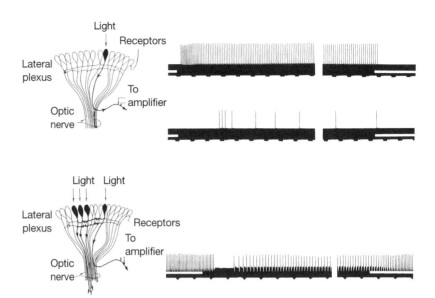

FIGURE 4.2 The interactions between retinal receptors in the *Limulus* eye are inhibitory. Intense illumination of a single ommatidium evoked a high rate of discharge of nerve impulses recorded by an amplifier from the nerve fiber of the ommatidium (upper recording on right). Decreasing light intensity by 10,000 greatly reduced the rate of discharge (second recording). Light is on during the darkened line (ticks mark 0.2s intervals). Illuminating neighboring ommatidia (darkened bar) inhibited the high rate of discharge of the recorded ommatidium (third recording). Diagrams of the cross-section of the eye on the left show the illumination of a single ommatidium (top) together with a cluster of ommatidia (bottom). The responses of optic nerve fibers from the neighboring ommatidia are not shown. *By permission from Hartline et al. (1956).*

FIGURE 4.3 Contrast affects our perception of brightness. Although the middle panel reflects exactly the same amount of light throughout its length, the left-hand side of the panel appears dimmer because light areas border it. The light areas evoke inhibitory retinal signals that reduce the response of neighboring neurons decreasing their output to the brain. Occluding the top and bottom panels confirms the uniformity of the central one.

neural systems. It led to a comprehensive description of the neural code the eye sends to the brain, as discussed later in this chapter.

Hartline's discovery of lateral inhibition provided important clues about human vision. Our vision is highly sensitive to borders and edges. We enhance the contrast between light and dark areas in the visual field—a phenomenon known as simultaneous contrast (Figure 4.3). In 1865 Ernst Mach had hypothesized that this ability of the human vision system could be explained by mutually inhibitory interactions in the retina (Ratliff, 1965). Physiological support of Mach's idea waited many years: Hartline found it in a visual system far simpler than our own.

Vision Is a Dynamic Process

Extending the elegant steady-state analysis of Hartline and Ratliff to the temporal domain significantly increased the complexity of the theoretical formulations. A convenient method for collecting and analyzing large numbers of impulses was required; and in 1961 Hartline, then at The Rockefeller University, purchased a state-of-the-art digital computer that nearly filled an entire laboratory. It had only 16K of memory and no software, but those of us (R. B.) lucky to be in the lab thought it was the most magnificent piece of equipment in the world. Thousands of those computers together could not come close to equaling the performance of today's laptop computer. With much diligence, Hartline and his colleagues succeeded in analyzing precisely patterns of optic nerve activity in response to dynamic patterns of illumination of the *Limulus* eye (Ratliff and Hartline, 1974).

Excising the Eye Decreases Its Sensitivity

The next epoch in *Limulus* eye research began in the 1970s, when Barlow decided it was time to study the eye without removing it from the animal. In the hands of Hartline and his colleagues, the excised *Limulus* eye yielded mountains of information about retinal function; but it had an annoying property—a slow but constant decrease in photosensitivity over the course of an experiment that usually lasted several hours. Hartline and Barlow had agonized over this problem, and both agreed that someday the eye must be studied in the animal with its blood supply intact. But this meant developing a new technique that Barlow, who was deep into his thesis research on the excised eye, was reluctant to consider. He pleaded with Hartline "to let me finish my dissertation, graduate and tend to the problem later." Hartline concurred.

Ehud Kaplan, Barlow's first graduate student at Syracuse University, became interested in the problem, and together they developed a successful technique, one that yielded long-term recordings—up to 6 days —from the same optic nerve fiber without decreases in sensitivity. But they were in for a big surprise: the intact eye appeared to be about 1,000 times more sensitive than the excised eye (Barlow and Kaplan, 1971). They were stunned and thought their "light-tight" cage or optical stimulator might be leaking light. Kaplan decided to take the place of the horseshoe crab in the cage and look for light leaks. With no outside air supply to the cage, Barlow kept tapping on its wall to make sure Kaplan was still alert (and alive!). He was, and he found no light leaks. They were forced to conclude that intact eyes were intrinsically more sensitive and later found that removing the blood supply impairs a regenerative mechanism that amplifies single photon events so the brain can detect them (Barlow and Kaplan, 1977). In an earlier study Dowling (1968) recorded regenerative potentials from the excised eyes of baby crabs. Apparently the eyes of baby crabs are more resistant to the effects of blood loss.

Discovery of Circadian Rhythms in the Eye

Even though the intact lateral eye is highly sensitive, Kaplan and Barlow occasionally detected decreases in its sensitivity immediately after severing the optic nerve. Aware of the report of efferent fibers in the optic nerve trunk by Fahrenbach (1973), they were suspicious that signals from the brain may modulate the eye's sensitivity. Challenged by Barlow to test this idea, two graduate students at Syracuse University, Stanley Bolanowski and Michael Brachman, recorded the ERG from the

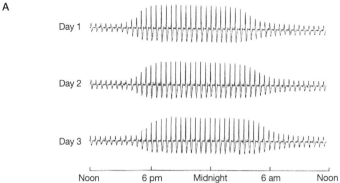

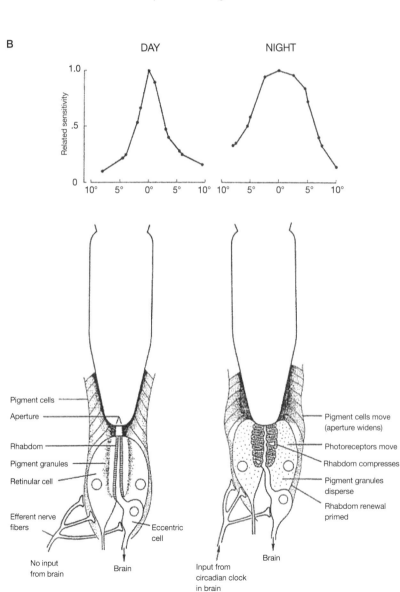

FIGURE 4.4 (A) Electroretinographic (ERG) recordings show the endogenous rhythm of increasing and decreasing sensitivity of the *Limulus* eye when an animal is kept in complete darkness. The ERG was recorded in response to a brief flash every 30 min. over a 3-day period. (B) Ommatidia show striking changes from day to night. The aperture that restricts the entry of light into the photoreceptors (retinular cells) during the day widens at night, pigment granules that absorb stray light disperse, and the rhabdom (light-sensitive membranes) folds and moves distally to enhance the capture of incident photons at night.

lateral eye, as Hartline had done years earlier. They were indeed astonished to see the amplitude of the ERG change with time of day: it increased at night and decreased during the day while the animal remained in constant darkness (Figure 4.4A). They had discovered a robust circadian rhythm in retinal sensitivity. But where was the endogenous oscillator that controlled the rhythm? In the eye, the brain, or some other organ?

Because Barlow and Kaplan detected no rhythmic changes in the lateral eye with the optic nerve cut, they suspected that the oscillator was in the brain and it sent information out to the eye to mediate the circadian rhythm. The students disagreed. Bets were taken on the effects of cutting the optic nerve. Barlow was the only one betting that severing the optic nerve would eliminate the rhythm. He won.

Neural Signals from a Clock in the Brain Increase the Eye's Sensitivity

These experiments showed clearly that a circadian clock located in the *Limulus* brain transmitted neural signals via efferent optic nerve fibers to the lateral eyes at night (Barlow et al., 1977; Calman and Battelle, 1991). Beginning at the time of dusk and ceasing at the time of dawn, the clock-generated neural signals increase lateral eye sensitivity as much as 1,000,000 times over daytime levels! This astonishing increase in sensitivity about equals the 1,000,000-fold decrease in ambient light intensity after sundown. Does the animal's visual system compensate so well for the dim light at night that it can see as well as if it were daytime? This appears to be the case, as described later in this chapter.

Such a pronounced circadian rhythm in retinal sensitivity should not come as a surprise, even in a creature as ancient as *Limulus*. Rhythms are widespread among both invertebrate (Barlow et al., 1989; Barlow, 2001) and vertebrate (Manglapus et al., 1998) visual systems. In most all cases, visual sensitivity appears to be under the joint control of a circadian oscillator and light. Why do the visual systems of some animals need to anticipate changes in light intensity rather than respond directly to them? Studies of *Limulus* suggest some answers.

How Can the Sensitivity of the *Limulus* Eye Change So Drastically from Day to Night?

The number of day-night changes in the eye's anatomy and physiology detected thus far is truly remarkable (see Table 4.1). *Limulus* appears to

TABLE 4.1 Circadian rhythms in the *Limulus* lateral eye.

Retinal Property	Day	Night	Retinal Property	Day	Night
Efferent input	Absent	Present	Screening pigment	Clustered	Dispersed
Gain	Low	High	Aperture	Constricted	Dilated
Noise	High	Low	Acceptance angle	6°	13°
Quantum bumps	Short	Long	Photomechanical movement	Trigger	Prime
Frequency response	Fast	Slow	Photon catch	Low	High
Dark adaptation	Fast	Slow	Membrane shedding	Trigger	Prime
Lateral inhibition	Strong	Weak	Intense light effects	Protected	Labile
Cell position	Proximal	Distal	Visual sensitivity	Low	High

have discovered "every trick in the book" to increase its retinal sensitivity at night (Barlow et al., 1989).

The first in the list, the efferent input, mediates the others. All except photomechanical movements and membrane shedding exhibit endogenous circadian rhythms; that is, they continue unabated when the animal remains in constant darkness. They all combine with mechanisms of dark adaptation to increase lateral eye sensitivity at night, the last in the list.

Although efferent fibers from the brain's circadian clock innervate all retinal cell types, the major physiological rhythms in Table 4.1 appear to originate in the photoreceptor cells, the retinular cells. Ehud Kaplan, George Renninger, Takehito Saito, and Barlow inserted glass microelectrodes through a hole in the cornea into underlying ommatidia, hoping to impale single retinular cells. They succeeded and discovered circadian changes in gain, noise, and quantum bump shape originated in the retinular cells (Kaplan and Barlow, 1980; Barlow et al., 1987; Kaplan et al., 1990). Remarkably, the circadian clock in the brain reaches out to these cells, the most distal cells of the visual system, influencing the earliest events in vision.

Kaplan and Barlow (1980) were surprised to discover the clock's ability to reduce photoreceptor noise at night. The photosensitive retinular cells have long been known to be spontaneously active when the animal is kept in darkness. They generate quantum bumps, the building blocks of the photoreceptor potential, in the absence of light during the day. Kaplan and Barlow found that at night neural signals from the clock lower the rate of spontaneously evoked quantum bumps and thereby lower the eye's background noise at night. The photoreceptor noise appears to result from thermal isomerizations of the light-sensitive molecule rhodopsin, and the clock appears to reduce the noise by stabilizing rhodopsin (Barlow et al., 1993). Also at night the clock prolongs quantum bumps, thereby increasing the eye's ability to sum photon events and trigger a signal to the brain (Kaplan et al., 1990). This also reduces the eye's response to rapid changes in illumination (Batra and Barlow, 1990). As a consequence of these circadian changes, the crab sacrifices temporal resolution for increased visual sensitivity at night.

Anatomical changes further increase the eye's sensitivity at night, and they do so at the expense of spatial resolution (Figure 4.4B). The clock's efferent output to the eye exerts both real-time and delayed changes in the anatomy of the retina. The delayed changes of membrane shedding and photomechanical movement are primed by the nighttime efferent signals but are triggered by daylight. Membrane shedding is a vigorous transient breakdown of up to 80 percent of the rhodopsin-containing membrane of the photoreceptor cells (Chamberlain and Barlow, 1979). This major metabolic event occurs once each day in every photoreceptor cell and can be blocked by eliminating the efferent signals during the previous night. Its function is not understood. The real-time changes in cell position, aperture, and screening pigment combine to increase photon catch and acceptance angle of each ommatidium, adapting the retina for dim light vision at night (Barlow et al., 1980). The aperture works much like the iris of vertebrates. During the day it constricts to reduce the photon catch by underlying photoreceptors, and at night it dilates to increase their photon catch. Changes in cell position cause the photosensitive region of the ommatidium, the rhabdom, to move closer to the aperture, maximizing the chance that a photon will be absorbed by a rhodopsin molecule. These nighttime anatomical changes expand the fields of view of adjacent ommatidia (Barlow et al., 1984) slightly blurring the crab's vision.

Such dark-light trade-offs in temporal and spatial resolution are widespread in the animal kingdom. For example, we switch from cone to rod receptors in response to dim light. Densely clustered in the fovea, cones provide high acuity but require moderate light intensities to function. Foveal cones also provide high temporal resolution, but cannot sum their inputs: each has a direct connection to the brain. Densely distributed in the periphery, rods provide poor spatial and temporal resolution but readily detect single photons. Several rods can sum their photon responses to trigger an optic nerve output to the brain. Although most vertebrates shift between their rods and cones in response to changes in ambient illumination, some anticipate them (Manglapus et al., 1998). *Limulus,* whose retina possesses only one type of photo-receptor, adapts to diurnal changes in illumination via a biological clock. The clock changes the receptor's properties in anticipation of changes in ambient illumination caused by the earth's rotation.

Octopamine, released by efferent terminals in the retina (Battelle et al., 1982), triggers many of the anatomical and physiological changes in Table 4.1 (Kass and Barlow, 1984). It does this by activating adenylate cyclase via a G-protein coupled receptor to synthesize cyclic AMP, which serves as a second messenger. Elegant studies by Barbara-Anne Battelle and her colleagues (Battelle and Evans, 1984; Edwards and Battelle, 1987; Battelle et al., 1982, 1998, 2000; Battelle, 2002) have uncovered a number of the biochemical mechanisms mediating the eye's circadian rhythms.

Circadian Rhythms Increase the Eye's Sensitivity, but Can the Animal See Better at Night?

Maureen Powers believed that the large day–night changes in the sensitivity of the retina ought to be detectable in the animal's visual behavior. Many scientists have been frustrated trying to study *Limulus* behavior in the laboratory. Powers and Barlow (1985) found that illumination of the lateral eyes evoked reflexive movements of the tail and that the probability of tail movement correlated with circadian changes in the amplitude of the ERG. Moreover, both the probability of tail movement and ERG amplitude are higher at night and lower during the day. Using a classical conditioning paradigm that paired light with an aversive stimulus, we measured the dimmest light intensity necessary to evoke the gill response at various times of the day and night under conditions of constant darkness. We found that the animal responded to lower light intensities at night (Powers and Barlow, 1985), indicating that the circadian rhythms of the eye (Table 4.1) do indeed affect the animal's ability to detect light.

What Does *Limulus* See?

Discovery of the robust circadian rhythms in the lateral eyes intensified a long-standing question: What does a horseshoe crab see? Hartline often joked that he had spent many decades "studying vision in a blind animal." After hearing about the newly discovered circadian rhythms in the late 1970s, he reminded Barlow that no one had succeeded in uncovering a role for vision in the animal's behavior. Intent on finding one, Barlow donned SCUBA gear and observed the animals in

Buzzard's Bay, a natural habitat near Woods Hole, Cape Cod. After many cold and lonely nights at the bottom of the bay, he learned only that the animals avoided overhead shadows he cast on their eyes by shifting his underwater clipboard in the downwelling moonlight. The response appears designed to avoid predators. Near the end of a frustrating summer, his MBL colleague Colleen Cavanaugh mentioned that *Limulus* often mate at night and that males seem to search for females. It was well known that every spring the animals migrate to beaches along the eastern coast of North America. Males and females pair off and build nests in shallow water. The mating activity is greatest during the nighttime flood tides of the full and new moons (Barlow et al., 1986).

To test whether vision has a role in *Limulus* mating behavior, Len Kass, Len Ireland, and Barlow made cement castings of female shells and other shapes (cubes and hemispheres) that were the same size as a female. The objects were painted black, gray, or white to see whether contrast affected the crab's behavior. We placed the cement objects along the water's edge near Woods Hole in the late afternoon and waited for the evening high tide. The experiment was a long shot. Standing on Mashnee Dike waiting for high tide, we hoped no one would wander down the beach to investigate. The last thing we wanted was to try to explain why serious scientists would be attempting to fool male horseshoe crabs with dummy females. Amazingly, as males came in with the tide, they began swarming around the castings of the female carapace, especially the castings painted black (Figure 4.5; Barlow et al., 1982). We were ecstatic, running up and down the beach like a bunch of kids, yelling to each other about how many males were around the submerged objects. It was truly a night to remember! There was no question that the female castings were the most attractive objects and the cubes the least. The black female castings, in fact, evoked the entire mating behavior of males: approach, mounting, and sperm release. As the tide receded, the males did not leave the immobile female castings and thus risked death by dehydration on the tidal flats. We took pity on these tenacious, love-struck males, detaching them from the castings and placing them in the outgoing tide.

The great attraction of the males to the cement castings proved that chemical cues (pheromones) were not involved. Moreover, males blinded by eye coverings did not approach the castings, much less try to mate with them. These experiments provided the first concrete evidence of vision's role in the animal's behavior—males use vision to find mates!

To measure how well *Limulus* can see, Powers, Barlow, and Kass

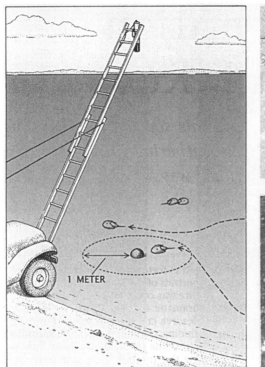

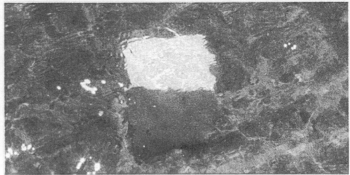

FIGURE 4.5 Video camera and image intensifier mounted high overhead record the response of adult male horseshoe crabs (left) to a cement casting designed to resemble a female *Limulus.* A casting (top right) evokes the complete mating response (approach, mounting, and release of sperm) when a male passes within sighting distance. Male crabs approach other objects (bottom right) briefly, but they do not remain in their vicinity or attempt to mate with them.

(1991) observed the behavior of males in the vicinity of a female-sized object using an overhead video camera. They fitted it with an image intensifier for nighttime observations and found that males oriented toward the black object at an average distance of 0.94 m during the day and 0.88 m at night. Females avoided such objects, as did juveniles (Ridings et al., 2002). It was clear that males can see a female-like object nearly as well at night as during the day (Powers et al., 1991). Apparently the large circadian increase in sensitivity of the lateral eyes compensates almost completely for the decrease in lighting after sunset.

To better understand the animal's visual performance, Erik Herzog, Powers, and Barlow investigated the ability of males to detect objects of different contrast. They found that males must be about 0.1 m closer to a low-contrast object to see it as well as a high-contrast one of the same size (Herzog et al., 1996). This was true both day and night. Their contrast sensitivity appears to remain about the same, regardless of the daily changes in illumination in their natural habitat.

How many ommatidia does a male use to see a female? The coarsely faceted lateral eyes enable crabs to see over a wide field but with very low resolution. At a distance of about 1 m, a male sees a female with very few ommatidia, perhaps as few as four. Males may well be operating near the optical limits of their lateral eye to find mates. At night they sacrifice what little acuity they have to increase their visual sensitivity. It is indeed surprising that they can see so well under water with only starlight.

What Does the Eye Tell the Brain?

Forming images of females with less than 1 percent of the eye's receptors raises the question of just what information the eye sends to the brain. Unfortunately, it is not feasible to record the responses of all ~1,000 optic nerve fibers from the lateral eye of a behaving animal. Fortunately, the *Limulus* eye contains one of the few neural networks whose properties can be modeled exactly. Extending the pioneering work of Hartline and coworkers described earlier in this chapter, Barlow and coworkers Christopher Passaglia and Fred Dodge constructed a realistic, cell-based model of the eye, one that is capable of computing the entire ensemble of neural activity the eye sends to the brain (Passaglia et al., 1998). No such realistic model exists for any other neural network of similar size or complexity in any animal.

Their strategy for deciphering the neural code underlying visually guided behavior was to record the lateral eye's view of its underwater world with a crab-mounted camera (CrabCam; colorplate 9). At the same time they recorded the response of an optic nerve fiber from an ommatidium looking in the same direction as the CrabCam. Upon returning to the laboratory, the team, together with Erik Herzog and Scott Jackson, played back the videotaped scene to the computer model and computed the ensemble of optic nerve responses, or "neural images," to the scene. They found that the response recorded from the single nerve fiber was a good match for that computed for the equivalent receptor of the model. They then examined the computed neural images for putative neural codes of potential mates.

Incredibly, the horseshoe crab eye appears to be "tuned" to detect other crabs. The eye responds vigorously to crab-size objects that move across its visual field (Figure 4.6; from Barlow et al., 2001). Moving retinal images of female-like objects can be generated by a female swimming past a male, a male passing a female, or both swimming by one another. Movement, not the sex of the crab, is the key. Male crabs are also attracted to males swimming by, but stationary retinal images evoke little response. Detailed studies show that the eye's spatial and

FIGURE 4.6 Computed responses of the *Limulus* eye and brain to moving mate-like objects day and night. The low and high contrast objects (top and bottom panels, respectively) approximate the size (0.3m diameter, 0.15m high) and range of contrasts of adult female crabs. They move across the field of view of a lateral eye at a distance of 0.6m where most visual detection occurs (Herzog et al., 1996). "Visual stimulus" (left column) are CrabCam images of the high and low contrast objects after sampling by the eye's optical apparatus. The arrays of pixels indicate the light intensities incident on the 16 x 16 array of ommatidia viewing the videotaped scene. "Neural image" shows the ensembles of optic nerve activities computed by the retinal model in response to the visual stimulus. The arrays of pixels in the neural images give the computed firing rates of optic nerve fibers mapped onto a gray scale. They represent snapshots of the responses of the 16 x 16 array of ommatidia to the visual stimulus. "Temporal integration" shows the computed neural images of synaptic activity in the brain after integrating the retinal neural image with an integration time of 400ms. The synaptic activities are mapped onto a gray scale. The fourth column displays the computed neural images of synaptic activity in the brain after "Spatial integration" within the excitatory centers of the presumptive receptive fields of laminar cells. Note that at night visual noise obscures the neural images of the visual stimuli, but temporal and spatial integration partially recovers them.

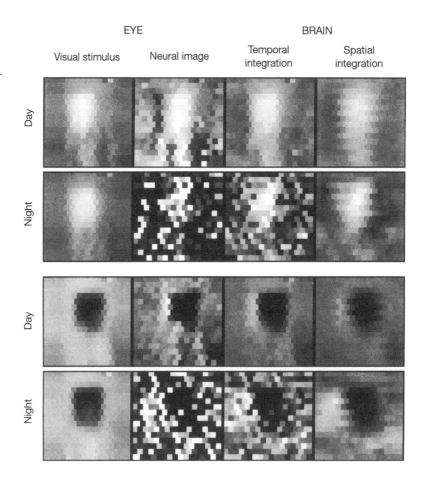

EYE BRAIN

Visual stimulus | Neural image | Temporal integration | Spatial integration

Day

Night

Day

Night

temporal properties are optimized for detecting moving, crab-like objects. Such properties readily account for the animal's ability to see high-contrast objects but not low-contrast ones.

How do male *Limulus* see females who have recently molted and possess low-contrast shells? Studies in the ocean suggest that males can better detect such low-contrast objects when their underwater world is illuminated with the flickering light that is a prominent feature of the shallow nesting areas. Wind-driven overhead waves act like lenses, creating beams of light that strobe the underwater scene at frequencies of ~2–6 Hz (Passaglia et al., 1997). Strobic light reflected off the low-contrast objects evoke coherent bursts of optic nerve impulses from neighboring ommatidia, enhancing the neural signals evoked by the object. Such reflective low-contrast objects activate retinal receptors in their paths and leave adapted ones in their wakes. Without image motion, the

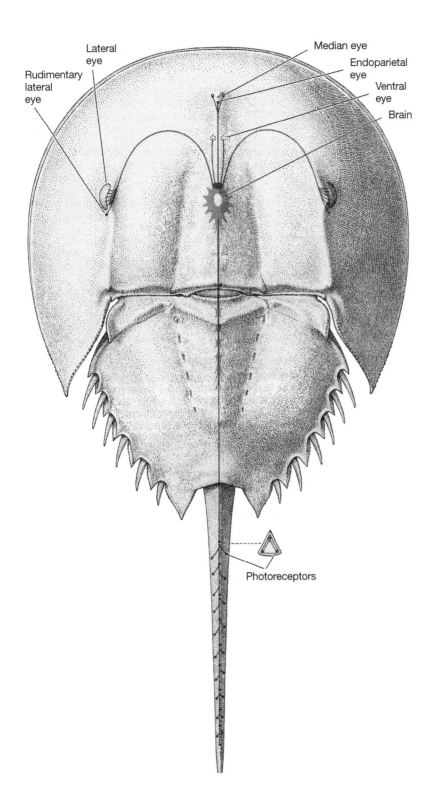

Lateral eye

Rudimentary lateral eye

Median eye

Endoparietal eye

Ventral eye

Brain

Photoreceptors

FIGURE 4.7 The eyes and photoreceptors of the horseshoe crab. Nine eyes transmit signals along optic nerves to the optic lobes located in the brain. At night a circadian clock in the brain transmits signals along the optic nerves to increase the sensitivity of the lateral compound and median eyes. Photoreceptors throughout the length of the telson transmit signals to the brain that synchronize the clock in the brain to daily cycles of light and dark. At night signals transmitted from the ultraviolet-sensitive median eyes enhance the sensitivity of the lateral eyes in accordance with the amount of ultraviolet light received from the nighttime sky.

eye provides no cues to the brain to help it distinguish flickering light reflected off an object from that reflected off sand.

The lateral eyes of this so-called living fossil are not as primitive as one might expect. They are elegant in design, incorporating many of the integrative mechanisms found in more complex vertebrate eyes. They possess universal excitatory and inhibitory mechanisms for processing the spatial and temporal features of visual information. They "tune" the lateral eye so it can transmit robust signals to the brain about mate-like objects both day and night (Figure 4.7). Circadian mechanisms of adaptation enable the eyes to operate over wide ranges of light intensity. Even on the darkest, most overcast moonless nights, they can tell the brain about potential mates (Atherton et al., 2000). Under such conditions *Limulus* can see what we cannot.

Limulus Has Ten Eyes!

We have focused on the most prominent visual organs of *Limulus,* the lateral eyes. But there are eight others (Figure 4.7) whose functions are not well understood. Two are the median ocelli that lie next to each other near the midline of the carapace. They are most sensitive to the ultraviolet light from the moon and stars, sending signals to the brain that increase the strength of the clock's efferent output to the lateral eyes and thus making them even more sensitive than they would be in total darkness (Herzog and Barlow, 1991). Why the animal evolved such complex interactions among its median and lateral eyes is not known. The ocelli and lateral eyes are the only eyes of the crab that have lenses and form images.

The remaining six "eyes" are primitive photosensitive organs. Three of them, the rudimentary lateral and the endoparietal eyes, are prominent in tiny crabs less than a year old and may provide young *Limulus* with an ability to orient to light before their lateral and median eyes are fully developed. Two of the remaining three eyes are the ventral photoreceptor organs that are located about 1 cm anterior to the mouth along the midline (Millechia and Mauro, 1969). They can tell the brain about the light intensity under the animal (Behrens and Fahey, 1981), and they certainly can convey information about whether the animal is upside down—but one would imagine there are better cues for this condition. The tenth eye is in the tail. Photosensitive organs distributed along the tail have an important role in the animal's life: their output to the brain helps synchronize the circadian clock to the day-night light cycle (Hanna et al., 1988). In short, if the horseshoe crab travels to another time zone, the tail "eye" will help it get over jet lag.

How Does the Brain Interpret the Information from the Lateral Eyes?

The lateral eyes encode and analyze visual information and send highly processed neural images to the brain. They tell the brain about moving crab-size objects as well as other dynamic stimuli. How does the brain decipher and discriminate them? A possible answer is suggested by the temporal properties of brain cells that receive the eye's input. These are neurons of the lamina, the first synaptic layer of the brain. They integrate optic nerve inputs with a time constant of ~400 ms, enhancing the neural images of moving crab-size objects relative to flickering light reflected off sand (Passaglia et al., 1997). Brain cells respond best to objects moving within the range of horseshoe crab speeds and respond poorly to fast-moving objects and rapidly flickering light.

Such temporal integration (see Figure 4.6) yields response envelopes that represent coherent changes in the mean firing rates among ommatidia viewing moving crab-size objects. The brain's neural code for a "mate" appears to be a characteristic pattern of activity among clusters of cells. The neural code is about the same whether a male passes near a female with a high-contrast shell or a low-contrast one. Model calculations confirmed by nerve recordings show that image motion binds together the responses of clusters of ommatidia in space and time, sending robust signals to the brain. Such changes in mean firing rate suggest that the *Limulus* eye transmits information to the brain with a rate code.

In addition to temporal integration, brain neurons integrate information from localized regions of visual space. Responses from a cluster of retinal receptors that view a small region of visual space converge on neurons in the brain. The cluster of retinal receptors forms the receptive field of a specific neuron. Although the dimensions of receptive fields in the brain have not been mapped with precision, we assumed for preliminary calculations that each neuron sums optic nerve inputs from a 3 x 3 array of retinal receptors. Figure 4.6 shows that spatial integration significantly improves the signal-to-noise properties of neural images computed for the nighttime state of the eye. The circadian increases in retinal sensitivity in combination with spatial and temporal filtering in the brain can yield detectable visual signals even in the very low nighttime levels of illumination. The circadian and neural integrative mechanisms may help explain why *Limulus* can see so well at night.

What Is the Neural Basis of the Crab's Behavior?

Do we know enough about how the visual system works to predict the animal's behavior? Do we know how the system receives visual information, processes it, and generates a behavioral response? No, we do not. Making some simplifying assumptions about spatial and temporal coding mechanisms, we simulated an animal moving underwater in the vicinity of a crab-size black object and found the computed estimates of brain activity could account reasonably well for the animal's visual performance (Passaglia et al., 1997). The underlying neural mechanisms—uncovered by more than 70 years of research—provide important insights not only about how *Limulus* sees but also about vision in general. The exquisite research initiated by Hartline has focused on a fundamental question in neuroscience: How is sensory information coded? Because of the relative simplicity of the *Limulus* lateral eye, we could answer this question by determining the entire ensemble of nerve activity the eye sends to the brain and how the brain begins to interpret the information. Can we take the next step and answer another fundamental question in neuroscience: What is the neural basis of behavior? For this we must probe deeper into the brain to find out how visual inputs control motor outputs. Perhaps then we will know how the animal sees.

GROWING UP TAKES ABOUT TEN YEARS AND EIGHTEEN STAGES

CHAPTER 5

Carl N. Shuster, Jr., and Koichi Sekiguchi

In many ways the life cycle of a horseshoe crab is similar to that of other hard-shelled, joint-legged arthropods. When it is different, it is markedly so. The life of a horseshoe crab begins in the intertidal zone of a beach. This is a rarity among marine invertebrates. Perhaps equally astounding is the change that occurs on about the sixth day of development—a transparent membrane lifts off the surface of the embryo and swells up to about twice the original diameter. From then on, the developing embryo can be seen rotating within this sphere.

The sequential account of the life cycle of *Limulus* began in Chapters 2 and 3 with discussions on nesting, spermatozoa, and the fertilization of the eggs. We continue here with descriptions of the egg, the mechanics of egg laying, and experimental fertilization of the eggs, before describing embryonic development and the later stages in the life cycle (see Exhibit 5.1 for information on how we became interested in these topics). We have used the term *stage* interchangeably with *phases* of embryonic development and with *instars,* the stages in the life cycle between two successive molts.

A Typical Arthropod Egg

Horseshoe crabs produce typical arthropod eggs. They are centrolecithal with a large densely packed yolk (lecithin) surrounded by a peripheral layer of egg protoplasm. The egg yolk of *Limulus* is usually bluish gray. The unfertilized egg is enclosed within a thick membraneous envelope, the chorion (also called the outer egg membrane), composed of distinct outer and inner layers (Sekiguchi and Yamamichi, 1988). The outer layer, or the basement lamina (5 microns thick), originates from the ovarian epithelium and the inner, or vitelline, envelope is formed by the egg itself (Dumont and Anderson, 1967).

Bennett's (1979) study on the chemical composition of the eggs determined the distribution of proteins, lipids, and carbohydrates in the unfertilized, mature *Limulus* egg. The envelope, cortical region, and yolk have high levels of 1,2 glycols, with the envelope containing fewer 1,1 glycol groups than the other components. Neutral mucopolysaccharides have been found in the cortical region and yolk, but only the cortical region has sulfated mucosubstances that are partly glycoprotein and glucose-6-phosphatase. Proteins are in all components of the egg. At first the protein in the egg envelope consists of sixteen amino acids; but, after fertilization, the envelope of a developing egg contains seventeen amino acids. Structural integrity of the chorion is due to electrovalent and S-S covalent linkages (disulfide bonds) between protein chains. Other constituents of the chorion include lipoproteins, unsaturated lipids, and fatty acids. Neutral lipids, unsaturated lipids, phospholipids, and fatty acids are in the yolk, while DNA is concentrated in the cortical region and the yolk.

How Are the Eggs Laid?

The process of egg laying was a mystery until fiber optics was used to film beneath a nesting female horseshoe crab (Natur Cine Pro Co., Ltd., 1993; directed by Dr. Sekiguchi). When she thrust her legs up and down in the sand beneath her, she created a slurry of sediment and water. The eggs, soft and sticky and covered by a jelly-like substance, were extruded from the oviducts in masses and dropped down into the slurry; spermatozoa attached to the egg coat when the eggs were laid. Newly laid eggs are more or less round and tend to clump together with beach material. Their surfaces soon harden after contact with seawater. Because the eggs are extruded from oviducts only a few millimeters apart (Figure 5.1), their proximity enhances the possibility that they will clump while they are dropping. The female uses her last pair of legs to push the eggs down into the slurry of sand and water beneath her. The whorl of spatulate blades at the tips of the legs aid in pushing the eggs closer together. Thus, beach material, usually sand and pebbles, is mixed with the eggs within a clump. The depth of the clump is dependent on the size of the female and how deeply she burrows into the sand; most nests are 15 to 20 cm deep. We wonder whether this mass of sand and eggs is easier to push deeper into the slurry. Is it less likely to be excavated by wave action than are single eggs?

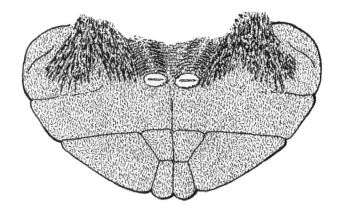

FIGURE 5.1 Posterior face of the female operculum showing the genital papillae (the tan-colored papillae are shown as white for contrast).

Experimental Fertilization of the Egg

In nature fertilization is external, with males in the vicinity of egg-laying females exuding copious amounts of spermatozoa into the water (Chapter 3). However, most laboratory studies on developmental stages of horseshoe crab embryos have been based on artificial fertilization. With this approach stages in embryonic development can be timed, beginning when fertilization occurs; this meets a basic experimental requirement. Apparently Osborne (1885), working at the Beaufort, North Carolina, laboratory of William K. Brooks, was the first to artificially fertilize *Limulus* eggs, in 1882. Now, the techniques described by Brown and Clapper (1981) and Sekiguchi (1988a) are used in culturing adults and juvenile specimens, in artificial fertilization, and in studies of the embryos.

Embryonic Development of *Limulus*

Details of embryonic development were among the earliest studies on *Limulus* (Packard, 1870, 1872, and 1885). Lithographic illustrations were prominent in descriptions of cellular, tissue, and organ development (Kingsley, 1892, 1893). When Patten (1912) described fifteen embryonic stages, he took advantage of the globular shape of the egg to create flat images from the three-dimensional embryos by using the Mercator projection, a map-making technique. These maps showed that the opisthosoma was the first main body part to grow around the yolk mass, followed by the prosoma. The segments, particularly of the opisthosoma, grow around the egg until their ends meet along the axial (dorsal) ridge and fuse. All features beyond the edge of the ventral plate become part of the dorsal aspect of the future animal.

In Japan, the culturing of all life stages of all four species of horseshoe crabs for research and resource management purposes has pro-

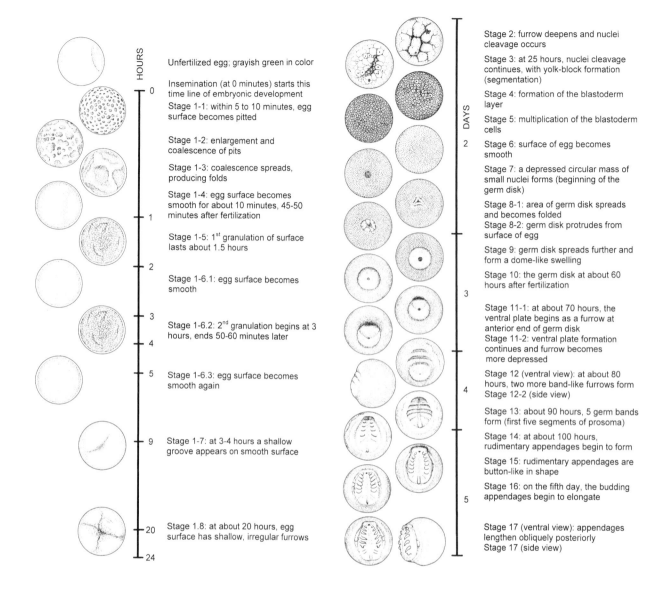

duced an impressive number of research reports. From these, we feature the observations on external changes during embryonic development of *Limulus*. The observations, made with a binocular stereomicroscope to see through the chorion, revealed successive stages in yolk segmentation and cell divisions during early development and subsequent tissue and organ formation (Sekiguchi et al., 1982). Details were enhanced by use of the vital stain technique developed by Oka (1943).

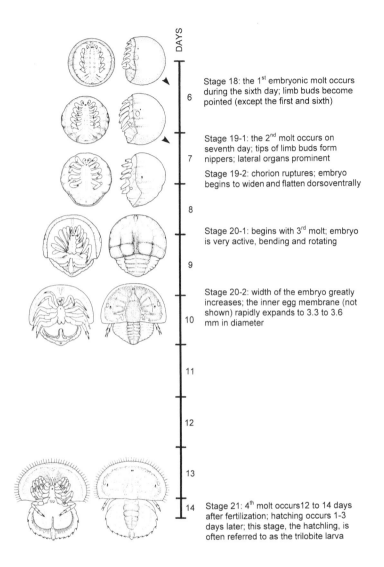

Stage 18: the 1st embryonic molt occurs during the sixth day; limb buds become pointed (except the first and sixth)

Stage 19-1: the 2nd molt occurs on seventh day; tips of limb buds form nippers; lateral organs prominent

Stage 19-2: chorion ruptures; embryo begins to widen and flatten dorsoventrally

Stage 20-1: begins with 3rd molt; embryo is very active, bending and rotating

Stage 20-2: width of the embryo greatly increases; the inner egg membrane (not shown) rapidly expands to 3.3 to 3.6 mm in diameter

Stage 21: 4th molt occurs12 to 14 days after fertilization; hatching occurs 1-3 days later; this stage, the hatchling, is often referred to as the trilobite larva

FIGURE 5.2 Diagrammatic representation of the sequence of distinctive embryological developmental characteristics arranged along a time line *(Sekiguchi, 1988b)*. Note that the time line is adjusted (not to scale) to accommodate each stage, beginning in increments of hours (h) and then days, with development of each subsequent stage increasing in time.

Charting Embryonic Development

In an important innovation, Sekiguchi et al. (1982) and Sekiguchi (1988b) placed all of the developmental changes along a time line (Figure 5.2). The time line provides a convenient summary of the sequence of embryological events under laboratory conditions of 30°C and 34–35 parts per thousand (ppt) salinity. Studies on the embryology of *Tachypleus tridentatus,* the Japanese horseshoe crab, revealed that it dif-

TABLE 5.1 A comparison of the early stages of the four species of extant horseshoe crabs.

Species	Number of Eggs		Diameter of Eggs (in mm)		Hatching Time (in days)	Prosomal Widths (in mm)	
	Per Clutch	Per Nest	Newly Laid	Capsule		Larvae	First-tailed
T. tri.	20,000	200–500	3.0–3.3	6.2	45	5.3–5.7	7.9–8.2
T. gigas	8,000	400	3.5–4.0	7.8	32–35	6.3–7.0	10.9–11.0
C. carcin.	10,000	80–150	2.0–2.2	4.2	32–35	4.0–4.2	6.0–6.3
L. poly.	90,000	3,650	1.6–1.8	3.6	15	2.7–3.7	4.5–5.5

Source: Compiled chiefly from Sekiguchi and Nakamura, 1979; Shuster, 1982; Sekiguchi and Yamamichi, 1988.

Note: In general and depending on the population considered, adult size in the four extant species ranks as follows, from largest to smallest: *T. tridentatus* > *L. polyphemus* > *T. gigas* > *C. rotundicauda*. Diameters of the newly laid egg (New), the egg capsule immediately prior to hatching, and the prosomal widths are in millimeters. The times (in days), from artificial fertilization to hatching, are based on experimental data.

fered from *Limulus polyphemus* in two major ways: the length of its time line and its large embryonated eggs (62 days and 6.2 mm in diameter, respectively). Table 5.1 summarizes these and other differences in the reproductive capacity and embryonic growth of the four extant (living) species.

Developmental Stages

In addition to the details given in Figure 5.2, Figure 5.3 provides a summary of embryonic events. Significant stages in embryological development begin on the surface of the fertilized egg. The large sphere of yolk modifies the rate and direction of cell division. Embryonic growth begins on one side of the egg, using the large yolk as the source of energy. At first the development is concentrated, with the growing embryo spreading around the spherical yolk mass until it engulfs the remnants of the yolk. Growth is first seen in the future oral region and spreads from there with the various organs arranged in a definite order from this center.

Rapid cell division, at the apex of the developing trunk of the future body, results in a terminal infolding (the telopore), that becomes the primitive streak, a line of cells along the future body axis (see Figure 5.2, stage 13). The mesoderm arises in part from the cells in that shallow depression (the telopore) that overlies a confused mass of proliferating nuclei destined to form mesoderm, yolk cell, and endoderm.

Eight distinct stages occur during the first day of development. Within 10 minutes of fertilization, the originally smooth egg surface becomes pitted with many crater-like depressions. At first, these pits increase in number and then they coalesce and spread until folds are formed (stages 1-1 to 1-3). By stage 1-4, a gap appears between the chorion and the egg surface, which is again smooth, but becomes granular (stage 1-5) and then again smooth (stage 1-6) until, by stage 1-8, the yolk is divided by folds into blocks. Interestingly, all of these stages also occur in unfertilized eggs.

On the second day after fertilization, the yolk blocks are subdivided with rapid increase in nuclei (stages 2 and 3) and the resulting spherical cells containing nuclei divide into small cells on the surface of the yolk mass (stages 4 and 5). Finally, the egg surface becomes completely spherical and is covered with many fine cells (stages 6 and 7). The germ disk, the beginning of the embryo, appears in stage 7, about 45 hours after fertilization. It is generally round and depressed at the center. In 50 hours a depression forms in the center of the germ disk and blister-like folds appear around the depression (stages 8-1 and 8-2). By 60 hours after fertilization, the germ disk has spread further and formed a circular, smooth dome-like swelling (stage 9).

About 60 hours after fertilization, the germ disk rises remarkably and an anterior semicircular part of the germ disk becomes deep red when the vital stain of neutral red is applied (stage 10). By 70 hours, the anterior semicircular area becomes triangular and rises at its posterior portion. The embryonic area of this stage is called the ventral plate (stage 11). The depression (blastopore) in the germ disk has migrated slightly posteriorly and has become increasingly indistinct. The ventral plate has become elevated and prominent arc-like furrows form across it. About 80 hours after fertilization, a germ band forms with two or three sections. These are primordial segments and appear on the ventral plate in the germ band stage (stage 12). At about 90 hours, the segments of the germ band increase in number to five (stage 13) and then, about 100 hours after fertilization, the bulges of the future legs appear (stage 14). These rudiments of the first to fifth appendages become round and button-like in shape (stage 15).

After several segments have appeared the large plate that forms the carapace is surrounded by an ectodermal fold on the periphery of the germ band (stage 16). During this stage the appendages, especially the middle three, become elongated obliquely backward and their tips become more pointed by stage 17. Also during this period of time, the embryo repeatedly undergoes expansion and contraction under the chorion (original egg cover). This movement separates the inner egg membrane from the surface of the embryo. The first embryonic molt is difficult to see because it consists of very fine cuticle shed by the em-

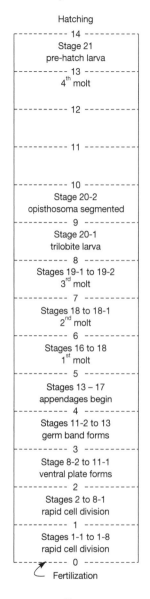

Time line of some
embryonic events

Days

FIGURE 5.3 Abbreviated time line of embryonic development in *Limulus polyphemus:* 14 days of development; read from the bottom upward *(based on Figure 5.2).*

bryo; it occurs on the sixth day (stage 18). By the end of this stage the egg swells to a diameter of about 2.0 mm. All of this movement is not depicted in the figure; but it is the most striking, easily observable aspect of early development—the cracking of the original egg coat (chorion) and the liberation of the embryo within a transparent spherical capsule (the inner egg membrane) that it had secreted. Uptake of water cracks the original eggshell and increases the space surrounding the embryo (by stage 16). From then on, until hatching, all stages of development are clearly visible, in contrast to the special staining and surgical procedures required to discern earlier embryonic development. The second molt occurs on the seventh day, at stage 19, by which time all limb buds are well developed. The lateral eye is the small black dot that migrates posteriorly in stages 18b to 19-2b. Closely associated with it is a lateral organ, the circular area, that may have functioned as a rudimentary eye, aiding young *Limulus* in orienting to light (Figure 4.9).

The third embryonic molt occurs on the eighth day, at which time the front and middle parts of the body (the prosoma and the opisthosoma) develop remarkably and increase in width (stage 20). The fourth and final molt occurs between 12 and 14 days after development began (stage 21).

Changes in the Egg Membranes

In an early stage of development, a very thin membrane forms on the surface of the embryo. Later another membrane is formed, with the material secreted from the whole surface cells of the embryo during stages 11 to 15 (see Figure 5.2). This inner egg membrane is thick and tough in the early stages and holds seawater in the perivitelline space between it and the embryo. The egg membrane or capsule gradually swells from an influx of seawater. The developing embryo is protected from the outer environment in this expanding spherical space for about 10 days, until it hatches. During this period, the egg capsule swells from about 1.7 mm in diameter to about 4 mm in diameter immediately before hatching (Sekiguchi, 1988b).

Trilobite Larva

While the larva of *Limulus* (see Figure 5.2, stage 21) is referred to as the trilobite stage, an earlier segmented stage (stage 20.1) would be more appropriately referred to as the trilobite larva (Sekiguchi, 1988b). Stage 20.1, the segmented embryo that emerges after the third embryonic molt, is more reminiscent of trilobite bodies, but the custom persists in designating the larva as the trilobite stage. Perhaps this can be excused as justifiable in the popular vernacular, because the horseshoe crab larva is the first stage in the life cycle that exists in the world beyond the egg and, therefore, is more comparable to a free-living trilobite. Molting within the egg capsule is a horseshoe crab trait; the molts are thin-walled and transparent (Figure 5.4).

Perpetual Motion?

A *Limulus* embryo begins rotating within its spherical capsule by stage 18, the stage after the first embryonic molting. From then on the appendages never seem to stop moving. In later stages, the movements are so strong that they propel the upside-down embryo to a righted position. Despite vigorous kicking, the yolk-filled embryo yields to gravity and falls back to its more prevalent upside-down position (see Figure 5.4). All this activity makes for fascinating observation under a dissecting microscope because these embryonic movements are precursors to the swimming behavior exhibited by the larvae and later stages (see Chapter 6).

Hatching

Over a century ago, just the discovery of newly emerged larvae of the horseshoe crab was noteworthy. Battey (1883) reported that while he and Professor Dwight of Vassar College were collecting shells at Martha's Vineyard on the second of August, Dwight discovered newly hatched *Limulus* larvae under a mat of seaweed lining the shore of one of the inlets. The individuals, about 4 mm in length, nearly filled a pint jar and, although out of water at the time, were moving in a lively manner. Possibly Dwight's discovery was the result of the mass hatching phenomenon described by Rudloe (1979). Along the shores of the northeastern Gulf of Mexico, *Limulus* nests are approximately 18 cm below the surface. When the larvae hatch beneath the surface, they are either distributed by storm waves that erode the beach or, in less turbulent water, they can move en masse to the surface and emerge on a spring high tide during nights of full moon. Apparently, this communal mass emergence from a nest does not often occur, if ever, on Delaware Bay beaches (Robert E. Loveland, 1998, Rutgers University; personal communication). Such observations reinforce our view that we cannot expect horseshoe crabs to act the same everywhere because their behavior probably responds to local environmental conditions.

Can Horseshoe Crab Species Be Hybridized?

Yes. Sugita et al. (1988) conducted artificial cross-fertilization experiments with the gametes of the four extant species. They used sperm from the four species to artificially fertilize the eggs of their own fe-

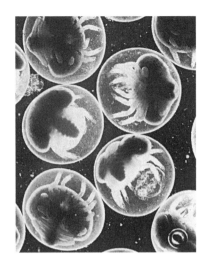

FIGURE 5.4 Growth and activity of a developing *Limulus* embryo within its transparent capsule can be observed for several days. These embryos are in late stages of development. Embryonic exuviae (shed skins) show clearly within the capsule in the lower right. *By permission from William H. Amos; from Shuster, 1960.*

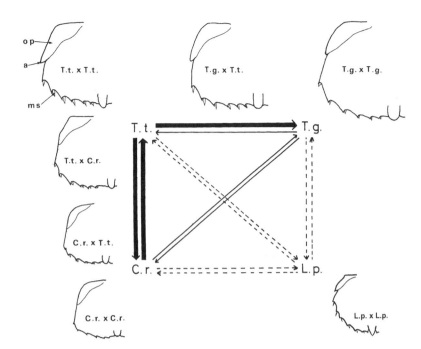

FIGURE 5.5 Hybrids produced by experimental reciprocal crosses with T.t. (*Tachypleus tridentatus*), T.g. (*T. gigas*), C.r. (*Carcinoscorpius rotundicauda*), and L.p. (*Limulus polyphemus*). The left margins of the opisthosomas of the larvae of the four species (corners) and the only three viable hybrids are all to the same scale. *From Sugita et al., 1988.*

males as well as reciprocal crosses. In marked contrast to an 89+ percent fertilization rate for crosses between the three Indo-Pacific species, no fertilization of eggs occurred in crosses between *Limulus* and those species (Figure 5.5). Only the reciprocal crosses of *Tachypleus tridentatus* and *Carcinoscorpius rotundicauda* produced viable larvae (female *T.t.* x male *C.r.*, female *C.r.* x male *T.t.,* and female *T.g.* x male *T.t.*). *T. gigas* gametes were the least compatible, with development occurring only when *T. Tridentatus* sperm fertilized *T. gigas* eggs. Interestingly, although the ranges of the three Indo-Pacific species overlap, the species apparently do not hybridize in nature.

Diagnostic characteristics include shape and size of the opisthosoma and the opercular pleurite (op), auriculate process (a), and the marginal movable spines (ms). Stout arrows represent development from fertilization to swimming larvae; slender arrows indicate the eggs that stopped development at the blastula stage; and dashed-line arrows indicate when fertilization did not occur.

Horseshoe Crab Chromosome

The demonstration of incomplete developmental compatibility between the horseshoe crab species in the hybridization study may be due to their differences in chromosome numbers. Iwasaki et al. (1988)

examined spermatogenesis (development of sperm cells) in the adult males. This was painstaking work as the tubules of the testes are intermingled with the greater tissue mass of the digestive gland. They determined that the diploid (2n) number of chromosomes for each of the four extant species was: *Tachypleus tridentatus* = 26; *Tachypleus gigas* = 28; *Carcinoscorpius rotundicauda* = 32; and *Limulus polyphemus* = 52 (in comparison, the 2n chromosome number in humans is 48).

The Environment Affects Development

Fieldwork has established that the development of the embryos may be as short as two weeks or as long as several months in the Delaware Bay area and, when they become larvae, may even overwinter within a beach, emerging and metamorphosing in the spring (Botton et al., 1992). This overwintering in the natural environment was mirrored by an early observation in the laboratory. When Lockwood (1870) stored viable eggs from the Raritan Bay area of New Jersey in jars, he was surprised to have larvae hatch as soon as 2 weeks and as late as a year later from the time the eggs were collected, even though the jars were subjected several times to freezing temperatures. One hundred years later at Woods Hole, Mass., eggs were refrigerated but not frozen and some of those eggs were brought up to room temperature and appeared to develop normally (French, 1979). In contrast, larvae and first-tailed stages died when subjected to the same treatment. When two batches of eggs were fertilized in April (1976 and 1978) and incubated at around 15°C, they hatched in 45 and 43 days, respectively. Batches of eggs fertilized in June of those years and incubated at approximately 23°C hatched in 25 and 28 days. The differences between the April and June hatchings strongly indicated a link between rate of development and temperature: hatching occurs sooner when incubated at higher temperatures. In nature, however, temperature and salinity vary throughout a tidal cycle.

Combined Effects of Salinity and Temperature

Due to the difficulties of investigating all variables that determine the rate of development or even survival of eggs in beach nests, many insights into what we might expect in nature have come from laboratory studies in the United States and Japan. In the United States, Jegla and Costlow (1982) examined the combined effects of salinity and temperature on development of *Limulus* embryos. Separate batches of 15–20 eggs, in an early embryonic stage, were exposed to 16 combinations of temperature and salinity. The time sequence, in which these stages oc-

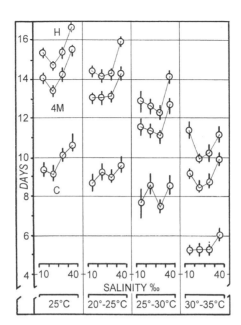

FIGURE 5.6 A graphic summary of the effects of temperature and salinity on the time in days for *Limulus* eggs to reach three developmental stages: chorion shedding (C), fourth embryonic molt (4M), and hatching (H). The circles show the mean times at which each stage occurred, and the vertical lines show the 95 percent confidence limits for each. *By permission from Jegla and Costlow, 1982.*

curred, varied according to the experimental design. The experimental matrix had four temperature ranges (25°C, 20°–25°C, 25°–30°C, and 30°–35°C) and four salinities (10, 20, 30, and 40 parts per thousand).

Development was slowest at the lowest salinity (10 parts per thousand [ppt]) and most rapid at the highest salinity (40 ppt) in any of the experimental water temperature ranges, while development increased as the water temperature increased (Figure 5.6). The shortest time between fertilization and hatching occurred in 10 days in 20 ppt salinity and 30–35°C.

Jegla and Costlow also compared the number of days between the hatching date of the larvae and when they molted. They noted that when embryonic development occurred in a beach habitat, molting occurred within 13 to 15 days; but, when embryonic development occurred in the laboratory, the larvae did not molt until 26 to 36 days after hatching. The results suggested that the environmental regimes within a nest are more conducive to future growth, at least to the larval stage, than controlled conditions in the laboratory. Further, the experiments provided information on the respective roles of temperature and salinity. In all examples molting occurred from 1 to 2 days earlier in the higher salinity. Overall, the length of time required for embryonic development increased from 20 ppt, 30 ppt, 10 ppt, to 40 ppt. The molting interval shortened, sequentially, in each of the temperature ranges (from 25°C, 20°-25°C, 25°-30°C to 30°-35°C).

Oxygen Levels Matter

When Palumbi and Johnson (1982) studied the responses of *Limulus* eggs and larvae to normal and hypoxic (low) levels of oxygen, they discovered that there was a deficiency in the amount of oxygen reaching the tissues under hypoxia. Because the oxygen content of sandy, intertidal sediments is often low, embryological stages may be subjected to at least periodic hypoxia. The larvae, however, are not normally subjected to such conditions. One of Palumbi and Johnson's experiments was designed to find out whether the physiological capabilities of one developmental stage in the life cycle of *Limulus* carried over to the next. In all the experiments, horseshoe crab developmental stages (embryos and larvae) showed remarkable adaptability and survivability. In fact, they had extreme tolerance when exposed to hypoxic conditions. If the tolerance of embryos to hypoxic conditions is also evident in the larvae, might this also carry over to older juveniles and to the adults (see Chapter 9)? However, under natural conditions this tolerance might not exist at beaches where toxic or inhibitory chemicals such as hydrogen sulfide could affect embryonic development or hatching success (Botton et al., 1988).

How Horseshoe Crabs Molt

Finding empty shells, complete in every detail, in beach wrack is a good sign that horseshoe crabs are in the area. These empty shells are not dead crabs; they are the molted skins (also called casts or exuviae) of juvenile horseshoe crabs. Like other arthropods, horseshoe crabs grow new shells when they increase in size.

Packard (1883) and Laverock (1927) described the process of molting but it was the naturalist Samuel Lockwood (1884) who colorfully and succinctly explained that a molting crab looks like "it is spewing itself from its own mouth." If you look closely at the under surface of a molted shell, you will see that the shell has split around the perimeter (rim) of the prosoma along the exuvial suture, or molt line. After the exoskeleton splits along this suture, the crab, with a soft new skin, begins to emerge. The split begins in the frontal area and, as the juvenile begins to emerge and expand, the split is forced open along the rest of the rim of the prosoma (Figure 5.7). The new shell that has grown below the surface of the old shell is soft, and the emerging animal

A

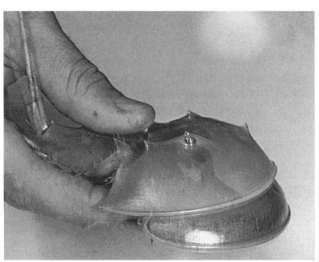

B

FIGURE 5.7 Juvenile *Limulus* in the process of emerging from their old shells *(Shuster 1955, 1982).* (A) When small (from a prosomal width of 28.5 mm, it reached 34 mm), the new animal has the consistency of foam rubber; (B) a larger (PW = 87 mm to 105 mm) juvenile has a pliant, firmer new skin. In marked contrast with a molted blue crab, *Callinectes,* which is soft and quiescent, *Limulus* can move if disturbed (note the fuzzy lines around the telson). *By permission of Alan A. Liss Co.*

increases in size by a rapid uptake of water that aids the animal in getting out of its old shell. This mode of molting—crawling forward out of the old shell and leaving the old shell intact—is another example of the uniqueness of horseshoe crabs among arthropods.

Getting Ready to Molt

Prior to shedding their old shell, juveniles generally burrow in wet areas. The larger the animal is, the longer it takes to emerge from its old shell. Larvae may completely emerge within 15 minutes after the exuvial suture has opened. The most difficult part of molting is withdrawing from the chitinous bars to which the book gill muscles are attached and from the leg casings. The new, premolt integument gets increasingly tougher at each successive molt, although it is still pliable at any juvenile stage. The new shell of small horseshoe crabs has the feel and look of foam rubber before it hardens. New shells of midsized juveniles are firmer and more rubbery at first. Several years later, the shell of a newly emerged large individual is somewhat leather-like before hardening. In both small and larger animals, the shell hardens by a tanning process that lengthens as the animals increase in size. Unlike crustaceans, which are virtually immobile during the molting process until their new shells harden, the usually quiet, molting horseshoe crab can be provoked into action despite the softness of the new shell (Figure 5.7). The length of time that elapses before emergence is complete is related to the size of the animals and environmental conditions.

An approaching molt is signaled by the yellowish borders that develop around the margins and along the molt line on the ventral surface of the carapace as well as by the turgid, plumpness of the body. In addition to the soft consistency of the new shell during molting, there are color differences between the premolt and the postmolt animal. The molted chitinous shell is thin and pale yellow in color in young, small animals. As the crab grows, the shell becomes thicker and more yellowish until, in larger animals, the shell is so thick that it appears brownish when molted. The color and other changes are best seen in half-grown animals. Just after molting, the new skin of these mid-sized animals is a bluish gray compared to the straw-color of the cast (colorplate 10). As the new shell tans and hardens, its color turns to a brownish yellow, often with a greenish tinge. That external coloration results from the blue-gray living tissue beneath the yellow chitinous shell.

Intermolt Periods and Duration of Molting

The size of an individual is correlated with the time it takes to complete the molting process, as well as with the interval between molts. Early growth stages occur rapidly but, as the individuals get larger, the time intervals between molts lengthen. Diet, rate of feeding, and environmental conditions are probably the major factors affecting the length of the intermolt period. The rapidity of growth to adulthood and the ultimate size of the adults appear to differ according to local environments, with temperature probably a prime controlling factor. In the northern portion of its range, *Limulus* may mature in 10 or 11 years and live up to 10 more, for a total of some 20 years. In the waters of Florida maturity may be reached sooner and the life span may be different (Anne Rudloe, 1979, Gulf Marine Laboratories, Inc.; personal observation). Except for studies on the hormonal regulation of molting, other physiological events underlying the mechanical changes have received little attention.

The interval of time required for an individual to emerge from its old shell depends in large part on the size of the individual. During their first year, horseshoe crab juveniles molt five to six times. Barthel (1974) observed the molting of a first-stage specimen from the Florida Keys and its next molt (instars 2 and 3). He kept the specimen in simulated natural conditions: instar 2, at an initial length of 6.2 mm, molted 10 days after capture to a length of 9 mm. Then, 23 days later, instar 3 molted again. It was quiescent for 5 to 6 days prior to each molting, staying under the sediment, and its feeding activities were much reduced or suspended. Molting occurred on the surface of the sediment. Postmolt inactivity lasted for less than 1 day.

The length of time to molt increases with the size of the animal until, in the larger stages, emergence may take up to 12 or more hours. After several molts during the first year, three may occur in the second year and two in the third year. After that, molting occurs once every calendar year. Over a year may pass between molts. When there is a minimum of disturbance by storms or heavy wave action, you can often see a sequence of casts appearing in the beach wrack over several months. Casts from older instar groups appear later and are larger than casts for younger instar groups. The largest, pre-adult juveniles molt as late as the fall of the year and, because they are usually in deep water on the continental shelf by then, their molts are not often found on beaches.

The Molting Hormone

Krishnakumaran and Schneiderman (1970) concluded that *ecdysterones* (Greek *ekdysis* = act of getting out—hence molting; *steros* = solid—hence the heavy-weight sterols) were the normal molting hormones of all arthropods. They came to this conclusion after demonstrating that they could induce molting in several crustaceans and chelicerates including juvenile *Limulus polyphemus*. The molting hormone in *Limulus* is likely to be 20-hydroxecdysone. Jegla (1982) and his students carried out an extensive series of studies on the histological and physiological aspects of molting in the larvae. Although neuroendocrine control was not determined, the larvae did synthesize a molting hormone, an ecdysone that appeared to be the major regulator of molting and development in the posthatch larvae and, presumably, also in the embryos.

The trilobite larva develops in a preprogrammed series of events that can be stopped at the beginning of premolt, at which time the animals can be stored for 6 to 8 months at 13–15°C. Upon return to an environment of 20°C the program restarts. The molt cycle has two distinct parts. The premolt period, during which the new integument develops, occupies over 50 percent of the cycle, followed by the sloughing (casting) of the old carapace.

Growing Up

After the hatchling leaves the comparative shelter of the beach nest, it undergoes further development: continued development of its digestive tract and an abrupt, stepwise metamorphosis into an adult in about 18 growth stages (instars). The stepwise growth is accomplished by growing a new shell underneath the old one and then shedding the old shell. This shedding process—molting (ecdysis)—and some examples of bizarre development are also described.

The larva, often called a trilobite larva or hatchling, undergoes two visible internal changes: the absorption of the remaining egg yolk and the completion of the digestive tract. Externally, its mid-piece has a barely protecting, nonmoveable terminal spine. This projection becomes larger in successive growth stages, producing a movable tail (Figure 5.8).

In the first-tailed stage that emerges from the larval shell during molting, the tail is longer but still fused to the body. However, by the next molt the tail is articulated and can be moved. Spines on the back of the small juveniles increasingly lengthen as the animals grow larger; but

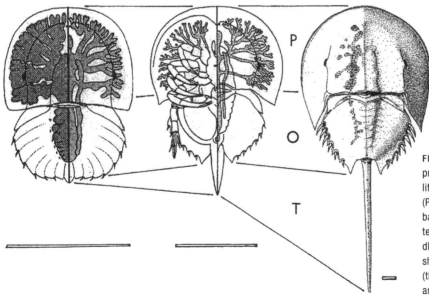

FIGURE 5.8 Comparison of the relative proportions of the bodies of stages in the life cycle of *Limulus* when the prosomas (P) are set to the same widths (each scale bar = 3 mm; 0 = opisthosoma; T = telson): larva (left), first-tailed stage (middle), and adult male (right). The figure also shows the extent of the digestive tracts (the shaded, branched areas) in the larva and first-tailed stages of *Limulus*.

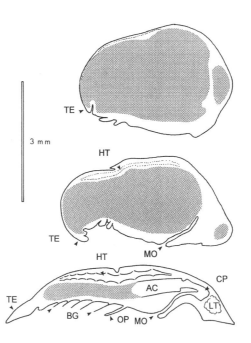

FIGURE 5.9 Diagrams of three developmental stages of *Limulus polyphemus*, showing the distribution of egg yolk (shaded areas) and the development of the digestive tract along the longitudinal axis (sagittal sections) of the body (ac = alimentary canal, bg = book gill, cp = crop, ht = heart, lt = liver [hepatopancreas], mo = mouth, op = operculum, and te = telson). (A) An embryo slightly beyond stage 19-2b *(based on Kingsley, 1892)*. (B) Embryo at stage 20-1 *(based on Kingsley, 1893)*. (C) Egg yolk fills more than half of the midgut of the premolt larva *(from Shuster, 1948)*.

after they are about half grown, these spines diminish in size and are blunt in the adult. Only the juvenile male undergoes sexual metamorphosis during its last molt. Most noticeable is the modification of his first pair of legs into a bulbous segment with a curved claw by which the male attaches to the edges of the female's shell (see Figure 3.1).

Development of the Digestive Tract

Throughout development, the embryo draws on the globular yolk mass of the egg, which flattens as the embryo develops around it until the remaining yolk is enclosed within the digestive tract (as depicted in the dorsal views, stages 20-2b and 21-2b and the longitudinal sectional views in Figure 5.9). The yolk is not solid but is composed of cells whose walls and nuclei are very distinct (Kingsley, 1893). The first external sign of the alimentary tract is the small depression that ultimately becomes the mouth (see Figure 5.2, stage 14). This depression shifts posteriorly in subsequent stages until it is just behind the first pair of legs. Concurrently, in the region of the future prosoma, broad sheets of tissues extend from the ventral plate into the yolk, dividing it into six pairs of lobes. These lobes become increasingly branched until they become a hepatopancreas, a mass of tiny brown tubules throughout the prosoma in juveniles and adults. Figure 5.8 shows the extent of the digestive tracts in the larva and first-tailed stages. In dorsal views of a larva (hatchling) stage, the segmental arrangement of the largely undigested yolk in a newly hatched larva is shown on the left; on the right side, just prior to molting, the system is nearly complete. The several lobes of the digestive diverticulae become more and more branched as the yolk is digested. In the ventral view of a first-tailed stage, the walking legs were pulled away from the axial region on one side to show the central position of the mouth (opposite the bases of the second and third pair of appendages); the dorsal view (right) shows the extent of the digestive system.

Ultimately the sheets of tissue become columns of muscles that are attached to the roof of the carapace and the bases of the legs. In the developing opisthosoma, the yolk decreases at the same time as the extensive muscular system of the gill-bearing appendages and large blood sinuses develop. The central, unsegmented part of the yolk that remains after the differentiation of the digestive gland (also called the liver, or hepatopancreas) becomes the alimentary canal. Yet, the digestive tract is still incomplete, half-filled with the greenish-brown yolk when the first instar larvae hatch (Kingsley, 1893). At this stage, larvae do not feed and, depending upon environmental conditions, they can subsist on the yolk for many months (Lockwood, 1870; French, 1979; Botton et al., 1992). As soon as the yolk is digested and the tract is fully developed, a larva is ready to molt into the first-tailed stage and then can feed (see Figure 5.8). At the first-tailed (second instar) stage, all components of the system are in place and the system is functional, although development of the midgut and digestive gland continue. All major internal parts of both larvae and the first-tailed stage are readily discernible under a microscope with the use of backlighting because the shell is virtually transparent. As an individual grows in size, the major change in its digestive tract is the increasing size of the digestive gland. No structural differences in the digestive tract have been reported between juvenile and adult horseshoe crabs.

Staircase Growth Pattern

Growing up is marked by abrupt changes in physical size. As in other animals with an external skeleton, a crab's shell develops under its current one, with the old shell acting as a mold for the new one. A new *Limulus* exoskeleton is soft and pleated in an intricate pattern of tiny folds. These shallow pleats unfold during molting when the animal emerges from its old shell. Figure 5.10 shows the staircase or stepwise pattern of growth on individual crabs. Such data collections, containing many stages in the lives of individual crabs, are rare due to the length of time required to raise crabs in captivity—possibly some 9 to 12 years.

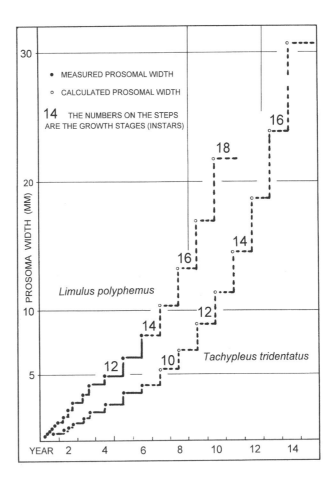

FIGURE 5.10 Growth in horseshoe crabs, as in other animals with an external skeleton, exhibits stepwise growth *(from Sekiguchi et al., 1988)*. The two steps contrast the ages and incremental increases in size at each molt for *Limulus polyphemus* and *Tachypleus tridentatus*. The solid circles represent actual prosomal width measurements (in centimeters) and the open circles represent widths based on the rate of growth calculations.

The figure also shows the time intervals between molts and the duration of the molting process. In such a growth series, each individual's growth can be determined directly by measurements of the old and new shells just after molting. The resulting increment of growth reveals not only the percentage of growth and relative sizes of the sexes but might indicate the condition or health of the animal. If aquaria-raised specimens are so rare, how has it been possible to determine growth stages, their number, and the increments of growth, in a population? Many specimens can be gathered, as Goto and Hatori (1927) demonstrated with the Japanese species, and then measured. The results usually define more or less bell-shaped curves for each sex, size, or age group (Figure 5.11). When groups of several sizes are collected, the tails of the curves usually overlap. We can also record the sizes of molting animals

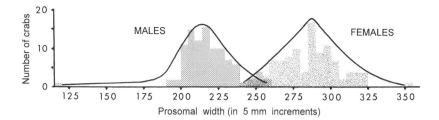

FIGURE 5.11 Size distribution (by prosomal widths) of males (open bars) and females (solid bars) from 100 mated pairs at Slaughter Beach, Del. *By permission from Shuster et al., 1961.*

(pre- and postmolt) and the sequential appearance of cast-off shells on intertidal flats and beaches. A composite of these separate observations helps us to determine the number of steps (instars) and their timing in the life cycle. Indeed, each approach can be used as a check on the other observations. A further refinement of this type of observation would be to determine whether juvenile males and females have different growth increments. The sex of juveniles, even the smallest ones, can be determined by examining the genital pores on the posterior surface of the operculum (see Figure 3.2).

Premolt/postmolt prosomal widths of juveniles taken from New England waters, at or just before molting, were measured by Shuster (1955, 1982). Incremental growth data also have been obtained from laboratory-raised *Limulus polyphemus,* from the first to the fourteenth instar (mean prosomal widths ranged from 3.30 mm to 81.2 mm, respectively) by Sekiguchi et al. (1988b). These data, even assuming different growth rates within discrete populations from Yucatán to Maine or even year to year within the same population, provide the best reference data thus far gathered. While suggestive, these data on incremental growth are too few to conclusively prove whether the females grow slightly larger at each instar than the males.

How Many Growth Stages?

Most studies have concluded that there are at least eighteen growth stages (instars) in the *Limulus* life cycle (see Figure 5.10). The evidence comes from measurements on many series of instars from nature and from individuals raised in the laboratory (Shuster, 1954; Sekiguchi et al., 1988b). With either approach, however, the series are incomplete due to either the difficulty of collecting the often far-ranging growth stages or problems in maintaining individuals in aquaria or outdoor tanks. Further, the number of growth stages may be different for the sexes because adult female horseshoe crabs of all the extant species tend to be appreciably larger than the males in the same population. This fact raises two possibilities: both males and females have the same number of growth stages, with the female having a greater increment at each molt, or females have more molts than the males. Carmichael et al. (2002) suggest that adult females may molt several times more than males.

A Gradual Metamorphosis

The hatchling is so similar in appearance to an adult, except for its lack of a tail, that the elongation of a telson in subsequent instars is an obvious change (see Figure 5.8). Subtle changes occur at each growth step; the body form changes from the free-swimming hatchling into an animal that can scuttle across the bottom and burrow. The second most obvious change is when a male matures and has the fist-and-thumb claspers used in mating.

Does Shape Change with Size? Lengths of the prosomal dorsal spines and lateral projections change with the age and the size of the animal. The spines increase in size with each instar until the midsized individuals and then gradually decrease in successive stages to the blunted spines of adult *Limulus* (Shuster, 1982). The length of these spines, at least in part, may be important for protecting the individual or preventing it from slipping during burrowing (Eldredge, 1970). Curiously, the projections may also be age-related. For example, the dorsal spines of an adult male (with a distance of 12 cm between the spines above the lateral compound eyes, usually termed the interocular distance) from the Florida Keys are blunt whereas the dorsal spines of an immature animal of the same size from Delaware Bay are sharp. Yet the terminal, lateral projections of the prosoma and opisthosoma of adult Florida males and the Delaware Bay juveniles are both sharper (form a more acute angle) than in the adult Delaware Bay adult male. The sharpness of these projections, as defined by the angle formed by the sides of the projection, is usually associated with the size of the crab—smaller animals have sharper projections. Thus, these carapace dimensions, when used to define different populations (see Riska, 1981), may include size-biased characteristics associated with same-size juveniles in other populations.

Do Adults Molt? The best answer for adult males is that they rarely molt, if ever. This conclusion is supported by measurements on 288 adult males comprising three age groups—young adults, middle-aged, and old—from Pleasant Bay on Cape Cod, MA (Shuster, 1955). Their similar size ranges and mean sizes indicated that molting did not occur in that population of adult males. Three other observations also suggest males do not molt: the obvious carapace deterioration over time, the

TABLE 5.2 Reports on large adult female *Limulus,* arranged from north to south.

Location	Comments	Observer
New York (north shore on east end of Long Island)	Specimen estimated to be at least 15 inches wide	Richard A. Azzaro, Esq.
New Jersey (Raritan Bay)	Very large specimen found in late 1940s; given to a visiting scientist from Europe by Dr. Robert Ramsdell	
Virginia (Kiptopeke Beach)	Found in April 1953 off Chincoteague by Mr. and Mrs. J. E. Snyder (Specimen #95748 in the collection of the Smithsonian Institution, National Museum of Natural History); prosomal width of 16 in (40.7 cm)	Jan Nichols, BioWhittaker
South Carolina	Prosomal width of 16 in (40.7 cm)	Dr. James Cooper, Endosafe

internal chitinous rods (trabeculae) that reinforce the adult carapace, and the age of the organisms living on the crab's shells (epibionts). Unless the internal rods are dissolved during a premolt phase, an adult male could not emerge from its old shell because its flesh would be enmeshed in the interstices of the trabeculae. In addition, large slipper snails, *Crepidula fornicata,* 5 to 6 cm in length, have been found on the carapaces of adult *Limulus.* Botton and Ropes (1988) determined that these organisms were at least 8 years old. Assuming that horseshoe crabs reach maturity in 9–11 years (Shuster, 1950), adults in the Delaware Bay area may live at least 17 to 19 years, This indirect method of aging *Limulus* appears to have general applicability. Grady et al. (2001) found that maximum *Crepidula* lengths can be used to estimate the minimum age of adult horseshoe crabs from Pleasant Bay, Mass.

For adult *Limulus* females, an extra molt explains why females are larger than males. The ultimate larger size of the female may result from additional growth stages, even after maturity has been reached (Carmichael et al., 2002). Gauvry et al. (2002) considered that this was certainly possible, based on the occurrence of four female molts with mating scars that were collected in the Delaware Bay area in 2001. Beyond additional growth stages, a possible explanation for the largest females observed (Table 5.2) is that these females experienced maximum or nearly maximum growth during each of the usual number of instars.

Reproductive Readiness

Adult males are easy to identify but not the females. Because a fine ovarian network mainly covers the dorsal surface of the digestive gland

(Makioka, 1988), an excised section of the prosomal shell usually has ovarian tissue attached to it. In a brief comparative study with Shuster (1955), who aged female specimens by their outward appearance, Osgood R. Smith examined shell/tissue samples from the same crabs, not knowing whether they were from juveniles or adults. This joint study indicated that external evidence was not as reliable as the absence of eggs in deciding which were juveniles. But, more important, Smith discovered that the adults had two and sometimes three sizes or batches of eggs. Thus, as soon as the batch (or clutch) of mature eggs is laid, the next clutch is already present, ready to begin maturation. Unanswered questions are: How soon will the next clutch be mature and ready to be spawned? How many clutches will a female produce during her lifetime?

When Shuster participated in two National Marine Fisheries Service survey cruises in 1986 on the continental shelf from Cape May to Cape Fear, he examined horseshoe crabs from among all the marine species collected by trawling (nine specimens in March and twenty-two in September). Although the numbers of adults collected were few, mature ova could be palpitated from female *Limulus* by stroking the area over the oviducts on the posterior face of the operculum. All of the males were molting, possibly due to changes in temperature and pressure as they were dragged from depths to the surface or stress when the trawl contents were dumped onto the sorting table. These observations of adults, combined with the fact that spawning extends over several spring and summer months (usually May and June but sometimes as early as April and as late as August in the Delaware Bay area), suggest that once the adults begin to produce gametes, they may be reproductively ready most of the year.

Monstrosities Invite Curiosity

Not all embryonic development of horseshoe crabs proceeds in a normal manner. Patten (1896, 1912) obtained many abnormally developed embryos by placing the eggs from several nests in running seawater for 8 to 10 weeks at the Marine Biological Laboratory in Woods Hole, Mass. After most of the larvae had hatched and swum away, 5 to 10 percent remained. Many of these were apparently sound and healthy, but among them Patten found considerable abnormality, in early and late stages of embryonic development, in endless variety and combinations. Some were pygmies or giants; others were legless, headless, or tailless, or consisted of fractional parts (halves, quarters, or smaller divisions).

There were also fused doublets and triplets. Patten surmised that the abnormalities were due to variable conditions in the eggs, because so many of the eggs had developed normally.

Only recently has anyone attempted to study the causes of such abnormalities. The catalyst was the drastic reduction of horseshoe crabs in Japanese waters during the past 30 years. The most obvious cause for the problem was that this was due to the reduction in spawning habitat, but the question remained whether there were any other contributing factors. Because other marine species in the coastal waters were also decreasing, Tomio Itow wondered if pollution could have contributed to their demise (1993). One way to examine this theory was to look at the developmental stages and count the number and types of abnormalities. During a visit to the United States in 1996, Itow exposed developmental stages of *Limulus* to seawater collected at several locations and obtained results similar to those in Japan; trace metals in seawater produce abnormal embryonic growth. The study was conducted in the laboratory of Robert E. Loveland at Rutgers, The State University of New Jersey (Itow, 1997).

Examples of Abnormalities in Large *Limulus*

Abnormalities in the telson of *Limulus* are the most frequently observed and reported of any structural malformations (Shuster, 1982). The slenderness of the long telson and its fragility relative to other body parts, are factors in accidental injuries. As the telson heals, the crab's propensity to produce small spines along the dorsal axis of the telson may be directed toward the site of injury, with an elongated spine as the result (Shuster, 1982). In almost all of the malformations of the telson, small axial spines are usually missing. Examples of these telsons can be lined up to form a series, from shorter to longer elongated spines at different sites, to a perfectly forked telson. Other telsons may be foreshortened and swollen; this condition also has many variations.

Shuster (1954) interpreted the occurrence of an extra notch among the marginal spines on the opisthosoma that was repeated from one stage to the next in six growth stages of one animal as an indication that once an injury or malformation occurs, it is continued thereafter in the life stages of the individual. This is consistent with the notion that the old carapace is the mold that the new integument faithfully copies during its formation.

Occasionally a specimen with normal-sized appendages has one or more small-sized ones. This may be an attempt at regeneration. Jegla (1982) and his students tested this possibility in a variety of experiments

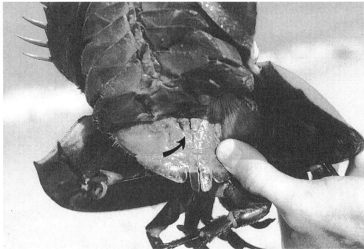

FIGURE 5.12 A hermaphrodite with (left) mating scars of a female (top arrow), skid marks (middle arrow; usually caused when an animal is pulled across the bottom by a trawl or dredge), and male claspers (bottom arrow). The genital papillae project like those of a male, but the openings are slit-like, more like a female (right, above arrow). *Photographs by C. N. Shuster, Jr.*

on the larvae. They found that, if a few of the limbs were cut off, the larvae would regrow them. This raises a question—could larger juveniles also regenerate lost or injured appendages? Apparently so—Schaller (2002) and Haefner et al. (2002) photographed adults with partially regenerated appendages. Further, because *Limulus* apparently does not have the capacity to drop an appendage by abscission as commonly seen in the common blue crab *Callinectes sapidus* and other crustaceans, the foreshortened appendages could be either embryological abnormalities or the result of regeneration after injury.

Hermaphrodite horseshoe crabs are found infrequently (Figure 5.12). The mix of male and female characteristics is not the same in all specimens. One had the general appearance of an adult male and had the mating scars of a female (Baptist, 1953). Its operculum bore a typical female papilla on the right, a male papilla on the left; internally, both spermatozoa and ova were present.

Some bizarre live animals have been exhibited at the National Marine Fisheries Service aquarium in Woods Hole, Mass. One specimen

A

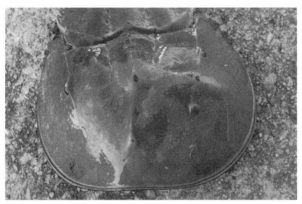

B

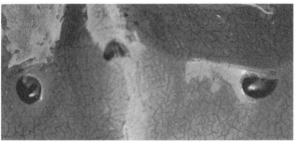

C

FIGURE 5.13 A three-eyed *Limulus.* (A) Dorsal view: the third compound eye is anterior to the normal lateral compound eye on the left side. (B) Oblique head-on view. (C) Close-up view: the anomalous axial ridge bearing a photoreceptor; the third eye is on the left, the normal eye on the right. Specimen was collected by the Woods Hole Marine Biological Laboratory supply department and given to the NMFS Aquarium by R. B. Barlow. *Photographs by T. L. Morris, Jr., National Marine Fisheries Service Aquarium, Woods Hole, Mass.*

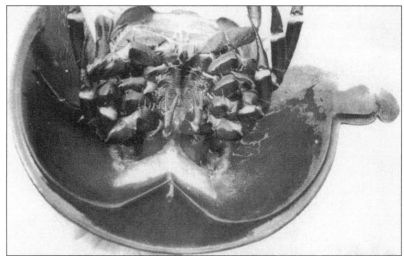

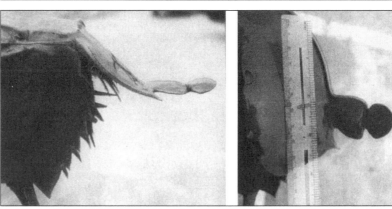

FIGURE 5.14 Adult male *Limulus* with an unusual nonarticulated projection from the right margin of its prosoma. Top: this oblique view of the ventral side of the prosoma shows that the rim extends into the base of the projection. Lower left: the thinness of the lateral projection is seen in this posterior view. Lower right, dorsal view: the projection was about 40 mm in length (millimeters are the smallest unit of measure on the plastic ruler). *Courtesy of T. L. Morris, Jr., National Marine Fisheries Service aquarium, Woods Hole, Mass.; photographs by C. N. Shuster, Jr.*

was a young adult male with three compound eyes. Its central prosomal axis was bent to the right near the position of the middle axial spine of the prosoma. Apparent malformed simple eyes were between the normally placed lateral and the more anterior third eye on the left side (Figure 5.13). Another young adult male had a nonarticulated lateral projection on its prosoma that superficially looked like the terminal portion of the paddle leg of an extinct sea scorpion (Figure 5.14).

Together, the authors have over 100 years of experience in the study of horseshoe crabs.

Koichi Sekiguchi: "My interest in horseshoe crabs began in the 1930s, when I assisted my teacher, Professor Hidemiti Oka, in his studies on experimental embryology using the Japanese horseshoe crab, *Tachypleus tridentatus.* Many times we went to see and collect horseshoe crabs at Oe-hama Beach (Okayama), a beach that the Japanese government designated as a natural preserve and breeding place for horseshoe crabs in 1928. That beach and neighboring areas no longer exist. They were reclaimed as land by the construction of a polder in 1970."

From that beginning, Sekiguchi established a method of artificial fertilization of *T. tridentatus* and produced a time line of the normal development of the horseshoe crab. "In the summer of 1974, I visited Professor Dr. John Costlow, Jr., the director of the Duke University Marine Laboratory. We used artificially fertilized *Limulus* and *Tachypleus* eggs to study their comparative development and larval growth. My interests have extended over several other fields, including studies on egglaying activity, natural habitats and distribution, and comparative biochemistry. On the basis of my studies and those of many collaborators, I proposed a new interpretation of the phylogenetic relationships among the four extant horseshoe crabs and a biogeographic theory of their distribution in *The Biology of Horseshoe Crabs* (Sekiguchi, 1988a). Carl Shuster became a close friend in the 1970s. I visited his home and we investigated *Limulus* mating behavior in Delaware Bay in 1989. In 1994, when Dr. Shuster and his colleague Dr. Mark L. Botton visited Japan, my colleague Dr. Hiroaki Sugita and I participated in a study on the mating behavior of the Japanese horseshoe crab at Imari Bay and Kitsuki Bay, Kyushu."

Carl Shuster: "My introduction to *Limulus* was a bottle of its eggs and larvae. This led to a thesis study of the development of the digestive tract in the early stages (Shuster, 1948). Then, in May 1949, I accompanied my mentor, Professor Thurlow C. Nelson, to Cape May, New Jersey, on a weekend trip to Delaware Bay to observe spawning for the first time. When we walked out of the Nelson cottage onto the beach just before dawn, the area was shrouded in a heavy mist. No wind was blowing. Only the gentle lapping of the water on the beach and a click-clack sound were heard. The entire setting was like stepping out of the present into some prehistoric scene. As we approached the water's edge, we saw groups of rounded stones. But these stones were moving—they were the nesting horseshoe crabs. The click-clacking came from their bodies bumping against one another. During that weekend several observations were impressed on my mind: the impact that changes in the amplitude of the waves striking the beach had on the size of mating groups; the mating scars on the females; and the incessant feeding by migratory shorebirds on the *Limulus* eggs that were strewn over the beach. Since then, I have observed spawning activity hundreds of times, often with other observers, at locations from Florida to Maine. Perhaps the most memorable scene and the one that continues to attract the most attention, from me and others, is the combined spectacle of masses of mating horseshoe crabs and thousands of shorebirds feeding on horseshoe crab eggs on Delaware Bay shores [see Chapter 1]. My interests in horseshoe crabs have included their natural history and ecology, morphology, morphometrics, functional anatomy, evolution of mating, serological relationships, the fossilized species, and the conservation and fisheries management of the species."

HORSESHOE CRABS IN A
FOOD WEB: WHO EATS WHOM?

Mark L. Botton and Carl N. Shuster, Jr.,
with John A. Keinath

Inquisitive beachcombers encountering us during the field season often ask how horseshoe crabs eat, and how they interact with other species in the estuarine food web. The location of the mouth isn't obvious, because it is located on the ventral surface and is largely concealed between the spiny bases of the walking legs. In fact, horseshoe crabs are classified in the group of arthropods known as the Merostomata, which includes animals with thighs (meros) surrounding the mouth (stoma). Nor is it easy for us to demonstrate that horseshoe crabs are surprisingly versatile feeders. Unlike lobsters and true crabs, whose formidable claws and powerful mandibles reveal obvious predatory behaviors, the horseshoe crab's feeding appendages don't seem capable of inflicting much damage.

Two of the authors have been fortunate to have had the opportunity to study the feeding ecology of *Limulus*. In 1949, Alfred C. Redfield invited Carl Shuster to the Woods Hole Oceanographic Institution to help investigate the predation of soft-shell clams by horseshoe crabs in the Barnstable Harbor area of Cape Cod. In that era, horseshoe crabs were suspected of causing so much damage to clam beds that Massachusetts paid a bounty for their destruction. Nearly 30 years later, Mark Botton began his doctoral research at Rutgers University under the direction of Harold H. Haskin (who, by coincidence, was a former doctoral student of Redfield's). Botton was struck not only by the massive numbers of horseshoe crabs spawning on Delaware Bay beaches, but also by their density on the intertidal flats at low tide. Surely such a concentration of large predators had the potential to have significant impacts on the benthic invertebrate community. In the late 1970s, marine ecologists such as Robert Virnstein and Sarah Ann Woodin were popularizing the use of manipulative field experiments, including exclosures and enclosures, to study the impacts of predation on soft-bottom benthic communities. Inspired by their approaches, Botton applied these techniques along with analyses of stomach contents and aquarium experiments to study the feeding ecology of horseshoe crabs in Delaware Bay.

In this chapter, we describe the feeding biology and digestive system of *Limulus*. We also consider the significance that horseshoe crabs play in the ecology of shallow-water marine communities, as predators, prey, and hosts to a diverse suite of epibionts that grow upon their shells.

How Do Horseshoe Crabs Eat?

Horseshoe crabs dig into the substrate in search of their usual food, clams and worms. Under the broad protective dome of the prosoma, the legs probe into the sediment, producing a slurry from which they can more easily pluck prey. The six pairs of book gills on the opisthosoma may indirectly contribute to food gathering; these appendages probably move pulses of water beneath the carapace that loosen the substrate, making it easier for the claws to grab prey.

You can view feeding under simulated conditions by presenting an overturned specimen with a food item, or by using a mirror under a glass dish containing the crab. When first observed, the legs and book gills seem to be writhing in a random and uncoordinated manner. To fully appreciate the complex feeding movements, you must study the appendages—their shape, musculature, position and arrangement of the bases, and the soft cuticle that surrounds their bases. Hydraulic forces developed during certain muscular actions probably enhance the appendages' movements (see Chapter 11). The elongated, heavy bases of the legs can roll from side to side and rock back and forth on their axes. Not only do the leg bases surround the mouth, but they are also canted toward it at an angle, placing their medial portions in close proximity (Figure 6.1). The tips of the legs have a more extensive excursion in several directions, and they sweep back and forth in virtually straight paths during walking (Manton, 1977).

The thirteen pairs of horseshoe crab appendages have overlapping roles in locomotion, burrowing, food gathering, and water flow. Of these, the first seven pairs have the most direct roles in feeding. The

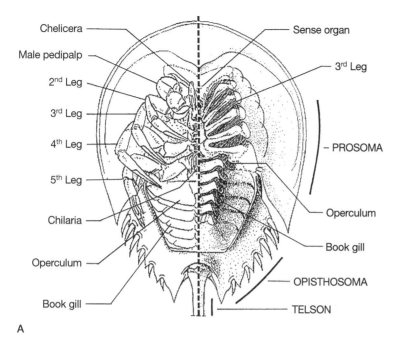

A

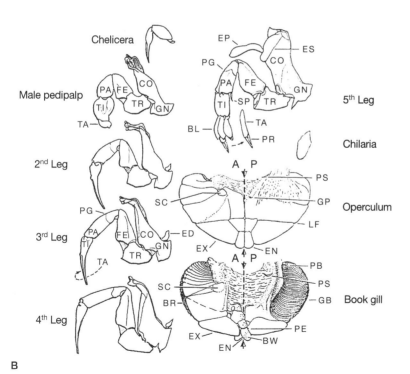

B

FIGURE 6.1 Thirteen pairs of appendages and a sense organ are on the ventral portion of the body of *Limulus (based on Shuster, 1955).* (A) In this divided drawing, the appendages are in their natural positions on the left. On the right side, the appendages have been removed and their areas of attachment indicated by the elongated, shaded areas. The relative amounts of forward and backward rocking of the coxal joints of the legs are represented by the five arcs to the right of the leg bases. (B) Details of the appendages. Only the third book gill is shown because all are similar. The segments of the legs, from base to tip, are coxa (CO), trochanter (TR), femur (FE), patella (PA), tibia (TI), and tarsus (TA). Appendage details: third leg (ED = endite, GN = gnathobase, PG = patellar groove), fifth leg (ES = epipodite suture, EP = epipodite, SP = patellar spine, BL = blade, PR = pretarsus), operculum (PS = parabranchial stigmata, GP = genital papilla, LF = line of flexion, EN = endopodite, EX = expodite), and book gill (PB = proximal gill leaflets, PS = parabranchial stigmata, GB = book gill, PE = penultimate spine, BW = brachial wart, EN = endopodite, EX = expodite, BR = branchiae [all of the book gill leaflets], SC = sclerite). A = anterior and P = posterior sides of the operculum and the third book gill.

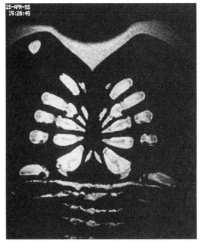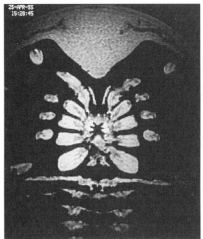

FIGURE 6.2 Two magnetic resonance images (MRIs), or "slices," through a live inactive adult female horseshoe crab show the medial sections of the coxal segments, longitudinal sections of the trochanters, and small parts of the other leg segments (see Figure 6.1). The gnathobases (endites) on the coxal joints radiate from the mouth area, and the nearer they are to the mouth (bottom view), the more they look like "teeth." *MRIs courtesy of E. M. Haacke, President, Magnetic Resonance Innovations, St. Louis, Mo.*

claws (chelae) on the walking legs have no crushing power to speak of; rather, their role is to move food to the gnathobases (the spiny projections of the basal joints) surrounding the mouth. These heavy projections, particularly on the last pair of legs, serve like rudimentary external molars that can easily crush the shells of small mollusks. Each basal segment of the five pairs of walking legs can rock side-to-side along the long axis of the basal segments (coxae; Figure 6.1B) as well as tip up or down from its dorsal articulation with the prosoma. Thus, the position of the bases of each leg imparts a slightly different role.

Two basic orientations of the five pairs of walking legs facilitate feeding—the angle at which they are attached to the ventrum and the angles at which the projections of their basal joints approach the mouth area (Figures 6.1 and 6.2). The biomechanics of walking and feeding are quite distinct, although the same appendages are involved. The swinging motions of the bases of the legs during locomotion are at right angles to the "biting" movement in feeding (Manton, 1977). Despite the biting implication, there is not much tearing or shredding of the food material. Rather, the gnathobases macerate the food while it is moved toward the mouth.

Watching the gnathobases in action is like viewing opposing rows of spike-covered fists aligned above a table (the sternal region of the prosoma)—each pair of spiked protuberances alternately dipping (grasping the food) and rising (releasing their hold on the food). The coxa usually follows an oval path, dipping and rocking toward the mouth and then rising and rocking away during feeding. The oval motions move toward the median line of the body as food is pushed toward the mouth. When rejecting food, the process is reversed. Nine muscles move the coxa but the reverse action of only two of these produces either feeding or egesting (Wyse and Dyer, 1973). This activity results in a peristaltic-like movement of the food toward the mouth. The food disappears so quickly at the end of the action that it appears to be sucked into the esophagus.

The paddle-like chilaria aid in keeping food from passing posteriorly and in pushing food anteriorly, while the small pincer-like chelicerae assist in placing food particles back over the mouth. Soft-bodied prey and smaller clams (less than about 20 mm in length) are usually crushed by the gnathobases and shoveled, shell and all, into the mouth. A different feeding technique is used when a crab is feeding on large bivalves of about 40 mm (Botton and Haskin, 1984). The walking legs hold the ventral margin of the shell (the edge opposite the hinge) against the gnathobases. The shell is worked along the gnathobases, creating chips and fractures. Eventually, enough of the shell is broken so the crab can remove the meat using its chelae.

How Do Horseshoe Crabs Digest Food?

The digestive tract of the horseshoe crab is a simple, thin-walled tube that is differentiated into six discrete sections: mouth, esophagus, proventriculus (a combined crop and gizzard), intestine, rectum, and anus (Patten and Redenbaugh, 1899; Shuster, 1948; Lockhead, 1950). Chapter 5 describes the tract's development; see colorplates 11 and 12 for the anatomy. Both ends of the digestive tract are lined with chitin and shed with each molt. A longitudinal section through an adult *Limulus* shows the placement and dimensions of each part of the digestive tract. The mouth, esophagus, and proventriculus lie entirely within the prosoma; the intestine extends from the prosoma to the opisthosoma. Schlottke (1934) and Shuster (1948) have described the histology of the various regions of the gut. The mouth is centrally located, between the bases of the legs, and just posterior to a lip-like protrusion, the labrum. The esophagus (a short, flexible, and slightly muscular tube) leads slightly upward and anteriorly from the mouth, passing through the crab's brain (a large nerve ring) before entering an enlarged section at the base of the proventriculus. The proventriculus is a muscular, elongated, pear-shaped organ occupying the frontal area of the prosoma. It is near the anterior-most part of the alimentary tract. Its base is extensible and sac-like and is sometimes referred to as the crop.

Ingested material passes from the crop upward into the gizzard lined with several parallel, notched ridges of thickened chitin extending along its length. Within the gizzard, food is fractured into smaller particles by muscle action that grinds the prey with other engulfed particulates such as sand, pebbles, and shell fragments. The gizzard mashes food material into a pulp that can be squeezed through the pyloric valve, a posteriorly bending, fleshy, conical protrusion at the end of the proventriculus, into the intestine for enzymatic digestion. Sometimes, virtually intact small bivalves may be passed into the intestine (see colorplate 12). Two pairs of tubules lead from the intestine into brownish-colored masses of multibranched midgut diverticula, or hepatopancreas, which occupies much of the prosoma (Lockhead, 1950). Fecal pellets are encased in mucus and may be several centimeters long in an adult animal. A short, chitin-lined rectum leads to the anus, located by the muscular, ventral base of the telson.

Schlottke (1934) and Yonge (1937) emphasized the homologies of the digestive process of *Limulus* with that of arachnids. Preliminary digestion in the intestine is extracellular, using enzymes secreted by the hepatopancreas. The hydrolysis of dipeptides is intracellular within the many tubules of the hepatopancreas. The alimentary canal of the Indo-Pacific horseshoe crab, *Tachypleus gigas,* has a number of digestive enzymes, including acid and alkaline proteases, esterase, amylase, invertase, and cellulase (Debnath et al., 1989). The demonstration of cellulase activity is particularly interesting, as it lends support to prior speculation that plant detritus might be of nutritional value to horseshoe crabs (Botton, 1984a).

Fat, glycogen, and protein are stored in yellow connective tissue between the tubules of the hepatopancreas (Makioka, 1988). At times, these absorptive cells pass considerable quantities of calcium phosphate into the lumen of the hepatopancreas. From there, crystals of this material pass into the alimentary canal, adding to the feces (Lockhead, 1950).

How Do Horseshoe Crabs Find Food?

Turner (1949) and Baptist et al. (1957) have observed that horseshoe crabs aggregate in areas of abundant food supplies (Figure 6.3). Horseshoe crabs almost certainly make primary use of chemical cues to detect prey. Wyse (1971) found that extracts of fish or clam elicit a chemoreceptory response by the chelae, and the spines on the gnathobases also contain chemoreceptors (Barber, 1956). The latter chemoreceptors are relatively unresponsive to solutions (sweet, sour, bitter, salty) but are strongly activated by juices from bivalve mollusks. Stimulation of the sensory cells elicits the feeding reflexes described by

FIGURE 6.3 A pair of horseshoe crabs and a single animal feeding at the edge of a blue mussel bed. Typical water-filled excavations show the extent of previous activity. *Photograph by C. N. Shuster, Jr.*

Manton (1977). Collectively, the chemosensory signals from chelae (three million receptors) and gnathobases (one million receptors) comprise a major sensory input to the *Limulus* brain (Wyse, 1971). Considering the dorsal-lateral placement of the compound eyes and the ventral location of the mouth, vision may be less important to a horseshoe crab during feeding than chemical and tactile cues.

Is there an explanation for the tendency of horseshoe crabs to aggregate in areas of high prey density? Does chemoreception work over long distances, or are crabs cuing in on particular bottom types that they have learned to associate with good feeding areas? By whatever mechanism, there is little doubt that horseshoe crabs can locate prey, once they are in the vicinity. Smith (1953) interpreted the pattern of crab excavations around planted plots of soft-shell clams *(Mya arenaria)* as nonrandom, suggesting a possible means by which a crab could locate and perhaps remember where patches of clams were located. If indeed horseshoe crabs can locate and return to areas of abundant prey, their predatory behavior may be more sophisticated than we presently understand.

What Do Horseshoe Crabs Eat?

Horseshoe crabs are opportunistic foragers that can take advantage of a wide range of locally available prey. Early reports noted that the food of horseshoe crabs consists of two major categories, marine worms (Lockwood, 1870; Fowler, 1907; Shuster, 1950; Lockhead, 1950), and bivalve

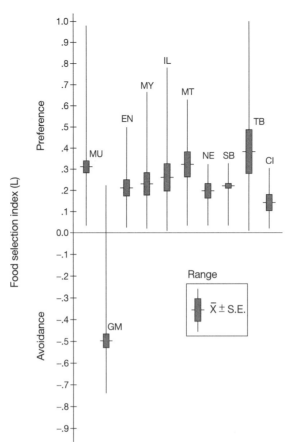

FIGURE 6.4 Selectivity indices for ten of the most frequently occurring gut items in horseshoe crabs from Delaware Bay in 1979 (n = 68). The food selection index, L, compares the relative abundance of the food item in the gut with its availability in the environment. Positive values of L indicate selection, and negative values indicate avoidance. Horizontal lines represent the mean, shaded rectangles the standard error, and vertical lines the range of the calculated L. Species are abbreviated as follows (reading left to right): MU = *Mulinia lateralis*, GM = *Gemma gemma*, EN = *Ensis directus*, MY = *Mya arenaria*, IL = *Ilyanassa obsoleta*, MT = *Mytilus edulis*, NE = *Nereis* sp., SB = *Sabellaria* sp., TB = *Turbellaria* sp., CI = *Cirrepedia* sp. *By permission from Botton, 1984a.*

mollusks such as *Gemma, Macoma, Mya,* and *Ensis* (Shuster, 1950; Smith and Chin, 1951; Smith, 1953; Smith et al., 1955).

Botton and colleagues have studied the feeding ecology of horseshoe crabs in Delaware Bay and the middle Atlantic continental shelf. Botton (1984a) examined the diet and food preferences of adult horseshoe crabs feeding on a Delaware Bay tidal flat. Food items belonging to 42 taxa were identified from 96 gut samples. Bivalves were the most important prey, both in terms of number of food items and frequency of occurrence. A few animals had horseshoe crab eggs in their gut, probably ingested incidentally, and others had large volumes of sand with little recognizable food. Botton (1984a) compared the frequency of occurrence of prey items in *Limulus* guts with their occurrence on the tidal flats where crabs were foraging. While the most abundant potential item on the flats was the small (< 4 mm) hardshell gem clam, *Gemma gemma,* this species was avoided relative to other food items (Figure 6.4). In aquarium feeding experiments, horseshoe crabs that were offered 2,000 *Gemma* without alternative prey ate an average of 417 individuals per day. However, when crabs were offered a choice between *Gemma* and larger, thinner-shell dwarf surf clams *(Mulinia lateralis)* and/or soft-shell clams *(Mya arenaria),* there was a consistent avoidance of *Gemma* (Figure 6.5). Both *Mulinia* and *Mya* shells are easily crushed by a crab's gnathobases. There was no preference between *Mulinia* and *Mya* of the same size range, but crabs selected *Mulinia* over similarly sized but hard-shell quahogs, *Mercenaria mercenaria.*

Subsequent studies of *Limulus* on the continental shelf by Botton and colleagues confirmed its generalist feeding strategy. At one station off the New Jersey coast (where 150 horseshoe crabs were found in a single 5-minute dredge haul), crab guts were literally stuffed with an average of nearly 400 blue mussels, *Mytilus edulis,* averaging 6.3 mm in length (Botton and Haskin, 1984). Elsewhere off the New Jersey coast, bivalves such as surf clams *(Spisula solidissima)* and tellinids, small brachyuran crabs, and polychaetes were most common. Botton and Ropes (1989) identified 50 prey taxa from 72 horseshoe crabs from the continental shelf. Bivalves comprised the vast majority of the ingested food during all seasons (Figure 6.6). As many as 465 nut clams *(Nucula proxima),* 166 surf clams, 230 razor clams *(Ensis directus),* and 230 dwarf tellins *(Tellina agilis)* were found in individual crabs.

Dietary studies of the Indo-Pacific horseshoe crabs are limited, although the similarity of the feeding appendages and internal anatomy suggests that all horseshoe crab species probably have similar food habits. Debnath et al. (1989) and Chatterji et al. (1992) found that *Tachypleus gigas* from India fed mainly on mollusks, crustaceans, and polychaetes. Both of these studies, as well as Botton (1984a), Botton and Haskin (1984), and Botton and Ropes (1989) found that guts sometimes contained considerable amounts of sand along with minute fauna such as foraminiferans, nematodes, plus plant material.

Collectively, the dietary studies of adult horseshoe crabs suggest that there may be three distinct feeding modes that reflect sediment type, prey density, and prey size.

Selective feeding on preferred prey. Horseshoe crabs may aggregate where preferred prey (such as *Mytilus, Mya, Spisula,* or other thin-shell bivalves in the range of 6 to 20 mm) are found in great abundance. Clams are probably handled individually and crushed by the gnathobases, with the ingestion of a minimal amount of ambient sediment.

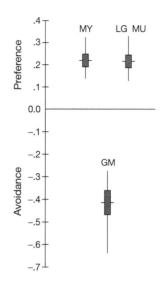

FIGURE 6.5 Aquarium food choice experiments. Adult horseshoe crabs were offered a choice between 50 soft-shell clams, *Mya arenaria* (LG MY), >10 mm shell length; 50 dwarf surf clams, *Mulinia lateralis* (MU), >10 mm; and 1,000 gem clams, *Gemma gemma,* 2 to 4 mm (n = 6 trials; see Figure 6.4 for an explanation of the data presentation). *By permission from Botton, 1984a.*

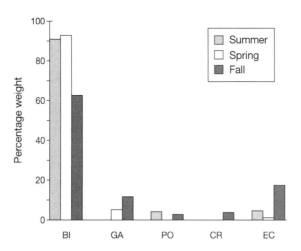

FIGURE 6.6 Taxonomic composition of prey identified in gut contents of horseshoe crabs during three seasons from the middle Atlantic continental shelf (n = 72). The primary food items were bivalves (BI), gastropods (GA), polychaete worms (PO), crustacea (CR), and echinoids (sand dollars) (EC). Results expressed as percentage of the total dry weight of ingested food of each principal dietary taxon. Weights include shells when present. *By permission from Botton and Ropes, 1989.*

Indiscriminate feeding on benthic fauna. Small mollusks (such as *Gemma* or *Nucula*), polychaetes, and other benthic infauna may be ingested along with plant detritus and sediments. Incidental ingestion of items such as foraminiferans, nematodes, and harpacticoid copepods probably occurs using this feeding mode.

Feeding on large bivalves. As described by Botton and Haskin (1984), horseshoe crabs can break open and consume bivalves up to about 40 mm in length by chipping away at the ventral margin using the gnathobases. Neither shell nor sediment is ingested.

To this point, there have been no dietary studies of juvenile horseshoe crabs in the natural environment, and this remains an important gap in our knowledge of horseshoe crab feeding ecology. Trilobite larvae do not feed and, depending on environmental conditions, they can subsist on the yolk for many months (Lockwood, 1870; French, 1979; Botton et al., 1992). In aquaria, second instars take a variety of foods, including bits of polychaetes, mussels, brine shrimp *(Artemia),* nematodes, and the alga *Enteromorpha* (French, 1979).

What Effects Do Horseshoe Crabs Have on Prey Populations?

Horseshoe crabs dig into intertidal sediments to avoid desiccation and/or search for food; these activities leave the sediments pock-marked with bowl-shaped depressions (Figure 6.7A). The mechanics of burrowing involve the coordinated movements of the prosoma, opisthosoma, and the walking legs and are described in detail by Eldredge

A

B

(1970) and Vosatka (1970). Burrowing by large juveniles and adult *Limulus* has a major effect on the stability of marine sediments. During an 11-month period of observation at Barnstable Harbor, Massachusetts, Rhoads (1967) observed that *Limulus* disrupted sediments to a depth of 3 cm. Burrowing on intertidal flats is frequently deeper, and in many cases the animals are barely distinguishable from the surrounding sediment. Often, the only indications are a low mound by the prosoma, and two small holes at the rear of the opisthosoma that mark the openings of the respiratory channels (Eldredge, 1970; personal observations). Kraeuter and Fegley (1994) compared caged and uncaged plots on a Delaware Bay tidal flat to gauge the magnitude of vertical sediment reworking by horseshoe crabs. Uncaged sediments were disturbed to a mean depth of 11.1 cm, much greater than the 3.2 cm depth of disturbance in areas protected from horseshoe crab burrowing. In New England north of Cape Cod, where horseshoe crabs are considerably smaller than in New Jersey, sediments were reworked to a depth of 5 to 8 cm deep (Baptist et al., 1957).

Horseshoe crabs act as predators as well as sediment disturbers, and these combined activities can have substantial impacts on benthic fauna. They also attract other benthic feeders (Figure 6.7B). The destructiveness of *Limulus* to bivalves in New England was first quantified by Turner (1949), and further elaborated in studies by Smith and Chin (1951) and Smith et al. (1955). In the latter report, *Mya arenaria* planted at a density of 20 clams per square foot in a Massachusetts estuary were reduced to a density of less than two per square foot by one week's feeding by a *Limulus* population. These areas were inundated with feeding depressions over the entire planted zone. A dramatic improvement in clam survival was noted when the clams were protected by predator exclosures. Juvenile hard clams in tanks were crushed in large numbers

FIGURE 6.7 (A) These typical excavations, made by adult horseshoe crabs on an intertidal flat in Barnstable Harbor, Mass., are evidence of the extent to which *Limulus* disrupts the substrate, whether dug to wait out an intertidal period or to feed. *Photograph by C. N. Shuster, Jr.* (B) When horseshoe crabs are feeding, they disturb the sediments, often excavating worms and sand shrimp that attract benthic fish. *Photograph by B. Luther, Jr., Fairhaven, Mass., taken with a flash camera at a 35 ft depth in Cape Cod Bay, Mass. From the collection of Col. E. S. Clark, Jr., of Sandwich, Mass., personal communication, 1963.*

by the casual movements of *Limulus* (Maurer, 1977; Maurer et al., 1978). Woodin (1978) sampled benthic invertebrates in pits caused by *Limulus* and blue crab *(Callinectes sapidus)* activity on tidal flats in Virginia and found reductions in numbers of species and individuals. She regarded the change as resulting from both predation and sediment disturbance, but did not distinguish between the magnitudes of the two causes. Other aspects of her study emphasized the importance of sediment stabilization to intertidal macrofauna.

Botton (1984b) used predator exclosures to examine the impacts of horseshoe crab predation on an intertidal sand flat in Delaware Bay. Areas protected from *Limulus* had more individuals, biomass, and species per core than unprotected sediments. The most dramatic effects were on the survival and growth of two bivalves, *Mya arenaria* and *Mulinia lateralis,* that were also shown to be preferred prey for adult horseshoe crabs (Botton 1984a). Few clams above 4 mm in length survived outside the predator exclosures, suggesting that grazing by horseshoe crabs "cropped off" clams as they reached this size.

Comitto et al. (1995) studied the recolonization of *Limulus* pits in the intertidal zone at Chincoteague, Virginia. Numbers of *Gemma gemma,* the dominant species, generally recovered within one day, via passive sediment transport. This small, shallow-burrowing bivalve has no way of avoiding sediment disturbances by horseshoe crabs, but its hard shell makes it resistant to both predation and disturbance. *Gemma's* ability to rapidly recolonize disturbed patches furthers its persistence on intertidal flats subjected to frequent disturbances caused by horseshoe crabs (Comitto et al., 1995).

Who Eats Horseshoe Crabs?

Most of the life of the horseshoe crab is subtidal and relatively concealed from us. Consequently, we know much more about predation on those stages residing on the beaches and intertidal flats, which we can study more readily. At various stages of its life cycle, *Limulus* is eaten by fish, crustaceans, birds, and marine reptiles. Much remains to be learned, however, concerning predation on subtidal juvenile and adult crabs, and about the overall demographic importance of predation to the population.

Predation on the Eggs, Larvae, and Early Juveniles

Horseshoe crabs are most vulnerable to predation at their earliest life history stages, the eggs and larvae. When spawning, female crabs deposit most of their eggs at depths below 10 cm (see Chapter 2). Those eggs that remain there are relatively safe from the feeding activity of the shorebirds and gulls, which congregate along the shoreline (see Chapter 1). Wave action, coupled with the continued digging activities of the crabs themselves, brings countless numbers of eggs to the surface, where birds eagerly consume them. However, birds are mainly eating eggs at or near the sediment surface, which would most likely desiccate and be lost to the population even if birds did not eat them. Beyond Delaware Bay, the significance of horseshoe crab eggs as food for birds has not been established.

Large numbers of minute invertebrates (meiofauna), particularly nematode worms, are often seen in association with *Limulus* eggs developing within the sand (Hummon et al., 1976). Although the nematodes do not appear to be predatory or parasitic, the exact nature of their interaction with the eggs is unclear. It may be that the sticky material that holds the egg mass together stimulates microbial growth, which in turn attracts the nematodes.

Several species of crustaceans and fishes eat *Limulus* eggs and trilobite larvae in the vicinity of beaches and salt marsh creeks. The sand shrimp, *Crangon septemspinosa,* consumed crab eggs in Delaware Bay (Price, 1962). Eels at the Kickemuit River, Rhode Island, ate the eggs as they were being laid. As described by Warwell (1897), "the eels . . . made a strange sight with their heads under the shell and their tails sticking out sideways. Sometimes 2 or 3 were under 1 horse-foot."

Perry (1931) described a similar behavior for catfish feeding on horseshoe crab eggs in Florida. Stomach contents of fishes in the Delaware estuary showed that *Limulus* eggs were a common food item from May through August in juvenile striped bass *(Morone saxatilis)* and white perch *(M. americana)* (deSylva et al., 1962). *Limulus* eggs have also been found in the stomachs of eels *(Anguilla rostrata),* striped killifish *(Fundulus majalis),* silver perch *(Bairdiella chrysura),* weakfish *(Cynoscion regalis),* northern kingfish *(Menticirrhus saxatilis),* Atlantic silversides *(Menidia menidia),* summer flounder *(Paralichthys dentatus),* and winter flounder *(Pseudopleuronectes americanus).* The feeding behavior of the Atlantic silversides during a 24-hour period along the shore of the York River at Gloucester Point, Virginia, at the time of the full moon also reveals something about the activity of *Limulus* larvae in the plankton. Using an index of relative importance, Spraker and Austin (1997) found that *Limulus* larvae were a dominant food item just after midnight (55 percent) but virtually absent in silversides collected during daylight hours. This suggests that the larvae were most abundant at night, and indicates that silversides and probably other fish might be an alternative method to sampling the distribution and occurrence of planktonic *Limulus* larvae. These findings agree with Rudloe (1979), who found that the greatest number of larval horseshoe crabs emerge from the sand on evening full moon tides, and that larvae are most active at night. These behaviors may be adaptive in minimizing losses to visual predators, notably shorebirds and gulls.

We know relatively little about predation on juvenile horseshoe crabs. Post-settlement trilobites and early instar juveniles are readily attacked by blue crabs *(Callinectes sapidus),* spider crabs *(Libinia* sp.), and hermit crabs *(Pagurus* sp.) (personal observation). Amphipods fed on larvae during pilot aquaculture studies in Israel (Kropach, 1979), and an unidentified species of fiddler crab appeared to be feeding on second and third instar *Limulus* on mud flats at Beaufort, North Carolina (Shuster, 1982).

Predation on Adults

As they grow through a series of molts, horseshoe crabs gradually move into deeper water (Rudloe, 1981). Predation may be concentrated on the newly molted soft-shell individuals. Beyond a carapace width of about 4 cm, corresponding to the tenth or eleventh instar, or about 3 years of age (Sekiguchi et al., 1988), crabs probably attain a size refuge from most predators except when they are molting.

Predation on the large juveniles and subtidal adults is poorly known and largely anecdotal. One remarkable case was reported from Florida, in which over a bushel (35 l) of adult *Limulus* was found in the stomach of a leopard shark *(Triakis semifasciata)* (Kirk, personal communication, 1953, cited in Shuster, 1955). Tiger sharks *(Galocerdo cuvieri)* also eat adult horseshoe crabs (Rudloe, 1981). American alligators *(Alligator mississippiensis)* have been observed eating adult *Limulus* in Florida's Indian River Lagoon system (Ehlinger, personal communication, 2002). In the Chesapeake Bay area, juvenile loggerhead turtles *(Caretta caretta)* migrate inshore during the warmer months and feed extensively on horseshoe crabs (Research Note 6.1).

Adult horseshoe crabs that are stranded on beaches are highly susceptible to bird predation. In Delaware Bay, both herring gulls *(Larus argentatus)* and great black-backed gulls *(L. marinus)* attacked live horseshoe crabs, causing substantial mortality to the population (Botton and Loveland, 1993). Debnath and Choudhury (1988) found that crows *(Corvus spelendens)* could overturn and then eat *Tachypleus gigas* on an Indian River beach, but the large gulls observed by Botton and Loveland (1993) only attacked crabs that had been stranded upside-down.

What Organisms Live on Horseshoe Crabs?

Adult horseshoe crabs are often heavily overgrown with sessile organisms—so much so that Allee (1922) referred to *Limulus* as a "walking museum." The majority of these interactions are best described by the term epibiosis, defined by Wahl (1989) as a nonsymbiotic, facultative association between the substrate organism (basibiont) and sessile animals (epizoans) or algae (epiphytes). Table 6.1 lists known epibionts found on *Limulus* (Figure 6.8). None of these species lives exclusively on *Limulus;* as shown by Wahl and Mark (1999), epibionts are typically colonizers of a range of hard substrates, and they are rarely species-specific in their choice of basibionts. Although some of the relationships in Table 6.1 are probably little more than chance hitchhiking, other interactions occur with such spatial and temporal predictability that they present opportunities for more detailed ecological study.

Attaching to horseshoe crabs provides potential benefits and risks to epibionts (Wahl, 1989; Key et al., 1996). For example, sessile organisms living on horseshoe crabs may have enhanced gene dispersal, and filter feeders such as barnacles, mussels (*Mytilus),* and slipper limpets *(Crepidula)* may benefit from enhanced water movements by growing on a mobile substrate. Conceivably, epibionts on moving targets might be less susceptible to predation by slow-moving predators (Wahl, 1989). On the other hand, horseshoe crab epibionts are exposed to the same wide range of environmental conditions as their hosts; in particular, epibionts are at risk from exposure to air, high temperatures, and considerable abrasion when the crabs are spawning.

Host horseshoe crabs may have adverse effects from heavy overgrowth by epibionts. For example, dense growths of *Mytilus* on the ventral surface may interfere with feeding, locomotion, and gill function (Botton, 1981), and crabs with heavily fouled lateral eyes have impaired vision (Wasserman and Cheng, 1996). Large *Crepidula fornicata* often grow on the rear margin of the opisthosoma (Botton and Ropes, 1988; Dietl et al., 2000), and in females this could interfere with a male's ability to attach during mating. None of the potential benefits of epibionts to hosts reported in other systems, such as camouflage, predator defense, or protection against desiccation (Wahl, 1989), is likely to be applicable to horseshoe crabs.

The development of the epibiont community on horseshoe crabs probably involves both larval choice of settlement sites and postsettlement processes such as competition for space. Dietl et al. (2000) related the distribution of epizoans on *Limulus* to the patterns of water flow over the dorsal carapace, and found that the epizoans tended to be most concentrated in recessed areas of the opisthosoma. Except for bryozoans, epizoans were rarely found on the anterior of the prosoma, where flow rates were higher and there was also a greater probability of being dislodged by abrasion as the crabs moved through the sediment. Both the study by Dietl et al., and a similar analysis of fouling in

TABLE 6.1   Epibionts associated with *Limulus polyphemus.* To Allee (1922), the adult horseshoe crab was a "walking museum"—in reference to the many species he found on their shells in the waters around Woods Hole, Mass.

Species	References
Algae	
Enteromorpha sp.	Botton and Loveland (unpublished)
Ulva lactuca	Botton and Loveland (unpublished)
Cnidaria	
Haliplanella luciae	Allee (1922)
Hydractinia echinata	Allee (1922)
Metridium dianthus	Allee (1922)
Obelia sp.	Allee (1922)
Podocoryne carnea	Allee (1922)
Platyhelminthes	
Bdelloura candida*	Lauer and Fried (1977), Sluys (1989)
Bdelloura propinqua*	Sluys (1989)
Bdelloura wheeleri*	Sluys (1989)
Syncolidium pellucidum*	Sluys (1989)
Bryozoa	
Bugula turrita	Allee (1922)
Membranipora sp.	Allee (1922)
Schizoporella unicornis	Pearse (1947)
Mollusca	
Anomia simplex	Pearse (1947)
Crassostrea virginica	Botton and Loveland (unpublished)
Crepidula convexa	Allee (1922)
Crepidula fornicata	Botton and Ropes (1988)
Crepidula plana	Pearse (1947)
Eupleura caudata*	MacKenzie (1962)
Geukensia demissa	Deaton and Kempler (1989)
Ilyanassa obsoletus*	Botton and Loveland (unpublished)
Modiolus modiolus	Allee (1922)
Mytilus edulis	Botton (1981)

TABLE 6.1 *(continued)*

Species	References
Odostomia sp.	Hedeen (1986)
Urosalpinx cinerea*	MacKenzie (1962)
Mercenaria mercenaria	Shuster (1955)
Annelida	
Filograna implexa	Dietl et al. (2000)
Hydroides diathus	Pearse (1947)
Polydora sp.	Botton and Loveland (unpublished)
Sabellaria vulgaris	Dietl et al. (2000)
Arthropoda (barnacles)	
Balanus amphitrite	Pearse (1947)
Balanus eburneus	Allee (1922)
Chelonia patula	Pearse (1947)
Semibalanus balanoides	Dietl et al. (2000)
Echinodermata (sea star)	
Asterias sp.	Botton, Loveland, and Shuster (unpublished)
Chordata	
Amaroucium constellatum	Allee (1922)
Didemnum sp.	Allee (1922)

*Indicates that egg cases or cocoons have been found as well as adults.

Tachypleus gigas by Patil and Anil (2000) found that barnacles were often concentrated in grooved portions of the prosoma and opisthosoma, which was probably a reflection of larval selection for such sites.

Extensive carapace fouling has only been reported in adult horseshoe crabs, which seldom (if ever) molt after reaching sexual maturity (Botton and Ropes, 1988). The aging of large *Crepidula fornicata* growing on *Limulus* allowed Botton and Ropes (1988) to deduce the longevity of adult horseshoe crabs. Juvenile and recently molted adult carapaces are usually smooth, glossy, and free of all macroscopic biofouling. A mucus, secreted onto the carapace by hypodermal glands, has antifouling properties (Harrington and Armstrong, 2000; Patil and Anil, 2000).

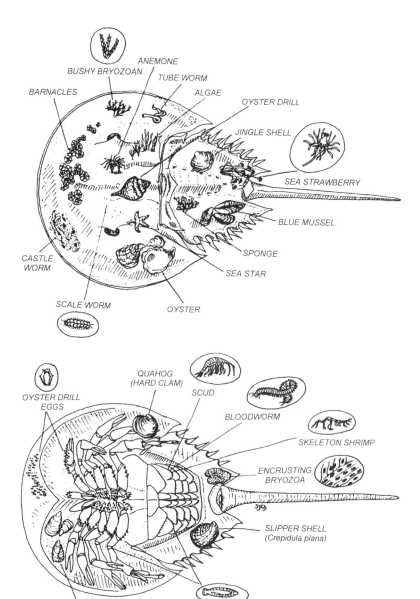

FIGURE 6.8 A composite of the many species that attach to *Limulus. Diagram by S. K. Draxler; modified from Grant, 1998, with permission of the* Underwater Naturalist, *American Littoral Society.*

What Is the Role of *Limulus* in the Food Web?

Horseshoe crabs are an important link in the food webs in many areas along the Atlantic coast of North America. Most of what we infer about their importance is based on studies from New England and the mid-Atlantic, especially the Delaware Bay area. As is often the case for marine animals, *Limulus* occupies different trophic levels depending on age and size. On the beaches, horseshoe crab eggs are directly consumed by shorebirds, gulls, and fish. The ecological importance of the horseshoe crab eggs for migratory shorebirds in Delaware Bay (see Chapter 1) has stimulated much of the recent controversy over the commercial horseshoe crab fishery (see Chapter 15). The considerable number of dead horseshoe crab eggs are linked to a variety of scavengers, such as talitrid amphipods ("beach hoppers") and ghost crabs. Beyond this, it has been demonstrated that the presence of horseshoe crab eggs stimulates meiofaunal productions (Hummon et al., 1976), and furthermore, the input of eggs significantly increases sediment organic carbon (Loveland et al., 1989). Horseshoe crabs are also hosts for a number of species. They are infected and infested by a number of microscopic and larger organisms and parasites (see Chapter 10), and the adult shell provides a suitable habitat for a number of mobile, sessile, and encrusting species (see Table 6.1).

Adult horseshoe crabs are important as "biological bulldozers," reworking sediments and affecting benthic communities through their feeding and burrowing activities. Where they are concentrated, horseshoe crabs can be important as predators of benthic invertebrates, especially mollusks. Adult horseshoe crabs themselves fall prey to few enemies while subtidal (with the notable exception of loggerhead turtles; see Research Note 6.1); but when they come ashore to lay their eggs, many adults are stranded and eagerly eaten by large gulls (Botton and Loveland, 1993). Finally, man has been a major predator of adult horseshoe crabs in fisheries that have harvested millions of animals for bait, fertilizer, animal feed, and a variety of other commercial uses (see Chapters 14 and 15).

PREDATION OF HORSESHOE CRABS BY LOGGERHEAD SEA TURTLES

John A. Keinath

As a group, sea turtles have remarkably diverse food habits. Green turtles are herbivores, leatherbacks specialize on gelatinous zooplankton such as jellyfish and salps, and hawksbills feed primarily on sponges. Of the five species of sea turtles along the Atlantic coast, the loggerhead *(Caretta caretta)* is the most abundant and the only one that feeds extensively on horseshoe crabs. During the summer in Chesapeake Bay, they eat a variety of benthic and pelagic invertebrates and scavenge nearly any other perceived food item, but it is the horseshoe crab that is the most abundant food item found in their stomachs (Lutcavage, 1981; Lutcavage and Musick, 1985).

Even though adult horseshoe crabs have a refuge from most potential predators due to their size, they are no match for a loggerhead. In Chesapeake Bay, most loggerheads are juveniles, but the largest specimens of these hard-shell sea turtles often reach 1,000 to 1,200 pounds in weight (Musick, 1979). Loggerhead turtles are well adapted for foraging on horseshoe crabs. Their thick beak, a keratinous jaw sheath up to 3 cm thick, along with very large jaw muscles account for their name. These jaws are ideal for eating hard-bodied organisms. When a crab is encountered, the turtle turns the crab upside down and uses its beak and claws to scoop out the legs, gills, eggs, and underlying structures.

Like horseshoe crabs, loggerhead turtles occur in many places along the east coast of the United States. They enter turbid estuaries during the spring when horseshoe crabs are abundant and use the tides to carry them across the bottom. Loggerheads are not dexterous in their movements, but rather slow and clumsy. They drift with the tide as they encounter food; presumably moving with the tides is energetically advantageous. Loggerheads prefer the river mouths and channel edges at depths below 3 m (Byles, 1988).

Prior to 1981, sea turtles occurring north of Cape Hatteras were considered waifs. In that year, Chesapeake Bay was found to be an important foraging area. The turtles inhabit Chesapeake Bay from May through November (Byles, 1988), when they migrate south to overwinter in warmer waters (Keinath, 1993). Subsequently, Long Island Sound was also found to be a foraging area. Loggerheads also occur in Narragansett Bay and Cape Cod Bay. In 1997, a study by Drexel University, using tangle nets, aerial surveys, and sightings, documented many loggerheads in Delaware Bay (six turtles per square kilometer), perhaps in higher abundance than in Chesapeake Bay (Byles, 1988).

So perhaps it is not surprising that loggerhead turtles aggregate in areas of Chesapeake Bay wherever horseshoe crabs are abundant. Both species arrive in Chesapeake Bay in late spring and adult horseshoe crabs comprise an important part of the diet of juvenile loggerheads. Based on aerial censuses, there may be 2,000 to 10,000 loggerheads in Chesapeake Bay during the summer (Keinath et al., 1987). Thus, predation on adult horseshoe crabs may be more intense than previously assumed.

CHAPTER 7

A HISTORY OF SKELETAL STRUCTURE: CLUES TO RELATIONSHIPS AMONG SPECIES

Carl N. Shuster, Jr., and Lyall I. Anderson

At first glance the clumsy and sluggish activity of horseshoe crabs spawning on a beach seems appropriate for their tank-like architecture. This, however, belies their activity in water. Much of their adeptness in their aquatic environment is due to their unique body design—a front part that is shaped like a helmet, a movable midpiece that tapers posteriorly and from which a rotating, spike-like tail projects (Figure 7.1).

The body shape of the earliest horseshoe crabs was similar to that of the trilobites and the sea scorpions in having freely articulating midbody segments (Exhibit 7.1). Once those segments were consolidated into a solid midbody part, the three-piece exoskeleton so familiar today was established. Because the concave, vaulted undersurface of the front and midparts is the major structural characteristic of most horseshoe crab species, it was a momentous evolutionary step, and all species have been consistent in this body design since—hence the reference to "living fossil." The second important attribute of the extant species is physiological: they are jacks-of-all-trades capable of living under a wide range of environmental conditions. If these anatomical and functional characteristics were exhibited by the extinct species, such a capability would seem to presage not only a successful but a plentiful group. Yet, paradoxically, few of the species ever coexisted. Despite their conservative anatomical history, this branch of animals has been found in aquatic habitats for about 450 million years. Thus their survival through geo-

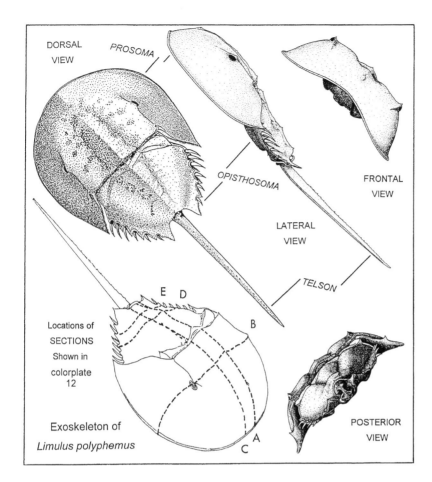

FIGURE 7.1 From all sides, horseshoe crabs have sleek lines. The morphology of the adult male *Limulus polyphemus* exoskeleton: dorsal, lateral, frontal, and posterior views (without appendages). The three-quarter view (lower left) shows the locations of the sections in colorplate 12. The prosoma is dome-shaped. Given that domes are architecturally strong, does the dome-shape strengthen the exoskeleton?

logic time seems due to the combination of relatively few changes in body form with an early evolution, and retention of wide ranges of tolerance to environmental changes. Discussion of the horseshoe crab body plan is central to this chapter; for a review of environmental and physiological considerations, see Chapters 8 and 9. Chapters 2 through 4 describe the most easily observed behavior, nest building and mating.

Structural Components of the Body

We begin with the description of the bodies of the extant (living) horseshoe crab because they are best known. First, we focus on the body plan of *Limulus* and describe some of the functions of the body parts; then we consider the three Indo–Pacific species. Fossilization, species relationships, and the changes and variations of the body plan of fossil species are reviewed, going back in geologic time to the early horseshoe crabs in the Silurian (some 430 million years ago).

Looking down upon *Limulus* provides clear evidence that its exoskeleton is a relatively simple mechanical structure. Only two joints are visible: a long hinge connecting the anterior part with the middle part and an almost universal joint at the base of the telson. These two body joints and those among the multiarticulated appendages aligned along the ventral axis of the body provide sufficient flexibility for several different activities. The legs function principally in locomotion, feeding, burrowing, and spawning; an operculum and book gills in respiration, locomotion, and spawning. The overturned horseshoe crab uses its telson to right itself.

Except for the vertically placed crop-gizzard, above the triangular frontal area of the prosoma, the major organs in the axial portion of the body are, from top to bottom (colorplates 11 and 12): the tubular heart, just beneath the axial ridge of the exoskeleton, fills a sinus, is about one-half the body length, and extends about equally forward and backward beyond the hinge. The tubular alimentary canal is below the heart. The mouth, at the midway point of the prosomal ventrum, passes into the esophagus that curves diagonally upward and forward through a central aperture in the brain to the chitin-lined crop-gizzard. From there a pyloric valve enters a long tubular alimentary canal, ending in a short chitin-lined rectum that exits at the base of the telson. A tough, prominent white shield (endosternite) covers the donut-shaped brain from which the central nervous system extends in all directions. These organs are embedded in connective tissue and their branches extend out into and intermingle with other tissues in the lateral umbrella-like portions of the prosoma and extend into the opisthosoma. The bases of the legs bracket the centrally located organs with columns of muscles of varying sizes and orientation that attach the bases of the legs to the inner, uppermost surface of the prosoma.

A chelicera and a chilarium are below and anterior and posterior to the mouth, respectively (see Figure 6.1). The brownish tissue in the lateral portions of the prosoma, below the crop and anterior, is mainly the digestive gland. Prominent structures are the coxal bases of the legs and cross-sections of the trochanters, clumps of the dorsal portion of the muscle columns between the bases of the legs and the roof of the carapace, and sections through the operculum and the branchial appendages (book gills). The cross-section through the opisthosoma (colorplate 12D) shows the depth of the vault containing parts of about three book gills, two vertical columns of muscle tissue, the posterior end of the heart, and the circular alimentary canal. Colorplate 12E shows the cross-section near the posterior end of the opisthosomal vault. The massive telson muscles occupy over one-third of the width of the cross-section and encircle the circular alimentary canal.

Forepart (Prosoma or Prosomatic Carapace)

The axial ridge of the extant species is higher than the ridges bearing the laterally placed pair of compound eyes. It forms the roof of the deep axial portion of the body that contains the central portions of the organ systems. Faintly visible external markings, between the furrows on each side of the axial area and the lateral eyes, indicate the internal positions of muscle groupings. On three sides, anteriorly and laterally, the edges of the carapace overhang the appendages. This provides a large recess (vault) within which the legs fold. While this became a universal characteristic of all subsequent horseshoe crabs, it was less common among trilobites, for which species variability in the vaulted carapaces has been linked to varying feeding strategies (Fortey and Owens, 1999). In any case, the serially arranged appendages were partially protected by the dorsal shield formed by the thoracic segments. This was true for the sea scorpions (eurypterids), in which adaptation to different ecological niches produced variation in the height of the vault. The swimming species, like *Baltoeurypterus,* were predominantly vaulted, whereas the larger plankton feeders such as *Hibbertopterus* were even more vaulted than *Limulus.*

Cutting a bowl in half and placing it over the front part of a lobster creates a cavity between the thin walls of the bowl and the central body. This resembles the cavity (vault) underneath the shell of a horseshoe crab that plays a significant role in the functioning of the horseshoe crab body. Within this vault, the actions of the appendages are usually unseen and largely protected from outside disturbances, whether animal or environmental. Eight pairs of appendages, with various segment shapes and kinds of joints (see Figures 6.1 and 6.2), are sequentially aligned along the ventral axial portion of the prosoma: chelicera, five pairs of legs, chilaria, and the operculum. The first six appendages are pincer-tipped. The first pair, the chelicera, has only two joints that move laterally and back and forth. The next four appendages are the walking legs, and each has five joints, beginning with the broad connection of the coxal base to the body, that bear gnathobases (protuberances on the bases of the legs that aid in mastication and moving food toward the mouth). In the adult male, the pincers of the first pair of walking legs are modified as claspers, used in mating (see Figures 3.1C and 6.1B). The sixth pair of appendages, the pusher legs, is not only the most robust but is also different in several respects. These legs lack spines on the gnathobases but have knob-like projections, sometimes called the nutcrackers, a spatulate bailer on the coxa, a whorl of blades on the tarsi, and two movable terminal spines. All seven pairs of these appendages function in feeding (see Chapter 6). The eighth pair is fused and forms a broad, muscular flap (the operculum).

The long hinge between the prosoma and the opisthosoma was an early modification. The hinge was narrow and restricted to the axial region of the body in early species, for example, the Upper Devonian *Bellinuroopsis.* When that early hinge is compared with that of *Limulus,* it looks as if the primitive connection between the two main body parts was jammed forward. This pushed the anterior segments of the midbody into the prosoma in the region of its central axis and formed a much longer hinge. Viewed from above, the jamming created the subcircular axial rings of the hinge joint and the lateral slanted wings (auricles) of the midbody. Beneath the carapace, two pairs of appendages of the ancient midbody—the chilaria and the operculum—wound up forward of the hinge. The primary evidence comes from analyses of *Limulus* embryos (Patten, 1912; Scholl, 1977; Sekiguchi, 1988) and examination of fossil species.

Midpart (Opisthosoma, Opisthosomatic Carapace, or Thoracetron)

Waterston (1975) postulated that the fusion of the abdominal segments in some ancient horseshoe crab was a significant step in evolution, especially because the creation of a single body part, with a deep, streamlined vault, increased the efficiency of aeration of the book gills. This fused body part is crucial to that and other functions in the extant species and no doubt was equally so in the fossil forms.

End Part (Telson or Tail)

When you see a horseshoe crab for the first time, you may wonder about the long spike-like end of the body and how it moves about. Is it a stinger? No. Will it hurt me? Certainly—if you step on it. A deep wound accompanied by gunk, including pathogenic bacteria from the Bay floor, is serious. Otherwise, the telson is harmless. However, you should not consider it to be a convenient handle to pick up a crab. Yanking on the telson could damage its muscles to such an extent that the crab might not be able to turn over. You can find information on the microscopic structure and functioning of the telson muscle in the series of studies conducted by Rhea Levine and her associates (Levine et al., 1982, 1985, 1988, 1989, 1991a, 1991b; Stewart et al., 1985). The telson moves up and down and laterally (Figure 7.2). It is essential in the righting process for all sizes of horseshoe crabs, but is used mainly by the larger, heavier animals to turn over after they have been flipped

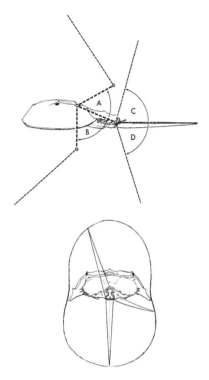

FIGURE 7.2 Two hinges join the three body parts of an adult male *Limulus.* Top: the axis of the opisthosoma (dashed line) and the axis of the telson (long line) and degrees of flexion (arcs A–D). Arc A is the extreme position of the opisthosoma and telson in the righting mode, and Arc C is the maximum position of the telson in the protection (enfolded) mode. Bottom: excursion of the tip of the telson, viewed from the posterior aspect of the opisthosoma (small arrow points show that the path traced by the tip of the telson can be in either direction).

A HISTORY OF SKELETAL STRUCTURE

onto their backs, usually while on a beach (Figures 7.3 and 7.4). Flexing of the opisthosoma extends the reach of the telson.

If the telson is too short, usually as a result of injury, the crab may not be able to turn over (Botton and Loveland, 1987; Penn and Brockmann, 1994). Watching an overturned animal struggle to right itself makes you wonder about the efficiency of the motions. How long would it take to turn over? Usually just a few minutes, but sometimes it takes hours. It all depends on a combination of factors such as the age, weight, and health of the individual; length and condition of the telson; firmness of the substrate; and air temperature. Sometimes an overturned crab will virtually anchor itself if it thrusts its telson down several inches into moist beach sand at an angle too near to the perpendicular.

Ancestral horseshoe crabs had a short, triangular-shaped telson but also had articulating opisthosomal segments. Once these segments fused, the tail spine trended toward the elongate, sharply pointed form it has in the extant xiphosurans. Presumably, overturning was not a great problem for an animal with articulating segments because, with the aid of the appendages, it could flex itself upright. The fusion of the segments was probably accompanied by an increased space for the extension of the muscle mass that operates the telson, evidenced by the formation of a fused box-like compartment, which is lacking in the freely articulating segments. An internal view shows the large muscle mass associated with the telson (see colorplate 12). This increased mass of muscle produces greater power and a wider range of movement of the telson (Fisher, 1981)—both essential in the activities of *Limulus*.

Parts Functioning Together

One of the more intriguing aspects of the functioning of the body is how different parts work together in varying ways, to move about or to plow into sand or mud. In doing so, the same parts may serve several functions. Legs, for example, are used by adult males in mating (see Chapter 3), in feeding (see Chapter 6), and in several modes of locomotion. The operculum bears the genital papillae (see Chapter 2), forms a protective cover for the book gills, and creates water currents by flapping.

Creating Water Currents

Examination of *Limulus* reveals long, narrow, diagonal apertures, or channels, between the lateral posterior projections of the prosoma and the anterior part of the opisthosoma. Functionally, these apertures are passageways. Water, drawn down through these channels by pulsations

FIGURE 7.3 This stranded horseshoe crab—by a combination of flexing its carapace, kicking its legs, and thrusting with its telson—has rolled over onto its right side.

FIGURE 7.4 Some overturned crabs make several attempts to right themselves. It is uncommon for a stranded, overturned animal to complete a 360° rotation after many unsuccessful thrusts of its telson. *By permission from B. L. Swan, Limuli Laboratories.*

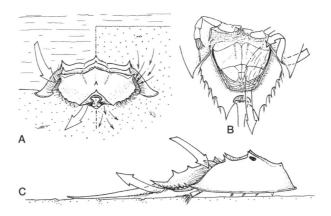

FIGURE 7.5 Movements of the operculum and branchiae produce water currents *(redrawn from Barthel, 1974; see Waterston, 1975)*. (A) Posterior—in water and under sand, (B) ventral, and (C) lateral views. Water enters the aperture between the prosoma and opisthosoma, flows across the book gills, and exits on either side of the telson. The flabellum may modulate the flow.

of the operculum and book gills, passes underneath the opisthosoma through the branchial chamber and exits on either side of the telson (Figure 7.5). Mechanically, the sequence of book-gill flexing movements begins at the fifth pair and proceeds anteriorly to the first. These movements are controlled by the rhythmic motor output of the branches of the ventral nerves (Fourtner et al., 1971).

The triangular projections that form the lateral portions of the hinge on the dorsal anterior end of the opisthosoma form contoured surfaces that partially shield the channels from above, while the curvatures of their ventral surfaces further define the channels that direct the flow of water under the carapace. These lateral projections, known as opercular pleurites, are recognizable in some but not all fossil forms. The operculum, a broad muscular flap underneath the body, extends posteriorly from the ventral part of the prosoma. When the animal folds up, as when overturned on a beach, the operculum provides a partial protective cover over the five pairs of book gills (branchial appendages) in the opisthosomal vault. These six appendages create the water currents that flow under the crab and perform several functions including aeration of the branchia, moving sperm-laden water over the eggs during spawning, and creating a jet stream that greatly facilitates their locomotion across the bottom of a bay. The deep-vaulted body enhances all these functions. The development of more efficient water flow occurred during the Paleozoic, particularly in the Carboniferous, when the species had a fused opisthosoma rather than a series of jointed segments (Waterston, 1979).

How Do They Burrow?

Locomotion across the bottom and burrowing dominate many aspects of the life of a horseshoe crab, including migrating, seeking food, excavating nests, and escaping from drying when exposed by an ebbing tide on an intertidal flat or on a beach. A dug-in position is the usual posture of *Limulus* during rest periods. The burrowing activities of horseshoe crabs (Eldredge, 1970; Vosatka, 1970; Barthel, 1974; Fisher, 1975b; Manton, 1977) pertain to the tunneling of the small, immature crabs as described by Guanzhong (1993) and digging by the larger juveniles and adults. Most horseshoe crab excavations—to avoid desiccation or to feed—are shallow (Botton, 1984). In intertidal areas, adult females rarely dig in below the level of their compound eyes (Kraeuter and Fegley, 1994). Yet this is not always the case. Recently Robert Barlow described a moon-like landscape of craters that he saw on the bottom while scuba diving many years ago during the wintertime (personal communication, 2000). This was at his favorite research area near Woods Hole, Mass. Upon probing under the depressed, flattened bottom of each crater, he found a single animal or even a mated pair.

Interest in burrowing arises not only from what *Limulus* does but how its digging might explain trilobite behavior (Eldredge, 1970). Ancient horseshoe crab burrows also have been preserved in trace fossils, as in a Jurassic trace from the northeast coast of England (Romano and Whyte, 1987). This behavior can be traced back to the Upper Devonian, where the xiphosuran *Kasibelinurus* probably produced the trace fossil *Protolimulus*. In addition to the basic maneuver of any arthropod, to excavate the substrate with its appendages, *Limulus* has the advantage of doing this under the caisson-like dome of its prosoma. There is no precise, short description of burrowing by *Limulus,* due to the range in its behavioral patterns. The size of the animal, kind and slope of substrate, and activity (resting, feeding, or spawning) are among the variables that are evident. It seems best, therefore, to use burrowing as a general term covering all kinds of digging.

A juvenile *Limulus* repeats a few distinct movements, substantially in sequence, until it stops pushing into the substrate (Eldredge, 1970; Vosatka, 1970). Adults do not appear to behave any differently. Essentially, an animal creates a depression beneath its body by moving sand centrally and backward with its first four pairs of legs while the hind legs push the collected material farther to the rear beyond the prosoma (Figure 7.6). When the body is flexed, the prosoma slips backward into the depression. Next the body is straightened and the backward thrust of the legs pushes the prosoma forward into the substrate. At various times during burrowing, the telson may swing from side to side, clearing the area behind the exhalant chambers. The whole procedure may be repeated several times, with all of the maneuvers occurring in varying degrees until an individual has reached a resting position. These rest periods and aeration of the branchia may also occur at various times

FIGURE 7.6 This stranded adult horse-shoe crab has begun to dig into a beach with a dry surface and a moist (darker) underlayer of sand. The forward movement of the crab can be gauged by the rim of sand around the front of the prosoma as well as the two swaths of moist sand that have been pushed backward by the pusher legs.

during the digging. Dorsal spines, more elongated in juveniles than in adults, may prevent backward slipping during burrowing.

In addition to the prosoma (especially the anterior arch), setae (primarily on the peripheral margins of the carapace and on the appendages) and the coxal epipodite (Figure 6.1B) are important anatomical features in burrowing activities (Eldredge, 1970). In all living horseshoe crabs, the leading edge of the prosoma is slightly arched—except in the adult males, in which it is even more concave (Figure 7.7, row D). We have seen examples of Synziphosurines, which possess this same characteristic from rocks of the Silurian, demonstrating the persistence in body plan, and even some trilobites show the same feature (Babcock, 1992). When the prosoma is flexed downward against a substrate, its leading edge is like a bulldozer blade, pushing into the substrate. This is in contrast to its position when it is crawling or skimming. At those times its forward edge, being arched, clears the bottom.

In burrowing, the legs, especially the hind legs, move the substrate backward and also push the prosoma farther into the substrate. The depth at which a *Limulus* burrows varies, but the most common stopping places are when the substrate rises to just below the level of the lateral compound eyes or when the animal is completely covered. According to Eldredge (1970), the marginal setae on the carapace and on the appendages are mechanoreceptors that enable *Limulus* to gauge the extent of burial. Even when completely covered, a horseshoe crab may maintain openings to the surface through which it pulls water down under its carapace across the book gills. Occasionally it may flush the water backward, clearing the intake channels. This is common in Japanese horseshoe crabs during spawning. Dr. Koichi Sekiguchi called our attention to this while observing *Limulus* spawning during one of his trips to Delaware Bay. When an egg-laying female is covered with several centimeters of calm water, back flushing produces spawning foam in two circles of bubbles above her, one for each intake of water (see Figure 7.5). It is as though quick movement of the branchia, perhaps modified by the coxal epipodite, creates a cavitation-like phenomenon, producing the bubbles or the release of gases trapped in the sediments. This back flushing also appeared to be common in juvenile limuli of either sex when they were observed burrowing into sand during a flume tank study (Luckenbach and Shuster, 1997).

Swimming Begins with the Embryo

Swimming may aid in the distribution of crabs. It may also help them hurdle barriers or escape from predators and anoxic bottom waters—the last of which may have been a problem for the fossil species in the freshwater and brackish lakes of coal swamps. When compared to skimming across the bottom in a scuttling manner, swimming is much less efficient for getting around.

How Do They Swim?

While the mechanics of horseshoe crab swimming have been studied (Vosatka, 1970; Manton, 1977), researchers have less insight into why they swim. Swimming may be inherent, learned while an upside-down embryo is still encased within its transparent egg capsule. As hatching time nears, the well-developed embryo is strong enough to turn somersaults, using both legs and book gills to create the movement (see Figure 5.4). Therefore, it is not surprising to see the newly hatched larvae swimming jerkily, upside down, while traversing a more or less diagonal path upward toward the surface of an aquarium. At the surface, they plane out and move slowly forward like a many-oared scow. If the water surface is disturbed, as by other larvae, they sink to the bottom, only to begin their upward swim again, immediately or after a short rest. Perhaps larval swimming is primarily a means of escaping the surf zone to an area where the shallow water is less turbulent. Because the duration of larval swimming lasts only a few days, it may be less important in widely distributing the larvae than in transporting them away from the beach to a quieter benthic habitat (Botton, 1999), because the larvae and later juveniles are abundant on the bottom in shallow-water, near-shore areas (Shuster, 1979; Botton, 1999). Interestingly, the larvae are active mainly at night as they are seldom captured by plankton nets except at night (Townes, 1938; Botton, 1999). They are also eaten by night-feeding fish (Spraker and Austin, 1997).

Crowded conditions apparently stimulate horseshoe crabs, especially the juveniles, to swim. This is easily verified in a small aquarium by crowding small juveniles: the greater the number of crabs, the greater the swimming activity. This behavior may be an attempt to escape crowded conditions. There is no way of knowing; but if the ancient horseshoe crabs grew up in calm waters, this avoidance reaction may have been successful in spreading out the population. Such avoidance may have been part of the normal behavior of all horseshoe crabs because *Limulus* and the other extant species do not appear to be social animals. It even may be linked to avoiding cannibalism because, in aquaria, larger hard-shell juveniles may prey on smaller soft, newly molted ones.

Before reaching the surface within the confines of a small aquarium, the smaller juveniles swim in almost any position. It is not unusual to see them swimming right side up or on their sides in addition to their usual backstroke method. Their takeoff from the bottom also varies. They may run up an incline, such as furnished by a rock. Or, while gliding across the bottom at a sufficient speed, a thrust of the telson against the sand spirals them upward at an angle into the water column. Then they slowly roll over onto their backs. If undisturbed, they usually move upward in a jerky, diagonal swimming path. No matter how they swim in the water column, some ultimately reach the surface and glide there. They appear to expend less energy once they plane out at the surface than when they are swimming diagonally upward.

For the smaller, more buoyant horseshoe crabs, swimming also appears to be an alternative to turning over in the water using the telson. When juvenile limuli are overturned in an aquarium, they occasionally propel themselves slightly off the bottom by using swimming motions before rolling over and righting themselves and then settling back on the bottom. The small size of the majority of fossil specimens from the Upper Carboniferous coal swamps suggests that they could have righted themselves in water in the same manner.

In nature, the water is rarely calm enough for swimming, even for large adults. Appearance of adult limuli in the surface waters of swiftly flowing tidal streams may be accidental. Horseshoe crabs generally move with currents and, therefore, may be drawn into a tidal stream that, due to its narrow streambed, may develop strong currents during a flooding tide. The turbulence of the water tumbles the crabs around and drives them upward. When the current is less swift and the surface is relatively smooth, they may actually swim at the surface. Twice, in several decades of fieldwork, we have seen adult limuli swimming at the surface in open water and only once has a horseshoe crab been reported swimming at the surface of an ocean. It was an adult *Tachypleus gigas,* about 1 m in length, captured in a large surface tow net. That was in the Andaman Sea, off the Burma coast, where the water depth was about 10 fathoms (Sewell, 1912).

Four Living Species

Three species—*Tachypleus tridentatus, Tachypleus gigas,* and *Carcinoscorpius rotundicauda*—are found in the coastal waters of southeastern Asia and the offshore islands, while the fourth species, *Limulus polyphemus,* is the only one inhabiting the coastal waters of the western Atlantic Ocean (Sekiguchi and Nakamura, 1979). See Chapter 8 for the geographic distributions of the four species.

Although the three Indo-Pacific species are closely related, they spawn in different habitats (Koichi Nakamura, personal communication, 1981); yet artificial hybridization is possible in some crosses (Sugita, 1988) (see Figure 5.5). We wonder, therefore, whether circumstances during earlier geologic times might have resulted in the production of hybrids. During Carboniferous times, up to four distinct species occupied the same habitats in the coal swamps, a far more restricted environment than that of the open marine setting (Anderson et al., 1999). Perhaps hybridization occurred among these forms.

Morphological and mating similarities among the extant species became readily apparent when we had the opportunity to compare aquarium specimens with the aid of the Japanese scientists who were the first to study all four species, alive and side by side (Botton et al., 1996). Specific characteristics of the adults of these species are given in Figure 7.7 and the place of these species in the animal kingdom in Exhibit 7.2. For an extensive, informative source on all four species, consult the book by Koichi Sekiguchi and his colleagues (1988).

The larvae of the four species are morphologically similar (see Figure 5.5), so it is not surprising to note that the adults are similar. Although each species exhibits relatively wide ranges in their adult sizes, this is usually a characteristic of discrete populations. *Tachypleus tridentatus* is generally as large or larger than *Limulus polyphemus,* and *Carcinoscorpius rotundicauda* is the smallest. The prosomal widths of the males are generally 70 to 80 percent that of the females. Differences among the females are in the shape and dimensions of the opercula (see Figure 7.7, row A—only *Limulus* has endites, at arrow, and the opisthosomas, row B). In addition to the overall dimensions of the opisthosomas, the shape and lengths of the female movable marginal spines are notable.

FIGURE 7.7 Morphological features of adults of the four extant species of horseshoe crabs *(based on Shuster, 1955, 1982; Sekiguchi and Nakamura, 1979; Sekiguchi, 1988).* The prosomal widths of the males (row E) were drawn to the same width and all other representations were drawn to the same scale to aid in making comparisons. Species are arranged in columns and body parts are shown in rows: (A) Anterior sides of the female operculum; (B) the dorsal aspect of the female opisthosoma (the first entapophyseal pit in *Limulus polyphemus* is triangulate, whereas those of all three Indo-Pacific species are elongate); (C) comparisons of the two distal segments of the second (2) and third (3) ventral appendages; (D) the frontal aspects of the males, drawn upside down to better illustrate their juxtaposition with the dorsal surface of the female opisthosoma; (E) the relative lengths of the telsons for the males in row F, with enlargements of the cross-sections; (F) dorsal aspects of the males.

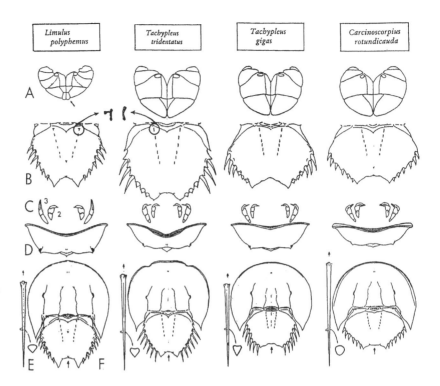

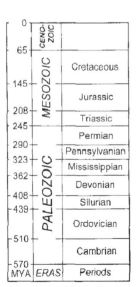

Geological Time Scale

FIGURE 7.8 A timescale of the divisions between geologic eras and periods in millions of years ago (mya).

The six pairs of entapophyseal pits on the dorsal surface of the opisthosoma become smaller from front to rear, corresponding to a decrease in the size of the branchial appendages (book gills). The first pair of pits in *Limulus* are triangulate, whereas those of the other three species are elongate, the same as the other pairs of pits in all four species (see Figure 7.7). If this pit had been preserved more clearly in the fossil of *Limulus coffini* (colorplate 13), there would have been no doubt as to which genus the species *coffin* would be assigned. Among males, the number of claspers (see Figure 7.7, row C) and the shape of the anterior portion of the prosoma, particularly the indentations, are distinct (rows D and E). *Limulus* has only one pair of claspers; the other species have two pairs. In cross-section, the telsons are triangulate except in *C. rotundicauda,* which has an oval one (row F).

Fossil Species

Arthropods with the horseshoe crab body plan have occurred at least since the beginning of the Silurian and in each of the geologic periods since then (Figure 7.8; Selden and Siveter, 1987). That is a long time. By tracking similarities as well as differences in body form down through

A HISTORY OF SKELETAL STRUCTURE

the ages, from the extant species to geologically older, extinct species, we can search for plausible connections to species at the roots of horseshoe crab history (see, for example, Størmer, 1952, 1955; Eldredge, 1974; Bergstrom, 1975; Fisher, 1984; Selden and Siveter, 1987; Sekiguchi and Yamasaki, 1988; Selden, 1993; Anderson and Selden, 1997; Dunlop and Selden, 1997). All extant species and those fossil species with the characteristic three-piece exoskeleton are in the Order Xiphosurida (Exhibit 7.2). Fossil species from earlier geologic times, Silurian back to the Cambrian, have segmented opisthosomas and are in the Order Synziphosurida (Figures 7.9 and 7.10).

This takes us from the present back to the Cenozoic era (65 million years ago, or mya), to the Mesozoic era (65 back to 245 mya) and then to the Paleozoic era (245 to 570 mya) (see Figure 7.8). Anderson and Selden (1997) subjected eighteen Paleozoic species and *Limulus* to a cladistic analysis, using twenty-six skeletal characteristics. Among these characteristics were those of the opisthosoma (extent of the fusion of segments, number of segments, extent of the first segment, discernible axis, moveable marginal spines, fixed lateral spines, transverse ridge nodes, and longitudinal ridges), the prosoma (prominence of the ridges bearing the compound eyes, cardiac lobe, ophthalmic spines, posterior margin, and genal angles), and prominence of the tail spine. Figure 7.9 shows diagrammatic outlines of sixteen of the species. The relationships determined by the cladistic analysis were plotted according to stratigraphic occurrence (see Figure 7.10).

While much about the external anatomy of the fossil species can be explained, assumptions as to their behavior and life style must largely be drawn from what is known about the four living species. These interpretations are based on the following reasoning: if the fossilized species were morphologically similar to the extant species, lived in similar environments and habitats, and had similar prey, their functions as well as their behavior were probably the same.

Observations on the utility of the body form in the living species provide the basis for imagining what activities the extinct species of horseshoe crabs may have been able to perform. The shape of horseshoe crabs also helps us to comprehend how they survived. It is a window to understanding the activity of other groups of arthropods, especially the extinct trilobites and the sea scorpions. Fossilized tracks give evidence of the ambulatory behavior of horseshoe crabs all the way back to the Silurian period (400 mya). Additionally, the vaulted carapace and the dorsal position of the lateral eyes are morphological characteristics pertinent to the benthic mode of life that can be traced back almost as far as the fossil record of the xiphosurans. Even predation has been recorded. Fossilized feces (coprolites) collected from English coal

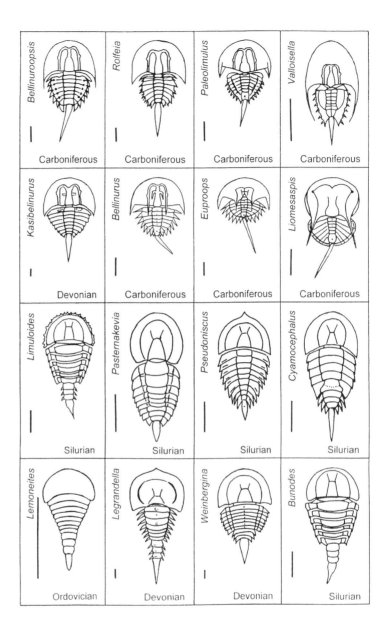

FIGURE 7.9 Diagrammatic reconstructions of early xiphosurans, Ordovician through Carboniferous *(modified from Anderson and Selden, 1997),* selected to illustrate variations between species and through geologic time. All views are dorsal; all prosomal widths are the same to contrast shape and extent (each scale line = 1 cm).

deposits contain the mangled remains of *Euproops.* Some fish had swallowed them whole (Anderson, personal observation).

Given the rarity of horseshoe crab fossils, how complete is the record? Not as good as we would like—there are many and sometimes long gaps in the fossil record. Nevertheless, the accumulated information has enabled scientists to map these species in geologic time and

A HISTORY OF SKELETAL STRUCTURE

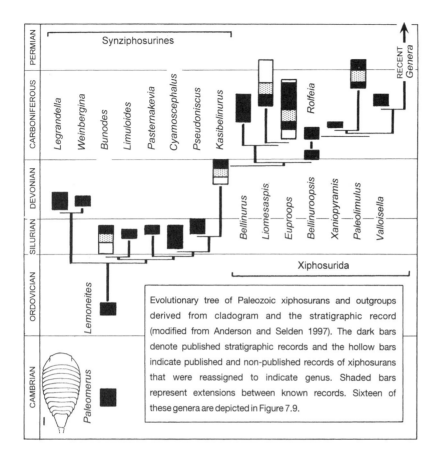

FIGURE 7.10 An abbreviated family tree, derived from cladistic analyses and the stratigraphic record (Cambrian through the Permian), showing the branches identified with the major fossil representatives of the Xiphosura *(Anderson and Selden, 1997)*. Legend for bars: vertical black bars = published stratigraphic records; unfilled = unpublished and published records of xiphosurans previously unassigned or redesignated as to genus; shaded = connections between known distributions of some of the genera. The horizontal brackets show sister-group relationships of several species within the Synziphosurines and the Xiphosurida.

space (Sekiguchi, 1988; Anderson and Selden, 1997). Although there is always the possibility that more fossilized species of horseshoe crabs might be discovered, we believe the present collection reveals most of the major steps in the body plan.

Chapter 5 outlines the anatomical variations, including a few aberrations. These variations could be troublesome in any interpretation of relationships between species, especially in the analysis of fossilized remains. Certainly substantial background knowledge is needed to avoid misinterpretations. Indeed, Waterman (1958) discovered one taxonomic error that was perpetuated by a draftsman who drew a mirror image of an anomaly. Some unusual aberrations may result during molting. In two cases, identical symmetrical abnormalities occurred in nature when soft-shell juvenile *Limulus* dug into a hard substrate (Shuster, 1955). Later an observation of a juvenile in a sand–bottom aquarium provided a plausible explanation. Pressure from sediments furrowed around the margins of the opisthosoma caused the margins of the soft-shell specimen to buckle. The juvenile in the aquarium was a hard-shell crab so the opisthosomal margins did not buckle when sediments covered the opisthosomal margins. This observation suggests that some reconstructions of fossil species may have been based on similarly buried crabs in which the prosoma was flattened but the full extent of the opisthosoma was not preserved. This may have occurred during the fossilization of *Austrolimulus fletcheri,* a crab of a Triassic species found near Sydney, Australia, the only species on record that had a greatly exaggerated horseshoe crab body.

Exoskeleton and Fossilization

Varying amounts of detail are preserved in fossil horseshoe crabs, depending on a variety of factors. These include the rapidity of burial, the kind of sediment (such as layers of rock, as shale, slate, or limestone, formed), and whether the fossil was formed from a shell cast off after a molt or from an entire animal. For the most part, because the internal organs were soft, they rarely created any imprints that were fossilized. As a result, most of the fossils of horseshoe crabs show features of either the outer surface or the undersurface of the exoskeleton or a combination of both. In some cases, the appendages were preserved, but not as distinctly as the features of the dorsal surface of the carapace and ridges on the ventral surface. Sometimes the tracks of the crab were also preserved.

The exoskeleton of *Limulus* is composed of a chitin, a cellulose-like biopolymer (Austin et al., 1981), similar to the protein matrix (conchyolin) in the shell of mollusks and the material in our fingernails. The exoskeleton is 28 percent chitin; the rest is proteinaceous with a fraction of some of the quinno (tanned) proteins that give it its toughness (Austin, personal communication, 1979). It is thin, compared to the exoskeletons of other animals such as the Crustacea, whose shells are reinforced with calcium carbonates. Further, as Hock (1940) demonstrated, chitinoclastic (from Greek *klastos* = broken, taking apart) bacteria readily destroy the *Limulus* carapace (see Chapter 10). These characteristics of the horseshoe crab shell, along with the manner in which embedding sediments harden into rock, may be some of the reasons why fossilized specimens are almost always flattened. The trilobites, whose exoskeletons were calcified though generally not as vaulted, tend to retain more of their three-dimensional shape as fossils. Additionally, silica replacement of calcium carbonate through processes associated with burial further reinforced the fossil remains against compaction. The unmineralized chitin in the horseshoe crab exoskeleton

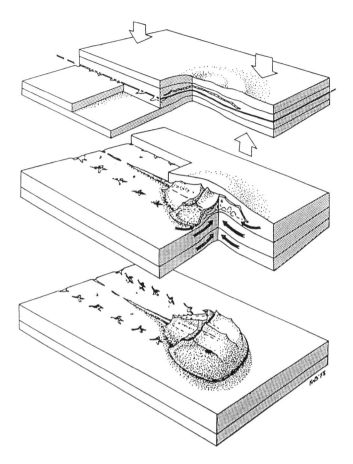

FIGURE 7.11 Diagram of limulid preservation and the formation of pseudo-undertracks on the Solnhofen lithographic limestone (Upper Jurassic), based on track sections and a trace by the tip of the telson (Barthel, 1974). Note that the trail consists almost entirely of pusher leg imprints. This is characteristic of a horseshoe crab skimming over the bottom.

offered less resistance and the shell was flattened. A different pathway involving the phosphatization of soft tissue through bacterial action sometimes has preserved incredible detail in fossil horseshoe crabs.

The flattening of fossils is due to compaction of sediment into rock. The highest relative degree of compaction occurs when muds are transformed into mudstone or shale. In this case, pore water is expelled, further decreasing the total volume of sediment (Figure 7.11). Sandstones are less prone to compaction and, as such, fossils preserved in them are less squashed. The situation is complicated, however, when the depositional environments, which these sediments represent, influence the type of fossils that may be preserved. Muds are deposited under relatively quiet water conditions, which are favorable for the preservation of whole exoskeletons. Anaerobic conditions below the surface of muds tend to prevent bacterial degradation and dissuade scavengers. Horseshoe crabs are usually better preserved, in their natural shape, when solid concretions form around the decaying organic matter early in the history of the sediment. These concretions are less prone to compaction as they are better cemented and can preserve fossil remains in three dimensions. Good examples of concretions are the Cretaceous of Colorado (Reeside and Harris, 1952) and those that formed in the strata above coal seams, particularly in the Mazon Creek area, IL (Baird, 1997a, 1997b) and the Lancashire Coalfield (Anderson, 1999).

Babcock and Chang (1997) experimented with preburial and postburial changes in recently dead juvenile *Limulus*, prosomal widths from 4 to 6 cm, as a guide to interpreting various stages in the disarticulation of the unmineralized exoskeletons of ancient arthropods, including fossilized horseshoe crabs. The freshly dead animals became bloated and floated for a few hours to 2 days. After a short period of rigor mortis, the body became flaccid and its parts flapped about loosely. The internal organs decayed within a week in oxygenated artificial seawater at 25°C. During days 5 to 7 they were covered with a bacterial-fungal coating that soon disappeared. Depending on tumbling, to simulate transport in the natural inshore environment, the exoskeleton disarticulated in stages. First the book gills were lost, and then the other appendages were broken within 30 to 65 days, followed by loss of the telson. The last to separate were the prosoma and opisthosoma. The movable marginal spines started to fall off after the book gills. In anoxic conditions decay of internal organs also occurred within a week, but afterward the specimens remained nearly intact.

While in no way comparable to the complex forces at work during the process of fossilization (taphonomy), compressing the thin-walled, empty shell (cast) of a *Limulus* juvenile between two heavy sheets of Plexiglas demonstrates its pliability and the likely areas of weakness. Compressing the wet cast, from between two-thirds to one-half of the original prosomal height, produced four visible results: 1) the most marked was the spread between the projections of the opisthosoma, 60 mm versus 75 mm; 2) despite the lateral flaring of the prosoma, the overall length of the specimen remained unchanged; 3) the thin wall of the broad prosoma wrinkled in several areas; and 4) the dimensions of the opisthosoma were not altered.

Selections from the Fossil Record

We have selected a few species to illustrate changes that have occurred in the horseshoe crab body plan. Actually, if you want to start with the basic details, begin with *Limulus polyphemus* in the section on structural components of the body (see Figure 7.1) and the other extant species (see Figure 7.7).

Representative Mesozoic Species

Limulus coffini—Cretaceous of Colorado (some 80 million years ago)—was identified from a single fossilized fragment (see colorplate 13). It was given the generic name because it virtually matches the opisthosoma of *Limulus polyphemus*. Unfortunately the anterior portion of this specimen, especially the hinge and anterior entapophyseal pits, were not preserved. If even one of these pits (compare with Figure 7.7, row B, *Limulus/Tachypleus* inset) had been preserved, the taxonomic relationship to the extant species could have been ascertained with greater certainty. Although the marginal, movable spines were not preserved, the similarity of the escalloped opisthosomal margin to *Limulus* indicates they were present in the live animal. Likewise, the shape of the terminal bay strongly indicates a telson must have been present.

Other features can be interpreted from certain conditions that are occasionally seen in *Limulus*. A well-defined depressed line on the fossil specimen of *Limulus coffini* extends laterally from the sixth entapophyseal pit. This line of depression occurs in soft-shell *Limulus* specimens that have been out of water and dried for a period of time. It is possible that the specimen, at the time it was preserved as a fossil, had dried out slightly after it had molted. Or, perhaps the crease in the soft shell was caused by contraction of the telson muscles that are attached just below on the inner surface of the carapace and drying out was not a factor.

Mesolimulus walchi was fossilized during the Upper Jurassic (ca. 150 mya) in the Solnhofen formation, Bavaria, Germany (Figure 7.12). This

FIGURE 7.12 An excellent positive-negative impression of *Mesolimulus walchi* from the extensive private collection of H. Leich. This specimen shows the characteristic axial ridge, ophthalmic ridges, and the broad ventral rim of the prosoma (prosomal width = 5.5 cm). *Photograph by C. N. Shuster, Jr.*

species is invaluable for studying anatomical features because there are relatively many specimens of different sizes, some of which are fairly large. This is a good place to respond to a pertinent question—how similar was *Mesolimulus walchi* to the extant species in body plan? So similar that if *Mesolimulus* could miraculously be brought back to life, it probably would qualify as a fifth extant species, as near in appearance to the four living species as they are to each other.

Detailed studies on the functional anatomy of selected extinct limulids by Daniel C. Fisher have expanded the understanding of those species and were one of the reasons we examined specimens of *Mesolimulus*. Fisher (1975a) interpreted *Mesolimulus* as being much flatter than *Limulus* (about 40 percent of the prosomal height). His experiments on the hydrodynamics of carapace models yielded convincing data that a flatter horseshoe crab, as he reconstructed *Mesolimulus,* would be a better swimmer than *Limulus* but less effective as a burrower. We certainly agree, if *Mesolimulus* was so flat.

An examination of *Mesolimulus* and associated fauna in museums and private collections in Germany makes strikingly clear that many other thick-bodied animals were also flattened during fossilization. This could be ascribed to the strong compaction that took place in the Solnhofen limestone, as described by Buisonjé (1985) (see also Figure 7.11). These observations and knowledge of the anatomy and activities of *Limulus* lead us to believe that *Mesolimulus* was a deep-bodied animal, similar to the living horseshoe crabs. In reconstructing an image of *Mesolimulus,* the chief criteria that we used were the elevation of the axial ridge above all other features, the depth of body accommodating the tiers of organ systems, the vaulted carapace, and the fact that it hardly made sense for an animal that was a benthic feeder to spend much time swimming. These criteria are just as important in interpreting the three-dimensional body form of other fossil species.

Of the living species, *Mesolimulus* most resembles the Indo-Pacific species—especially *Tachypleus (Carcinoscorpius) rotundicauda* in the robustness of the ventral rim, its smaller size, and the relative flatness of the carapace; and *Tachypleus tridentatus,* in the long, even-length, movable marginal spines of the opisthosoma. One question is whether *Mesolimulus walchi* and other fossil species such as *Limulus vicensis* might be more correctly aligned with the genus *Tachypleus* and, therefore, might reasonably be assigned the generic name of *Mesotachypleus.* This name would agree with the determination made by Yamasaki (1988) that *Carcinoscorpius rotundicauda* should be named *Tachypleus rotundicauda. Mesolimulus* deserves a more detailed comparison with the extant species.

Limulus vicensis was a species from the Triassic of France (215 mya) (Bleicher, 1897; see Fisher, 1990). This fossil had a prosomal width of about 6 cm and, like *Mesolimulus,* was very similar in body plan to the extant species. The posterior portion of the opisthosoma and the telson are missing as well as any details of the marginal areas.

Representative Paleozoic Species

By the beginning of the Mesozoic era, some 245 million years ago, horseshoe crabs had essentially the appearance of the extant species. Further back in geologic time, into the Paleozoic era (245 to 570 mya), the species are still recognizable as horseshoe crabs, but they were much smaller and varied more from the extant body plan. Two major groups have been recognized: the Xiphosurida, in rocks of the Carboniferous and Permian; and the Synziphosurines, which range in age from the Ordovician to the Devonian (see Figure 7.10). The major difference between the two is whether the midpart of the body was essentially a solid piece of fused segments, as in the Xiphosurida, or was composed of articulating segments, as in the Synziphosurines. In function, the most obvious difference was the greater control of various functions within the deeper vault of the solid piece versus the ability to roll up in the geologically older group.

Paleolimulus avitus has been found in a Lower Permian formation in Kansas (Dunbar, 1923; Babcock et al., 2000) and the Mazon Creek Carboniferous assemblage of fossils of Illinois (Baird and Anderson, 1997). Although tiny (only 2.5 cm [1 in] long), compared with adults of the extant species, this species had features common with both living and extinct species. The outline of the body is typically the shape associated with the more recent horseshoe crab species (Figure 7.13). Five pairs of slightly raised oval areas in the extracardiac region of the dorsal aspect of the prosoma indicate the size and location of the major muscle bundles of the legs. The distal portions of those legs were preserved, more or less in place on the ventral surface of the fossil; they appear to be more robust than those of the living species. The relative distance between the compound eyes and the prosomal width, as well as the hinge length, suggests that the prosoma was somewhat flattened during fossilization. Our reconstruction would re-create a more realistic water-flow channel between the prosoma and opisthosoma.

Because the axial region of the opisthosoma has the six well-defined segments associated with the six pairs of entapophyseal pits and the six pairs of marginal movable spines, *Paleolimulus* would have had these structures also. The extracardiac region of the opisthosoma had two features; one each of these can be seen in two other species, *Mesolimulus walchi* and *Tachypleus tridentatus. Mesolimulus walchi* also had

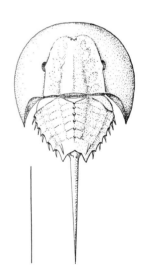

FIGURE 7.13 Although *Paleolimulus avitus* was a small animal (scale line = 5 cm), its external appearance was very similar to *Limulus polyphemus. Redrawn from Dunbar (1923), by C. N. Shuster, Jr.*

FIGURE 7.14 A species of *Euproops* from two concretions. Top, the dorsal view of an internal cast (left side) and its negative image (right side), showing the extent of the opisthosoma vault. Bottom, the positive impression (right side) illustrates the segmentation in the fused opisthosoma. This specimen had a prosomal width of 58 mm. *Photographs by C. N. Shuster, Jr., courtesy of the U.S. National Museum of Natural History, Smithsonian Institution, Washington, DC.*

prominent ridges bearing two or three pairs of blunt spines (or bosses) in the midregion of the extracardiac area on each side of the opisthosoma. Each ridge had a long terminal spine, which reached almost to the terminal bay. In *Tachypleus tridentatus,* two short spines are located on the rim of the terminal bay in positions that nearly match the location of the tips of the two *Paleolimulus* spines.

Euproops danae (Figure 7.14) existed in the Middle Pennsylvania, ca. 300 mya; it was discovered in Mazon Creek, IL (Meek and Worthen, 1865), and has been found in coal deposits in England (Anderson, 1994). Reconstructions of the animal revealed a three-dimensional structure not unlike later limulid body forms (Fisher, 1977a, 1977b; Anderson, 1994). Fisher concluded that the elongated prosomal spines were hydrodynamically suited to a free-fall when the animal stopped swimming and sank in an enfolded posture, avoiding a predator. When

A HISTORY OF SKELETAL STRUCTURE

enfolded, the animal was protected by the spiked edges formed by the long prosomal spines, the serrate margin of the opisthosoma, and the telson.

Synziphosurines

These early horseshoe crabs comprised a loose grouping of relatively small chelicerate arthropods whose fossil record extends from the Lower Silurian to the Upper Devonian (see Figures 7.9 and 7.10). All possessed a vaulted horseshoe-shaped prosoma and—in a major characteristic that separates them from the Xiphosurida—had segmented opisthosomas usually differentiated into an anterior broad region (mesosoma) and a narrow posterior (metasomal) region). The telson tended to be broad and stubby. Eldredge (1974) revised this taxonomic suborder; several more species have been added since then (Anderson and Selden, 1997).

Weinbergina opitzi, Lower Devonian, Germany (Richter and Richter, 1929) (Figure 7.15) resembles *Legrandella* more than any other. The three known specimens are greatly compressed but, as reconstructed by Eldredge (1974), they possessed a vaulted prosoma. The prosoma was large, semicircular, and smooth. Tips of several pairs of nonchelate walking legs extend beyond the margin of the prosoma in the flattened fossils. The terminal segments of the legs appear similar to the distal part of the pusher legs of the living species (see Figure 6.1), suggesting that it might have been an exceptional burrower. These appendages probably would not have been seen except that the flattening process during preservation also squashed the specimen to one side, creating a sort of three-quarter bas-relief of the animal. There were eleven abdominal segments and a short telson. The first segment, forming the hinge between opisthosoma and prosoma, was probably pushed forward under the prosoma, and hence hidden due to the deformation during compression (Eldredge, 1974). The last three abdominal segments formed a post-abdomen.

Legrandella lombardii, Lower Devonian, ca. 390 mya, Bolivia (Eldredge, 1974), was a large synziphosuran some 60 mm in prosomal width (Figure 7.16). It had a deep, vaulted body. The fossilized remains of this specimen are in an enfolded position with the pre-abdomen tucked under the prosoma and the post-abdomen and telson probably extended beyond the prosoma. Three features dominated the prosoma: 1) a cardiac (axial) lobe that protruded high above, 2) a pair of narrow crescentic sensory areas (lateral eyes) that extended almost the entire length of the ophthalmic ridges, and 3) five pairs of broad, low ridges, corresponding to the areas of leg muscle attachments or leg articula-

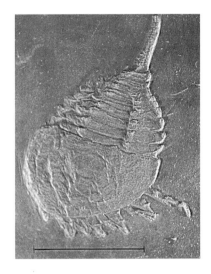

FIGURE 7.15 *Weinbergina opitzi.* The axis of the body, clearly marked by a series of knobs (identifiable as the series of light-colored areas running anteriorly from the base of the telson at an angle to the left), is flanked by two other series of nodular protuberances (the one on the crab's right side is mostly hidden). *By permission from Senckenberg Museum, Frankfurt-am-Main, Germany; photograph by C. N. Shuster, Jr.*

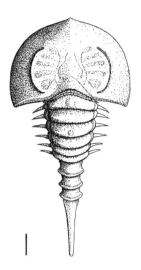

FIGURE 7.16 Although about 80 million years and body shape and size separate *Legrandella* (Devonian, South America) and *Paleolimulus* (Carboniferous, North America), their distant relationship is evident (see also Figure 7.13). The scale lines represent 5 cm.

tions in *Limulus,* that radiated from the cardiac protuberance. The extreme edge of the prosomal rim, extending around the ventral margin of the prosoma, is broken away, leaving a slight anterior projection that may be an artifact. A subcircular arch on the first segment of the opisthosoma apparently formed the only hinge that articulated with the prosoma.

The opisthosoma had eleven freely articulated segments and a prominent axial ridge, and was composed of two segmented parts. The anterior portion had eight hemicircular segments with lateral extensions (pleura). This portion comprised 65 percent of the length (3.8 cm) of the opisthosoma. The final three segments were smaller in diameter and somewhat hourglass in shape; each bore stout axial spines. Their articulation permitted side-to-side as well as up-and-down movement. Ventral appendages were not preserved in the only specimen. The telson, triangular in cross-section, was stubby but of undetermined length as it appears to have been broken off. Its excursion, in all directions, was enhanced by the motility of the three post-abdomen segments.

Tracks and Trails

Packard (1886) interpreted a trace fossil as an early horseshoe crab and he named it *Protolimulus.* But it was not a xiphosuran body fossil; it was produced by the activity of a xiphosuran. The trace maker has been identified as *Kasibelinurus* (Babcock et al., 1995). When Caster (1938) studied *Limulus* trails and tracks as models, he assigned fossilized tracks and trails in an Upper Devonian shale formation in Pennsylvania to a *Belinurus*-like limulid form, possibly *Protolimulus.* The tracks and trails had been made on a moist marine mud by specimens ranging in widths from less than 5 to more than 40 mm. Some trails were at least 4 feet (1.2 m) in length, until interrupted by the edge of the quarry where the fossils were found. Many of the wider trails ran in the same general direction—which, from the absence of resting impressions and carcasses, were thought to be shoreward migrations, much like the shoreward breeding migration of the extant species. Interestingly, there seems to be no essential difference between the fossilized tracks and trails, even those of a disoriented, dying *Mesolimulus* (see Figure 8.9) and those made by the extant species (Figure 7.17).

Especially on muddy but even in wet sandy places, parallel telltale tracks with other marks between them indicate the passage of a crawling horseshoe crab as well as the structures involved. When crawling, the animal usually leaves a trail of three parallel grooves, like a three-railed subway train. The outer grooves are outlined by the rims of the exoskeleton, with the telson usually leaving a wavy middle groove. Either side of the area between the rims and the telson may be punctuated by the whorl of four blades near the end of the pusher legs (see fifth leg, Figure 6.1, BL). These marks are also seen in fossilized trails (Caster, 1938). Trace fossils of walking or crawling are referred to as *Kouphichnium* (Chisholm, 1983; Malz, 1964) whereas resting traces are referred to as *Selenichnites* (Romano and Whyte, 1987).

Examples are known worldwide from the Upper Carboniferous, and they are often associated with brackish-water lake deposits. Some trails were straight; some moved side to side, leaving long, undulating curves in the substrate; and others were aimless wanderings like that of the fossil species *Mesolimulus*. Pusher leg imprints dominate records of animals crawling or skimming across the bottom; while the tip of the telson, functioning as a rudder or balancing pole, may also leave a trace. Depending on the roughness of the bottom or water velocity, rim marks may be frequent, indicating places where a skimming animal was tipped to one side until it regained its balance (Luckenbach and Shuster, 1997). Some xiphosuran trace fossils are undertracks (Goldring and Seilacher, 1971). These are traces, which form at different levels in the sediment, not necessarily in contact with the animal. At deeper levels in the sediment, only the most heavily impressed traces are preserved. These tend to be the pusher imprints. Interestingly, xiphosurids and their traces are well represented in the fossil record (Guanzhong, 1993). So far, published records have revealed at least one of these fossils, animal or trail, from all the major landmasses except South America and Antarctica.

What about the Aglaspida?

Aglaspids have been cited as primitive xiphosurans that lived in the Cambrian, but we do not consider them to be that closely related to the horseshoe crabs. A sketch of one, *Paleomerus*, is shown in Figure 7.10. They lacked several defining characteristics of Xiphosura. They had 1) no ophthalmic ridges, 2) far more opisthosomal segments than any known member of the Xiphosura, and 3) no medial opisthosomal axis; furthermore, 4) they showed evidence that the cuticle was of a different composition, perhaps more phosphate-rich and with differing ornamentation. Even though there was no evidence for them in the fossil record, *Aglaspis spinifer* apparently did have chelicerae, as determined in a restudy by Briggs and Fortey (1989).

Interpreting the Body Plan

One night while Shuster was working on this chapter, he was startled by a noise in one of the large plastic trays in which he was temporarily

FIGURE 7.17 The trail of an adult horseshoe crab, released near the high tide line, making its way down a beach slope to the water's edge of Delaware Bay shows one variation of the so-called three-rail trail. Marks made by the appendage movements punctuate the areas in between the ridges of sand formed by the rims and the telson.

keeping horseshoe crabs. Upon investigation, he found an adult male trying to scramble up a corner of the tray. When Shuster bent down and looked straight at the front of the animal, he was struck by how fierce it looked with its legs outstretched and waving. This uncommon view of the crab was reminiscent of reconstructions of ancient sea scorpions at the Museum of Natural History at the Smithsonian Institution. Their legs were out in the open; how fearsome they must have appeared to other aquatic creatures. The wonder is that horseshoe crabs have persisted while the often-gigantic sea scorpions and the more numerous, complex, and ornate trilobites became extinct. Presumably, it was not only the functioning of the exoskeleton but the crabs' ability to live through changing environmental conditions that ensured their survival. Perhaps the evolution of a superior protective cover, the domed exoskeleton, was what provided an important step toward survival. The vaulted underside not only protected the appendages from enemies but also provided a sort of caisson under which the appendages could operate, in the main, within their own special environment, whether feeding, walking, burrowing, or egg laying. The horseshoe crab vault distinguishes it from the relatively flat-bottomed sea scorpions and trilobites, which had appendages that extended well beyond their carapaces.

When did the deep-bodied, vaulted body plan evolve? There may not be a clear-cut answer to that question. Perhaps, because fossils of Silurian xiphosurans—for example, *Limuloides* (Anderson and Selden, 1997) and *Cyamocephalus* (Anderson, 1999)—provide evidence of a vaulted carapace, this was a fundamental characteristic of the group back to their first emergence. The taxonomic group in the Lower Carboniferous to which *Bellinuroopsis* belongs provides another link to the limulids (Anderson and Selden, 1997).

The embryology of the extant species barely suggests an answer—the vaults are not clearly defined until the larval stage and, even then, they are relatively shallow (Shuster, 1948). Suture-like lines on the forward portion of the prosoma of the larvae give the appearance of a fusion of a frontal lobe with lateral projections. Perhaps most of the answer is in fossilized remains. Even though these are primarily flat impressions due to compaction of the sediments in which the animals died, there can be little doubt that many, if not all, species possessed the vaulted body plan, at least by the Silurian period (about 430 million years ago [mya]).

Other features of the exoskeleton also help us to understand the behavior and capabilities of the extant species and to visualize the three-dimensional appearance of the fossil species. The horseshoe-shaped rim around the edges of the undersurface of the prosoma has two parts of interest. The lateral portions of these rims are slightly concave and look and act like sled runners (Caster, 1938). This shape of the rim increases contact with the substrate and lessens sidewise slippage. Externally, the triangulate front portion of the rim provides a platform of contact with the substrate, especially during burrowing. In the adult male it is functional in mating. The ventral margin of the prosoma is curved like a horseshoe, hence the common name of the animal. The posterior projections of the prosoma curve slightly inward, enclosing the sides of the front part of the opisthosoma, creating the water channel discussed earlier. The functioning of this channel is also enhanced by the position and shape of the opercular pleurites that direct the flow of water under the carapace (see Figure 7.5) and the curved, spatulate basal structures (epipodites) on the fifth pair of legs (see Figure 6.1, EP).

Internally, muscles are attached either to small escalloped areas of the exoskeleton or to seven pairs of elongated rods (entapophyses) that extend downward from the roof of the opisthosoma and hinge area. An extensive maze of chitinous rods strengthens internal portions of the exoskeleton. Most of these internal markings are also visible on the outer surface of the exoskeleton, including entapophyseal pits and clusters of scar-like markings that disrupt the overall pattern of zigzag lines, particularly on the dorso-lateral areas of the prosoma.

The vault containing the respiratory branchia is the main feature of the opisthosomal ventrum of the living species. Besides functioning in respiration and water flow regulation, the branchias are the primary locomotor organs used in swimming and in gliding or skimming across the bottom. The large muscles of the operculum are attached mainly to a pair of large projections (entapophyses) on the anterior (prosomal) edge of the hinge. Six pairs of entapophyses on the dorsal undersurface of the opisthosoma provide attachment for the branchia. Formation of the branchial vault, through fusion of a number of segments, either was the immediate development of a relatively deep, walled space for the book gills or the start of a trend. If the opisthosoma had remained a structure composed of a number of freely articulated segments, the functions performed by a solid vault, as illustrated by Barthel (1974) and Waterston (1975), probably would not have evolved. It cannot be ruled out, however, that these functions were not already being carried out in the early xiphosurans with unfused segments. Perhaps they used the substrate surface to provide a quasi-ventral tunnel for feeding and aeration of the book gills. Persistence of the solid vault suggests that it was the more efficient form.

When studying horseshoe crabs, especially when comparing species, we typically examine a set of structural characteristics including

the axial region; the distance between the compound eyes (or the length of the hinge line) relative to the prosomal width; the claspers of the adult males; the fifth pair of walking legs (the pushers); the arrangement and size of the opercular pleurite; the shape of the first entapophyseal pit; the horseshoe-shaped ventral rim of the prosoma; and the operculum and the opisthosomal vault and differences in the placement of additional spines. The main difficulty in looking for all of these features in fossils is that, invariably, we have only the dorsal surface, usually an internal mold, to examine due to the way in which the fossil was formed and how the rock or sediment broke away from the surface of the fossil. Thus, ventral structures are rarely preserved in fossil specimens, a problem compounded by the rarity of these animals as fossils.

Limulid Evolution: Summation

Persistent is an apt, one-word description of the survival of limulids. This adjective characterizes not only their body plan but also their developmental stages, adaptability, behavior, and activities. Consider these observations:

1. Most of the anatomical features of horseshoe crabs appeared early in their geologic history. Several of these seem to have been important in the evolution of the modern body plan: consolidation of a number of body segments into a prosoma and an opisthosoma; the telson; a wide body with deep ventral vaults; the height of the prosomal axial ridge; the prosoma rim; and the pusher legs.
2. It was a momentous evolutionary step when limulids first laid their eggs in intertidal beaches. Not only were the eggs left in a unique incubator, but also the embryology probably included a new type of development. All embryonic stages, from the time the embryo secretes a capsule and molts four times before reaching the larval stage, are protected within a sphere.
3. After studying the body plan and behavior of horseshoe crabs, we agree with Nils Eldredge (1991). He succinctly characterized horseshoe crabs as ecological generalists whose hardy, jack-of-all-trades nature holds the clues to their conservative evolution and accounts for the few species that lived at any one time. Occasional observations may be misleading and may not be representative of typical circumstances, due to the differing responses by age groups to variations in environmental conditions. Horseshoe crab activities and behavior vary, locally as well as at different times and

places. It is necessary, therefore, to observe many variations in activity to understand any isolated situation that you happen to observe. Any of the foregoing may be reasons why horseshoe crabs often confound amateurs.

It would be an exhilarating and probably an educational experience to be able to study all available xiphosuran fossils, assembled in one building, with access to living specimens of all four extant species. Imagine how much more illuminating that experience would be if a team of experts in geology, paleontology, morphology, functional anatomy, behavior, and biochemistry could then examine the assemblage and describe each species and compare their relationships and habitats. There would most assuredly be one outcome—the relationships and taxonomy would be revised. Several of the fossil species probably would be found to be identical and the number of species reduced.

On finishing writing this chapter, we wondered how many more times this topic—the xiphosuran body plan—would be studied and new conclusions reached. As new specimens, methods of study, and insights become available, new reports will be written. This is the fate of most knowledge—to be recycled, added to, or amended, and to be explained with greater understanding. Thus, additional discoveries of new fossil species or more examples of those already recorded all have the potential of being very informative and valuable.

EXHIBIT 7.1

HORSESHOE CRAB LOOK-ALIKES

Two extinct groups of early arthropods, the trilobites and the sea scorpions, were contemporaries of the early horseshoe crabs. Similarities in body architecture are noted here for comparison.

The external morphology of an adult of the trilobite *Triarthrus eatoni* (Figure 7.18) has been redrawn from illustrations in the detailed analysis by Cisne (1975; see Levi-Setti, 1993).

The body was divided longitudinally into a head, thorax, and abdomen. Dorsally the main features were the pair of compound eyes and well-defined segmentation. Underneath, a pair of antennae, the position of the mouth cavity at the end of the hypostome (the anterior median lobe, also known as the labrum) and opposite the bases of the first pair of legs, the spiny gnathobases, and the appendages extending well beyond the margins of the carapace are of particular interest in a comparison with *Limulus.* The cross-section indicates the distance the body might have been elevated above the substrate in an ambulatory animal compared to when at rest or in a burrow. The uppermost, filamentous branches of the limbs were gills; the lower branch, a multisegmented leg, was setose but did not have pincers (chelae) like *Limulus. Triarthrus* apparently fed on minute particles. Its alimentary canal began at the tip of a broad hypostome. Food was moved anteriorly from the mouth through an esophagus to a large crop in the dorsal part of the head. A long intestine extended the length of the thorax to the anus.

Waterston (1979) reconstructed the external appearance of the eurypterid *Parastylonurus ornatus* and gave the first detailed analysis of the walking mode of long-legged eurypterids (Figure 7.19). This eu-

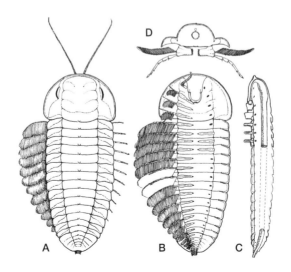

FIGURE 7.18 The external morphology of an adult trilobite, *Triarthrus eatoni.* (A) dorsal, (B) ventral, (C) longitudinal section with horizontal dashed lines connecting the front and end portions of the alimentary canal, and (D) cross-sectional views. Some of the details on the ventral side have been omitted to show underlying detail. A large specimen was about 4 cm long.

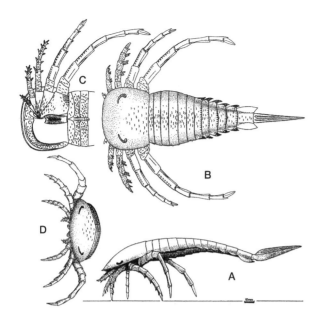

FIGURE 7.19 Four views of the eurypterid *Parastylonurus ornatus (from the reconstructions and interpretations of functions by Waterston, 1979).* (A) lateral, (B) dorsal, (C) ventral, and (D) frontal *(drawn by C. N. Shuster, Jr., based on A–C).* The heavy scale line under the telson is 1 cm.

rypterid was about 22 cm long (the short bar below the telson represents 10 mm). In (A), you see the lateral view of a walking position of this medium-sized, long-legged sea scorpion. The post-abdomen and the telson were specialized hydrodynamic structures that, by monitoring water movements, led to muscular coordination that maximized walking efficiency. The shape of the telson may have had an important function as a hydrofoil, balancing the walking animal in water currents. In (B), you see the dorsal aspect, and in (C), a partial view of the ventral aspect of the anterior area showing the prosoma, prosomal appendages, and the opercular segments of the pre-abdomen. The appendages on the right side of the animal have been removed to show the form of the underside (doublure) of the prosoma. Figure 17.19D shows a head-on view of the sea scorpion *Parastylonurus ornatus* in the walking position (by Shuster, based on the reconstructions by Waterston, 1979). Note that three pairs of walking legs support *Parastylonurus.* The small chelicerae are the central, innermost appendages, mostly hidden from view.

WHAT ARE HORSESHOE CRABS?

EXHIBIT 7.2

We are commonly asked about the kinds of animals to which horseshoe crabs are related. If not strictly crustaceans, as their common name implies, what are they? They belong to the vast array of joint-legged animals, the arthropods, that includes trilobites, sea scorpions, centipedes, insects, scorpions, and crabs. Because the relationships between horseshoe crabs and other arthropods have been confused in the popular media, a brief historical review may be helpful.

Centuries ago, naturalists identified these animals as crabs, based chiefly on their hard shell, pincer-tipped legs, gills, and their occurrence in coastal waters along with true crabs (decapods crustaceans). In 1881, the great English zoologist, Sir Ray Lankester, determined that horseshoe crabs are more closely related to scorpions and spiders than to any other living creatures, with a first pair of pincer-tipped feeding organs, the chelicerae (which are common to all arachnids—the scorpions, spiders, ticks, and mites—whereas mandibles are common to the crustaceans). Respiratory, alimentary, and nervous systems are also similar in horseshoe crabs and arachnids. Nevertheless, the term "crab" persists, in a variety of forms: horseshoe or horsefoot crab, pan crab (their empty shells have been used to bail out boats), helmet crab, and king crab (a real misnomer given there is a true crustacean with this name). Actually, king crab was an old common name for horseshoe crabs that was brought to America by English colonists. It is a name that persists today among many watermen who fish our coastal waters.

In recent years, the close relationship of crabs to arachnids has been questioned. Kraus (1976) has put forth an argument on a molecular basis that the Merostomata (horseshoe crabs = Xiphosura and sea scorpions = Eurypterida) should be co-equal with other arthropod groups: the crustaceans, arachnids, trilobites, and insects. Dunlop and Selden (1997) placed the horseshoe crabs (xiphosurans) in a separate group, assigning the extinct species of sea scorpions (eurypterids) to a taxonomic position closer to the scorpions and other Arachnida. Shultz (2001) also supports the position that, among extant chelicerates, xiphosurans and arachnids are monophyletic sister groups. In other words, the two groups are mutually exclusive—xiphosurans are not arachnids; they are probably as close to the ancient trilobites as any other group (Fortey, 1999). We consider that the Merostomata are an aquatic grade of Chelicerata, so we prefer it as the group name.

The scientific (taxonomic) classification of the extant species of horseshoe crabs, starting with the larger, more inclusive classification, down to the smallest taxonomic grouping, is as follows:

Phylum Arthropoda = Animals with jointed legs and hard exoskeleton.
    Subphylum Cheliceriformes = Chelicerate species have chelicera as the first pair of appendages.

Superclass Chelicerata

Merostomata = Animals with the bases of their legs arrayed around a ventral mouth (Greek *meros* = thigh, femur and *stoma* = mouth). The scientific names of the long-bodied eurypterids and the xiphosurans also have been derived from Greek words (*eurys* = broad, wide and *pteron* = wing; *xiphos* = sword and *oura* = tail).

Class Xiphosura

Order Xiphosurida = Those with the more characteristic horseshoe crab body plan, in contrast to the Synziphosurine species such as *Weinbergina, Bunodes,* and *Lemonelites* (see Figures 7.9 and 7.10).

Suborder Limulina = This suborder is characterized by the living species and fossil species; the oldest discovered genera are *Xaniopyramis, Moravurus, Paleolimulus,* and *Valloisella.* Suborder Bellinurina = other fossil species, such as *Bellinurus* and *Euproops.*

Superfamily Limulacea

Family Limulidae

Genus *Limulus* = From Latin *(limus),* meaning somewhat oblique, sideways, or askance—in reference to the position of the pair of compound eyes on the lateral ridges on top of the shell.

Species *polyphemus* = Polyphemus (in the *Odyssey* of Homer) was the monster with an eye in the middle of his forehead, in reference to the pair of single eyes that are centrally located at the base of a broad axial spine in the front part of the *Limulus* shell).

Family Tachypleinae = Greek *tachys* (swift) + Greek *plein* (to swim).

Genus Tachypleus = This genus subsumes *Carcinoscorpius* according to Yamasaki (1988), as reflected below. Although we agree with this redesignation, in the text we have retained the old classification, *Carcinoscorpius rotundicauda* (*Carcinoscorpius* = Greek *karkino* [cancer] and Latin *scorpio* [scorpion]; crab-scorpion), pending action on the reclassification of the genus by the International Committee on Nomenclature. It could be argued that, because *the* Tachypleinae are morphologically similar to fossil species of *Mesolimulus, Mesotachypleus* would be a more proper generic name.

Species *tridentatus* = Having three teeth, processes, or points (the three pairs of short marginal spines on the female opisthosoma; see Figure 7.7, row B).

Species *gigas* = *Gr.* for giant, large.

Species *rotundicauda* = Round tail, because the cross-section of telson is round (see Figure 7.7, row E).

COLORPLATE 1 Spawning horseshoe crabs, shorebirds, and members of the Delaware Bay Shorebird Workshop on a beach at Bombay Hook.

COLORPLATE 2 A Cape May, N.J., beach strewn with horseshoe crab eggs. The eggs, left on the surface after a spawning, were washed about during an ebbing tide. The wavy bands of eggs mark earlier water levels. Only a few horseshoe crabs have lingered at the foot of the beach; the majority of the spawners have already moved out into the intertidal area that will soon be bare when the tide completes the ebb cycle.

COLORPLATE 3 The four most abundant species of shorebirds along the Delaware Bay shore. From left to right, red knot, ruddy turnstone, three more red knots, semipalmated sandpiper, and sanderling. *Photograph by D. C. Twitchell.*

COLORPLATE 4 On some occasions, a major fraction of New World red knots can be seen in dense flocks roosting along the beaches of Delaware Bay. In the background, laughing gulls are lined up in a band along the water's edge at the foot of the beach. *Photograph by D. C. Twitchell.*

COLORPLATE 5 Ruddy turnstones and red knots feeding on horseshoe crab eggs along a Delaware Bay beach. *Photograph by D. C. Twitchell.*

COLORPLATE 6 Laughing gulls along the Cape May shore of Delaware Bay are opportunistic feeders that concentrate on horseshoe crab eggs during the spawning season. *Photograph by M. L. Botton.*

COLORPLATE 7 Horseshoe crabs nest
at Seahorse Key, Fla., along a narrow strip
of beach at the top of the high-tide line.
Photograph by J. Brockmann.

COLORPLATE 8 Keffer Hartline (1903–1983) sitting on his "think bench" in the woods behind his home near Baltimore.

COLORPLATE 9 A mini–video camera, CrabCam, mounted on an adult horseshoe crab records what the right lateral eye sees. A recording chamber mounted anterior to the eye contains a microsuction electrode (black cylinder on right) that records the response of a single optic nerve fiber through a hole drilled in the carapace. A white cap seals the recording chamber. The waterproof electrode and camera simultaneously record a nerve fiber's activity and the underwater scene as the crab swims near the cement objects shown in Figure 4.5.

COLORPLATE 10 Ventral views of mid-sized juvenile *Limulus* illustrating the characteristic yellow margins of the premold carapace (left) and the suture (split rim) through which a blue-gray specimen edged with pink is beginning to emerge from its old shell (right). The contrasts in color and size of the cast shell (left) and the newly emerged juvenile crab (right) are shown below. *Photographs by C. N. Shuster, Jr.*

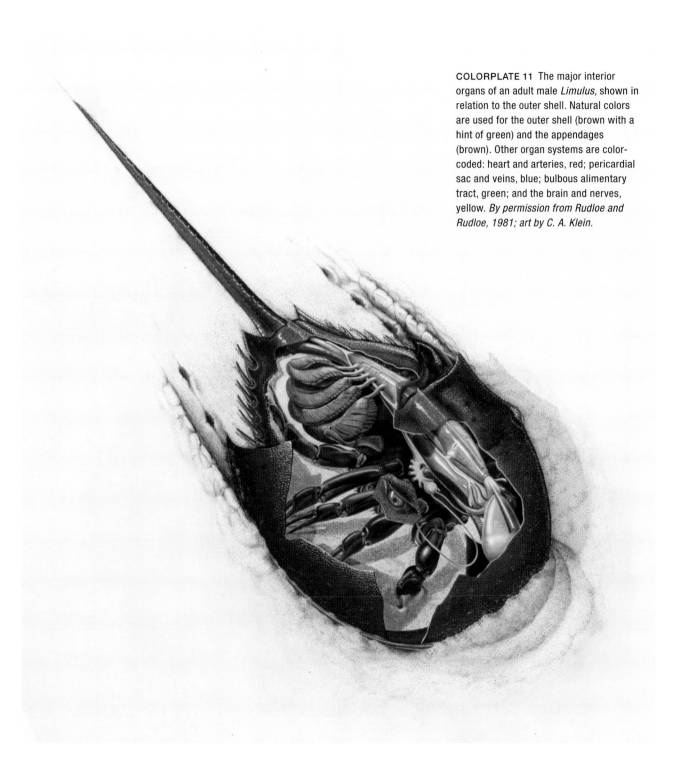

COLORPLATE 11 The major interior organs of an adult male *Limulus,* shown in relation to the outer shell. Natural colors are used for the outer shell (brown with a hint of green) and the appendages (brown). Other organ systems are color-coded: heart and arteries, red; pericardial sac and veins, blue; bulbous alimentary tract, green; and the brain and nerves, yellow. *By permission from Rudloe and Rudloe, 1981; art by C. A. Klein.*

A

B

C

D

E

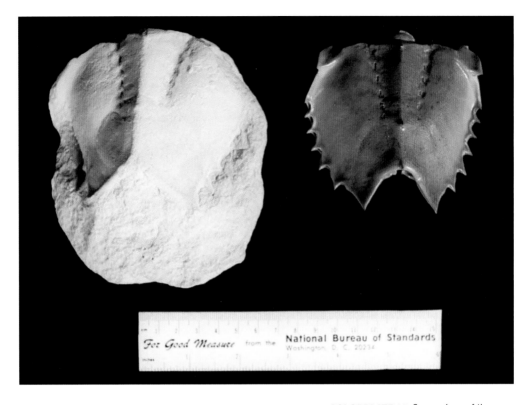

COLORPLATE 13 Comparison of the opisthosomal fragment of *Limulus coffini* (11 cm wide) to a similarly sized part of *Limulus polyphemus*. The *L. coffini* specimen is a cast made of the fossil in the Museum of Natural History, Smithsonian Institution, Washington, D.C.

COLORPLATE 12 Thick sections through a frozen adult male *Limulus* (compare with Figures 7.1 and 7.2). (A) Longitudinal section along the midline of the body. This specimen had recently fed on small blue mussels *(Mytilus edulis),* about 5 to 7 mm in length, which fill most of the long alimentary tube. (B) Cross-section through the compound eyes. (C) Longitudinal section about 3 cm left of the midline. (D) Cross-section midway through the opisthosoma (compare with C). (E) Cross-section near the posterior end of the opisthosomal vault cuts through the large circular ring of muscles (pinkish color) that move the telson.

COLORPLATE 14 This scene of a nesting pair of limuli, in a patch of sand at the edge of a tidal marsh on the eastern shore of Chesapeake Bay, may have been common at other times and places when extensive sandy beaches did not exist, as when marshes invaded the shoreline. Patches of sand may have been all that were available to earlier species of horseshoe crabs surviving changes in shorelines.

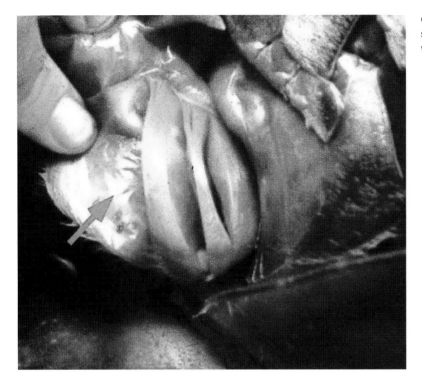

COLORPLATE 15 Book gill of the horse-shoe crab, *Limulus polyphemus.* Ventral view showing a "leaf" (lamella), at arrow.

COLORPLATE 16 Location of the urinary opening (nephropore) at the base of the fifth walking leg of *L. polyphemus. Photograph by D. W. Towle.*

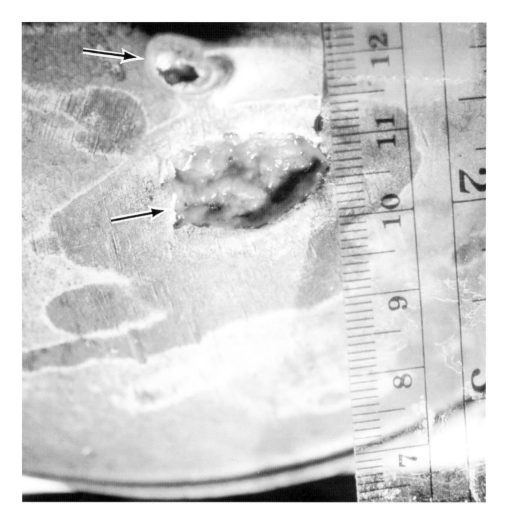

COLORPLATE 17 Healed wound, approximately 1 year after injury occurred. The lateral compound eye has been enucleated (top arrow), the tissue defect in the prosoma is not completely filled (lower arrow), and the internal viscera (digestive gland and gonadal tissues) are visible, but there is no evidence of infection. The animal has molted and the old exoskeletal tissue around the wound has been retained, protecting the wound during healing. The smallest units on the transparent ruler are millimeters.

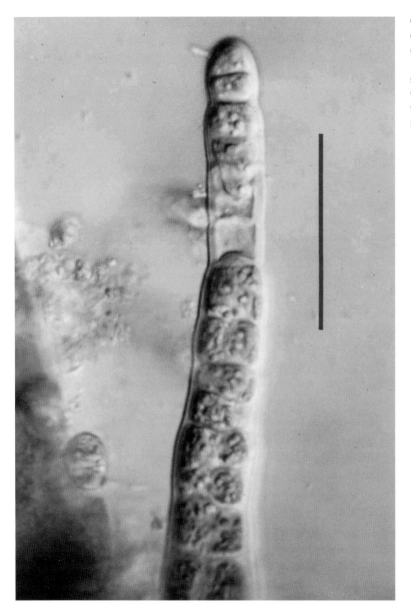

COLORPLATE 18 Microscopic view of a green algal filament growing from the exoskeletal surface outward (scale bar = 100 microns). The rounded distal tip and segmented walls contain green cells, which are distromatic (two cells thick) at the base of the filament. This arrangement is characteristic of the *Ulva* species.

COLORPLATE 19 Pathological changes associated with green algal infection of the large lateral compound eyes of an adult *Limulus*. (A) An early lesion: the light (white) circular lesion with green round circular spots in the upper-right area of the lateral eye. This area represents an early stage of destruction of the cornea and the deeper ommatidia beneath by invading green algae from the surface into the interior of the eye. Most of the remaining surface of the eye is normal ommatidia, with the occasional white spots representing single infected ommatidial units (scale bar = 1 cm). (B) A stained microscopic histological tissue cross-section of the early lesion in (A). Note the dense red mass containing spherical algal organisms in the central pink clear zone, which represents lysis of eye tissues by the organisms. The round black circular bodies surrounding the disease are normal units of the eye (ommatidia) (scale bar = 1 mm). (C) More advanced green algal infection of the eye. Note the large, irregularly shaped, dark ulcers extending over the lower half of the eye and the loss of symmetrical surface structures (scale bar = 1 mm). (D) Complete destruction of the soft tissues of the lateral eye by green algal infection. Note the empty orbit with a few small remaining colonies of green algae within (scale bar = 1 cm).

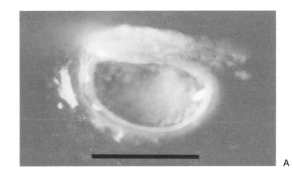

A

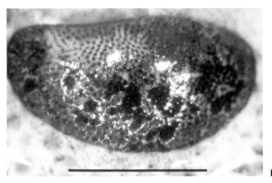

B

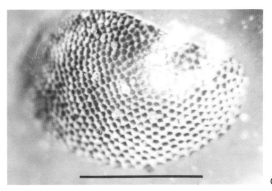

C

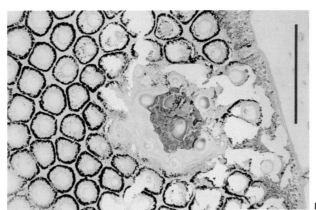

D

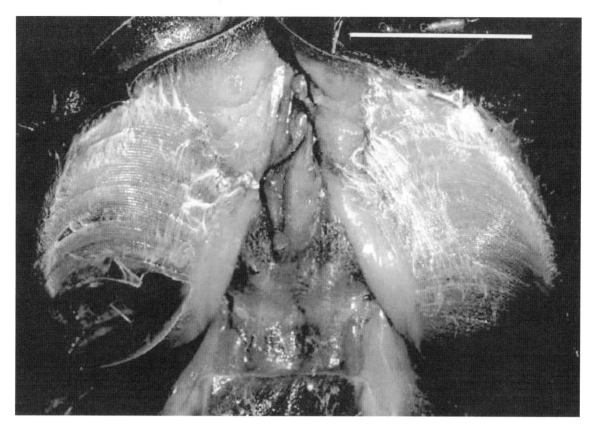

COLORPLATE 20 Early stage of cyanobacterial *(Oscillatoria)* gill infection of an adult *Limulus.* The trichromes have formed a blue-green, cotton-like covering on the lamellar surfaces of the gills (scale bar = 4 cm). The book gills have been pulled upward and the posterior end of the animal is toward the bottom of the picture.

COLORPLATE 21 Photomicrographs of blue-green *Oscillatoria* trichromes. (A) Direct smear of trichromes removed from the surface of infected gill lamellae (scale bar = 250 microns). (B) Blue-green trichromes entwined about a gill seta (scale bar = 1 cm).

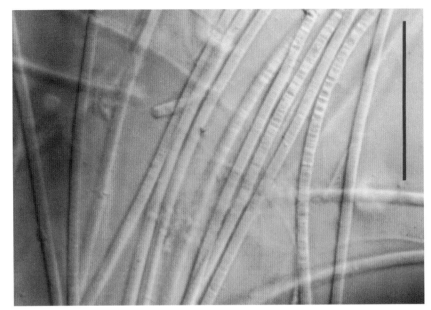

A

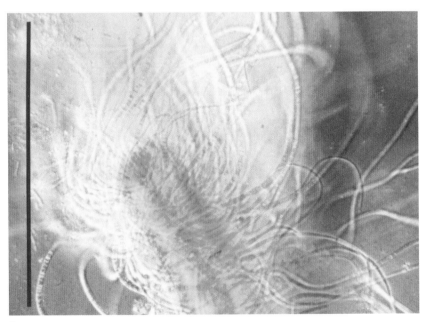

B

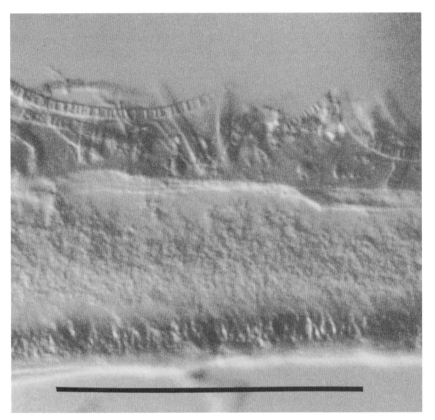

COLORPLATE 22 Stained histological cross-section of an early stage of cyanobacteria *(Oscillatoria)* on a gill lamella (leaflet). The attached trichromes (filaments) have covered the top of the gill lamella and have invaded its outer surface wall. The dilated vascular sinus (central portion of the lamella) contains coagulated hemolymph (scale bar = 0.5 mm).

A

COLORPLATE 23 Cyanobacterial infections. (A) Gross view of swollen gill lamellae due to a cyanobacterial infection (scale bar = 5 cm). (B) Histological section showing an early stage of infection by cyanobacteria *(Oscillatoria)* in an adult *Limulus*—swollen vascular sinuses filled with coagulated blood, detached internal epithelial layer, and the ruptured vertical internal supports of the vascular space (scale bar = 0.5 mm).

B

COLORPLATE 24 Late-stage *Oscillatoria* gill infection; ventral view of an adult female *Limulus*. Note the loss of gill tissue organization and displacement of remaining gill tissues (scale bar = 7 cm).

COLORPLATE 25 *Beggiatoa* infection of the branchial lamellae. (A) Gill lamellar surfaces covered with colorless filaments of *Beggiatoa* sp. Note the darkened gill tissue and clumps of white debris on the gill surface (scale bar = 2 cm). (B) Microscopic view of a direct smear taken from the infected gill surface. The filaments contain dispersed sulfur granules and gas bubbles surround the filaments (scale bar = 250 microns). (C) Ventral view of the blackened surface of infected gills of an adult *Limulus* (scale bar = 5 cm).

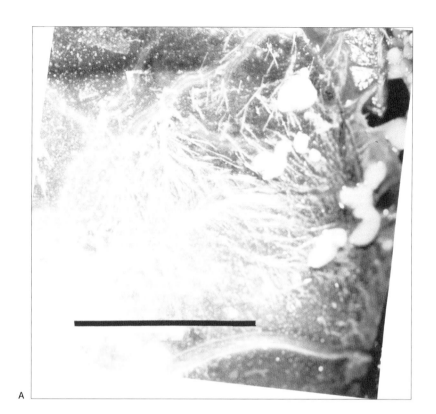

A

B

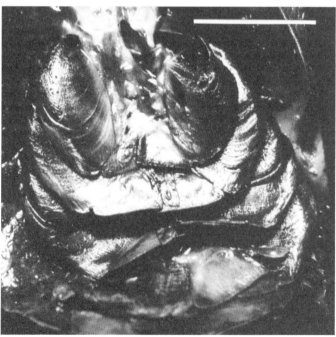

C

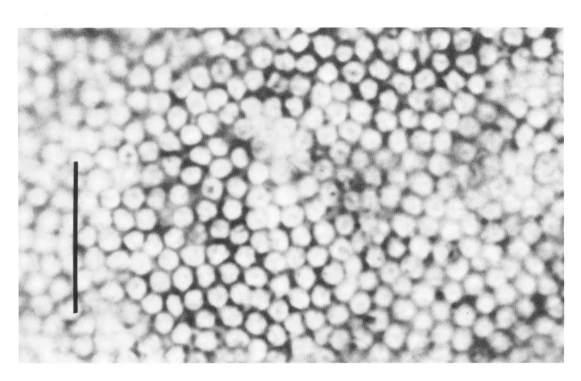

COLORPLATE 26 Invasion of the branchial lamellae of *Limulus* by a colonial cyanobacteria. Photomicrograph of the surface of an adult *Limulus* gill filament covered with a growth of colonial cyanobacteria (*Chroococcales* sp.). A fresh surface growth forms a golden mosaic pattern. As the organisms age, they become darker in color (blackened) (scale bar = 100 microns).

COLORPLATE 27 A book gill heavily infested with triclad turbellarid cocoons, with the surfaces of two gill leaflets exposed, showing the marked pathological gill changes associated with the triclad gill infestations. Many bean-shaped cocoons extend from the surface of the gill lamellae (leaflets), and a large oval ulcerative perforation is on the anterior surface of a gill lamella. Several white, irregularly shaped adult triclad turbellarids are evident on the posterior surface of a gill lamella. All the gill lamellae were deformed, wrinkled, and covered with hemorrhaging hemolymph (millimeter scale on right; scale bar = 1.5 cm).

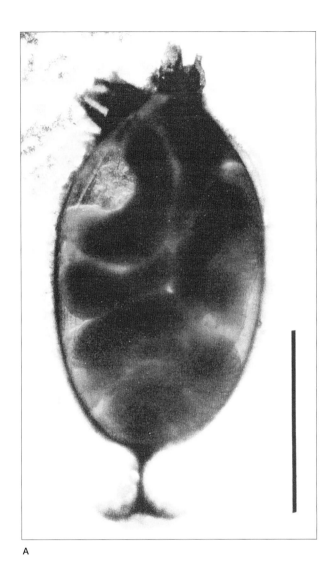

A

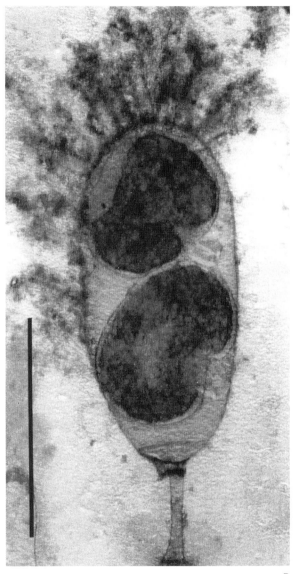

B

COLORPLATE 28 Photomicrographs of two species of embryonating triclad turbellarid cocoons attached to the surfaces of their respective gill lamellae (leaflets). (A) The cocoon of *Bdelloura candida* is larger, contains eight developing embryos, has a short pedicle at the base attaching the cocoon to the gill leaflet surface and ciliated protozoa at the top of the cocoon (scale bar = 1.5 mm). (B) The cocoon of *Syncoelidium pellucidum* is smaller, has an elongated oval form, two embryonating masses in the cavity of the cocoon, ciliated protozoans at the top, and straight walls of the pedicle (scale bar = 0.4 mm).

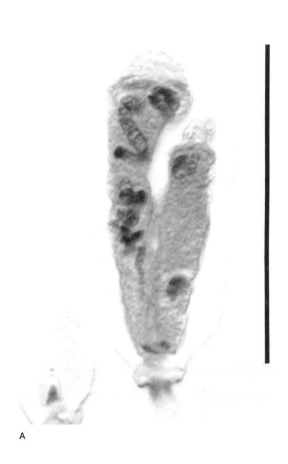

A

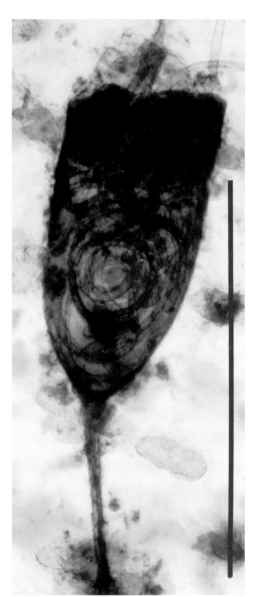

B

COLORPLATE 29 Commensals and parasites associated with triclad cocoons. (A) Photomicrograph of the fixed stained loricated ciliated protozoa (*Folliculina* sp.) revealing the internal structures and the outer protective case (lorica) (scale bar = 0.5 mm). (B) Nematodes (*Chromodaridia* sp.) have infested this cocoon of *Syncoelidium pellucidum* that was attached to the surface of a gill lamella. The nematodes (coiled organisms within the cocoon) have consumed the triclad larvae (scale bar = 1 cm). (C) A sexually immature parasitic copepod *(Nitrica bdellourae)* within a triclad co-coon, where it has consumed the triclad embryos (scale bar = 100 microns).

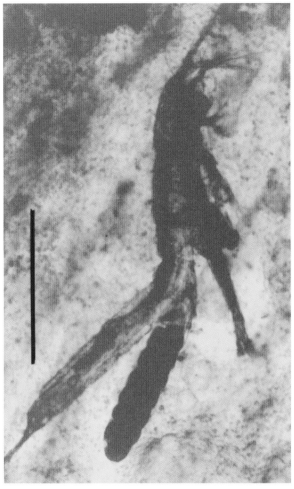

C

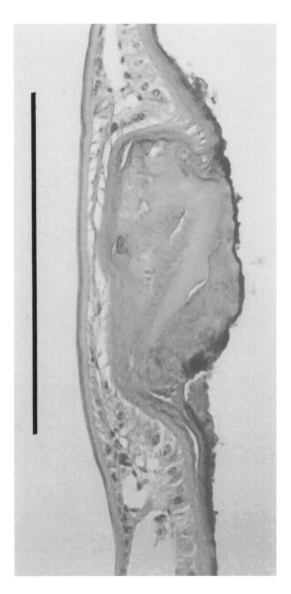

COLORPLATE 30 A stained histopathological cross-section of a healed triclad wound on the surface of the gill lamella. Note the large surface defect filled with coagulated hemolymph and reorganized epithelial tissue. The black surface on the lamella is a band of necrotic tissues containing bacteria (scale bar = 0.5 mm).

COLORPLATE 31 Photographs of metacercarial *(Microphallus limuli)* infestation of the compound lateral eye and muscle of a juvenile *Limulus*. (A) Lateral eye of a juvenile *Limulus* heavily infested with encysted metacercariae *(Microphallus limuli)*—the many small round white objects (scale bar = 7 mm). (B) Cross-section of muscle from a juvenile *Limulus* heavily infested with metacercarial cysts of *Microphallus limuli* (scale bar = 0.5 mm).

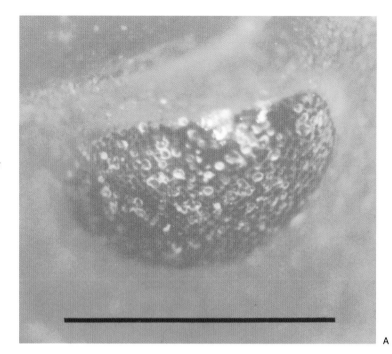

A

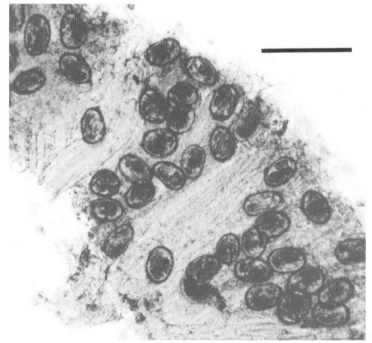

B

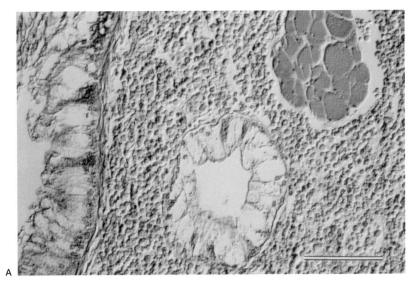

A

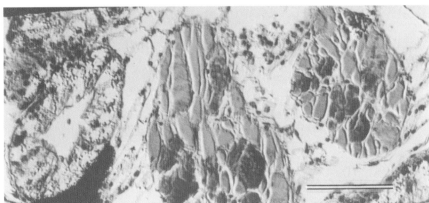

B

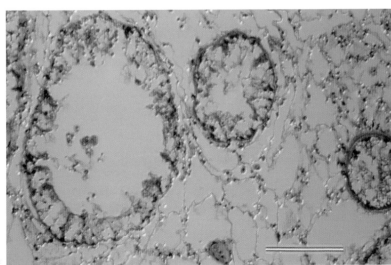

C

COLORPLATE 32 A comparison of the digestive glands of three adult limuli at different stages of starvation shown by stained, histological cross-sections (all scale bars = 75 microns). (A) Digestive gland of a freshly harvested adult *Limulus*. Note the dense number of secretory granules and well-defined cellular elements. The uniformly deep-stained muscle bundle, in the lower right, is also indicative of an active nutritional function. (B) Glandular tissues and muscle fibers of an adult held in the laboratory for approximately 3 months without food. Note the paucity of cellular elements and detail, and the irregular shapes and staining of the two muscle bundles on the right, suggestive of muscle degeneration and nutritional deficiency (Zenker's degeneration). (C) From a specimen that was kept in the laboratory for 6 months without food. Note the extreme cachexia characterized by the absence of secretory granules and sparse cellular elements.

COLORPLATE 33 Blue blood: after a horseshoe crab is bled through a syringe needle until the blood clots, it is replaced by another crab, and then by others, until the bottle is full. Synthetic substitutes may soon replace dependence on horseshoe crabs for blood. *By permission from Cambrex Bio Science Walkersville. Photograph by C. N. Shuster, Jr.*

THROUGHOUT GEOLOGIC TIME: WHERE HAVE THEY LIVED?

Lyall I. Anderson and Carl N. Shuster, Jr.

Nowadays, horseshoe crabs live only along the eastern coasts of Asia and North America; but even in these areas, their distribution is uneven. At first this discontinuity may seem puzzling, especially when the sites where fossilized species have been found are also considered. However, the answers to the distributions of all the species lie in the ecology of the living species and in the fossil and geologic records. Of these, the most important answer may be that the specific breeding behavior of the extant species that ties them to selective areas of the coastline probably did so throughout the whole geological history of the xiphosurans. This is what their distribution tends to suggest. It is also apparent that the abundance of *Limulus* on the continental shelf is directly related to the size of spawning areas and activities in nearby estuaries. Although some animals from nearby populations might intermingle and humans might transplant them even further, mixing of populations is probably not common enough to markedly change the morphological or genetic characteristics of any major population within a few decades.

The following discussion begins with a general consideration of long-term, large-scale geologic changes and a consideration of environmental factors, physical and ecological, affecting horseshoe crabs. This is followed by an account of the distributions of the extant species and examples of the habitat of fossil species.

Geologic Changes in Geographic Features

Coastlines change for a multitude of reasons and often quite rapidly, as by the erosional force of a storm. In contrast, plate tectonics, the movement of the earth's crustal plates, is a relatively slow process that accounts for various aggregations of landmasses and their separation over geologic time. At present rates of movement, tectonics probably would not push taxa to extinction through shrinking habitat. Hence, among all the alterations that have occurred in the surface of the earth during geologic time, what has mattered most is what has happened at the edges of the seas, at the interface between land and water. Although the physical and chemical attributes of those habitats undoubtedly have changed, the changes were sufficiently slow that available habitat persisted, furnishing a continuum of shallow water environments where Xiphosura could live. This is what occurred with the coal swamp forests of the Carboniferous age. In Europe and America, climatic change, and a little tectonic interference, shrunk the coal swamps. Instead of mass extinction evenly sweeping across plants and animals, the fossil record indicates that the inhabitants just shifted, with most to be found in the early Permian coal deposits of China.

Global maps of former continental position, as depicted in paleogeographical atlases, add little to an understanding of the former habitats of horseshoe crabs. However, they do illustrate the worldwide changes that have occurred in the sizes and distribution of landmasses and oceans from one major geologic era to the next (see, for example, Palmer, 1999; Scotese et al., 1979). Interpretation of these maps suggests that horseshoe crabs once existed in mainly tropical and subtropical regions despite where the fossils are now found.

According to the theory of plate tectonics, the crust of the earth is divided into rigid pieces (plates) that are moved by currents in the underlying semimolten mantle. Divergence of the plates forms ocean basins; their convergence, usually accompanied by volcanic eruptions and earthquakes, forms island arcs and mountain ranges. Collisions of the ancient continents also formed mountain ranges. Concurrently, erosion of the elevated land returned sediments back to the oceans.

Lengthy, submerged mountain ranges extend throughout the oceans. Material rising from the mantle along these midoceanic ridges is carried along the ocean floor by mantle currents toward the continents, where it sinks back into the mantle. These sinks are deep trenches that, with the oceanic depths bordering the continental shelves, are barriers to horseshoe crab migrations.

Besides the changes wrought by plate tectonics, the shape and extent of the land areas have been altered by the rise and fall of the sea level. Interestingly, continental shelves were once coastal plains, before the melting of extensive continental glaciers forced a global sea level rise and they were flooded. Today these underwater, gently sloping shelves are foraging areas for the extant species. Considering how modern-day crabs use the shelves, there has probably always been a correlation between the abundance and distribution of horseshoe crabs and the increase or decrease in the continental shelves. Narrow or missing shelves restrict crab populations.

Physical and Chemical Factors Affecting Distribution

Temperature and salinity are foremost among the environmental factors affecting horseshoe crabs; quiet waters, tides, currents, and water depths are also significant. These factors play important roles in Delaware Bay and on the continental shelf.

Temperature

The northward extension of the range of *Limulus* is probably limited by the shortness of the warm-water season rather than by cold temperatures (temperature zonation concept of Hutchins, 1947). Lack of sandy beaches may also be a factor as well as the possibility that the extreme tidal amplitudes in northern Maine are too hostile an environment for successful development of the eggs. The smaller-sized adults north of Cape Cod (Shuster, 1955) may be as much the result of short growth seasons as of the impact of cold, because horseshoe crabs have been seen under ice on Delaware Bay intertidal flats (Thurlow C. Nelson, 1952, Rutgers University, personal communication) and withstood chilling in laboratory experiments (Mayer, 1914). These observations indicate that *Limulus* from the northern part of this range is tolerant of freezing, but whether a short season of activity would affect reproduction is not known. Prolonged exposure at 40°C (104°F) may be lethal (Fraenkel, 1960).

Why *Limulus* is not found farther south than the broad continental shelf off Yucatán is not known. Warmer water temperatures should not be a problem, so it seems likely that the limiting factor is insufficient tidal amplitude combined with a lack of semidiurnal tides. Because the Indo-Pacific species are found both to the north and south of the equator, tropical temperatures do not restrict their range. If our interpretation is correct, the deep oceanic trenches and abyssal plains block further southward and eastward distribution of the Asiatic species. Their northernmost distribution may be limited, as for *Limulus,* by a shorter growing and reproductive season than elsewhere in their distribution.

Salinity

Coping with varying salinities is the main topic of Chapter 9 (the usual tolerance range for *Limulus* is from quarter-strength salinity [8 parts per thousand] to full-strength seawater).

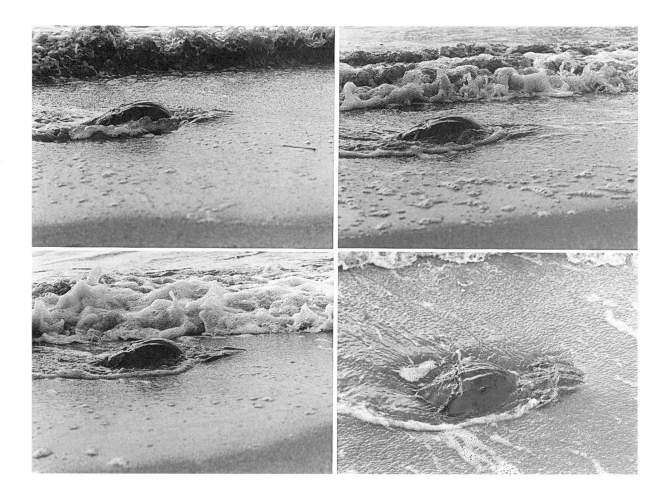

FIGURE 8.1 Spawning crabs often are oriented parallel to the wave front when waves obliquely hit a beach. In this case, the waves have driven all but the attached male from a nesting female that is almost completely buried and securely anchored (Cape May beach on Delaware Bay). *Photograph by C. N. Shuster, Jr.*

Importance of Still Waters

Relatively calm waters are a prerequisite to successful spawning. Waves over 30 cm (> 1 ft) in height generally wash the spawners from the beach. Increasing wave amplitudes diminish the number of males in a spawning group until only the attached male is able to stay with a nesting female. The remaining male may be buffeted with such force that, if he stays attached, the female is tugged from the nest and both are swept off the beach (Shuster, 1955, 1982). When waves strike the beach at an angle, some pairs are more or less aligned parallel to the wave front (Figure 8.1). Just what causes this behavior and whether it has a hydrodynamic advantage has not been reported.

Significance of Tidal Regimes

The lack of spawning populations within the western portion of the Gulf of Mexico is notable. There, high tides occur only once a day and the water level changes only a foot or less (Marmer, 1954). Such a tidal cycle and tidal amplitude may be factors limiting successful spawning because all known spawning areas are those beaches that are bathed twice daily by high tides with amplitudes varying, from site to site, from a few to several feet. Apparently the only beaches having suitable conditions for successful incubation of *Limulus* eggs are those where the tidal amplitude enhances the mixture of moisture, warmth, and oxygen (Shuster, 1982; Shuster and Botton, 1985; Penn and Brockmann, 1994). Besides being a key issue for both spawning and distribution, could tidal amplitudes enhance the potential for fossilization? Many of the Carboniferous fossils come from sites where sedimentation rates were consistently very high over a few thousand years. High sedimentation increases the chances of burial of a morbid or dead organism away from scavengers, and it places organisms in a physiochemical zone in the sediments where phosphatization of soft tissues and formation of siderite (iron carbonate) minerals can occur.

The first intensive survey of nine historic horseshoe crab sites has begun along the rocky shores of Maine, where there are pockets of sandy beaches (Schaller et al., 2001, 2002) and tidal amplitudes may exceed 13 feet. Crabs were tagged and their spawning activity reported. Maximum numbers of spawners during a single tide were a few hundred on these small beaches; the northernmost population was in Frenchman's Bay just north of Mount Desert Island.

Currents

The ability of horseshoe crabs to navigate in strong currents has not been fully investigated, but field observations in waters up to 1 m in depth indicate that they go with the tidal currents rather than buck them. In intertidal areas, horseshoe crabs do not become immediately active when first covered with the rising tide. When the water is 6 in or more in depth, they emerge from their shallow burrows in the flats and, after a few orienting adjustments, turn in the direction of the current and amble off. Much of the concentration and dispersal of horseshoe crabs may normally occur in the direction of prevailing and tidal currents because their innate behavior takes advantage of the direction of flow. A preliminary study on the activity of juvenile limuli in a flume indicates that they respond to rising water and current direction in the same way as the adults in nature (Luckenbach and Shuster, 1997). The main question is, How do they know which current to ride, ebb, or rise, or is their response nondiscriminant? A fossil record suggests a possible orientation of horseshoe crab burrows in sediments of Upper Carboniferous age from Bollington, Cheshire. Plotting the orientation of the long axis of burrows that were suspected to have been built by the horseshoe crab *Bellinurus,* against currents based on clues from sedimentary structures, suggests that the animals predominantly faced into the prevalent current.

Rudloe and Heernkind (1976) investigated the orientation of horseshoe crabs to the direction of surge currents on the Gulf coast of Florida by releasing adults at two sites. Only in one case did a crab head into the wave surge. All of the others moved almost directly with the surge and headed for deeper water.

Both the Yucatán Current and the Florida Current are prevailing northward-flowing, high-velocity currents next to narrow continental shelves. Is it possible that the combination of strong currents and restricted shelves deter the southward migration of *Limulus,* either from Georgia or from the Atlantic coast of Florida into the Gulf of Mexico? Further, the Yucatán Current intensifies each spring so that by May, maximum velocities are reached (Cochrane, 1966; cited by Nowlin, 1972). There are other similarities: (1) both currents sweep past the eastern edge of their respective peninsulas, (2) broad shelves extend westward from Florida and northward from Yucatán, and (3) their coastal tidal regimes are similar. The major portion of the offshore areas adjacent to horseshoe crab spawning habitats are the broad shelves off western Florida and northern Yucatán and the alluvial coastal area with smooth shorelines of sandy beaches off Alabama and Mississippi (Price, 1954). In these same areas, the range of the tide is small, not more than a foot or two on the average, with two high tides per day (Marmer, 1954).

Bathymetric Distribution

Limulus ranges from the intertidal zone of bays out onto the continental shelf of North America. The shelf is, except along the southern coast of Florida, an extensive underwater area that was the margin of the continent during the last great period of glaciation. Today it is a gradually sloping, shallow water plateau extending out to the 200 m (655 ft) depth contour, composed mainly of sand with regions of silty sand and clayey silt. *L. polyphemus* is most common on the northwestern Atlantic continental shelf, at depths in the inshore area up to 40 m (150 ft), although an errant specimen was once found at a depth of 1,097 m (6,558 ft) (Botton and Ropes, 1987). The shelf along Florida, from Miami Beach to Palm Beach (a stretch of some 102 km, or 64 mi) is approximately 10 km (6 mi) at its widest part. This narrow shelf and a minimum northerly flow of 2.7 knots of the Gulf Stream, 40 km (25 mi) offshore, may also limit the northward or southward migrations of *Limulus* through that stretch.

A comparison of the location of oceanic trenches (National Geographic Society, 1988–1999) with the distribution of the four extant species of horseshoe crabs suggests that great depths restrict distribution. As a corollary, the relatively shallow waters of continental shelves are avenues of dispersal and may have always been. Otherwise, the distribution of the extant species versus that of the extinct species becomes more difficult to explain. The progenitors of the extant species may have migrated eastward and westward at the time the Mesozoic and Cenozoic seas of Europe were becoming the Alps and the continent of North America was drifting away from the others (Ives, 1891; Sekiguchi, 1988c).

What may we conclude about the worldwide distribution of the extant species of Limulidae, restricted as they are to the eastern edges of the North American and Asiatic continents? In wondering about this distribution and after studying water depths on world ocean charts, we see that oceanic depths are a major deterrent to their further distribution. Nowhere have any of the extant species jumped an abyss to another continent, other than by human intervention.

Ecological Considerations about Distribution

Biological aspects of the ecology of horseshoe crabs have been described in earlier chapters, notably their breeding (1–3), life cycle stages (5), food webs (6), and symbionts, including diseases (10). Their breeding habitats are similar, mainly sandy beaches, although *C. rotundicauda* tends to spawn in muddy sand or mud banks (Sekiguchi and Nakamura, 1979; Shuster, 1982; Sekiguchi, 1988b; Debnath, 1992; Saha, 1989; Penn and Brockmann, 1994; and Botton et al., 1996).

Even though horseshoe crabs can live within a broad range of environmental conditions, in the absence of deleterious amounts of chemical and physical pollution, the establishment of successful breeding populations of the extant species depends on the favorable combination of contiguous habitats. Crucial habitats include those for breeding, a nursery area for the first-year stages, and nearby foraging grounds where the older crabs are more prevalent. Of these, suitable breeding beaches are the most limiting.

Although horseshoe crabs aggregate during the breeding season and when feeding, they do not seem to be gregarious otherwise. If they are truly nongregarious for most of their lives, we should consider this characteristic as a filter in interpreting the fossil record. Braddy (2001) illustrated a possible analogous scenario for fossil eurypterids (sea scorpions) in his "mass-molt-mate" hypothesis.

Horseshoe crabs are most evident where they spawn, on wave-protected shorelines. In this they are unique, because the crabs are the only large benthic marine invertebrate animal to spawn at the water's edge (see Chapter 2). At other times, foraging crabs of all ages are common where food organisms are concentrated. In contrast, dispersal appears to be the general rule, with the juveniles achieving an ever-expanding range as they grow larger (Shuster, 1955, 1979). Although these activities are common to all horseshoe crabs, they do not explain the geographical distribution of the species. Populations of *Limulus polyphemus* are unevenly distributed, locally and throughout their range. Basically, Limulidae are primarily estuarine species that show anadromous tendencies, at least in the extant species (Shuster, 1979). Exceptions can be expected, as in a spawning population in the Marquesas Keys (Mikkelsen, 1988, and personal communication).

In addition to the limits imposed by biotic parameters, certain physical and chemical conditions are known to restrict the range of Limulidae. These, listed in descending order of presumed significance, are temperature, salinity, and available oxygen; wave-protected beaches; food; tidal conditions; extent of shallow coastal water and abrupt changes in benthic topography; and velocity and persistence of benthic currents. All of these parameters have varying temporal and spatial effects—as notably evident in the seasonal abundance and distribution of *L. polyphemus* in relation to lunar phases (Powers and Barlow, 1985) and to latitudinal, salinity, and bathymetric gradients. How *Limulus* copes with environmental variability, especially in its adjustment to changes in temperature, salinity, and oxygen, is the theme of the following chapter. Here, we focus on environmental factors, distribution of the crabs, and habitat characteristics.

Ecology of Extant Species

Limulus thrives in diverse habitats over a wide latitudinal range. Its abundance and relatively large size are the basis of its important role in the ecology in several estuaries and near-shore marine localities. How this species and other horseshoe crabs react to their environment, throughout their life cycles, tells us a lot about the nature of the animal.

Spawning Activity and Suitable Beaches

A massive spawning event is the most obvious, most publicized activity of horseshoe crabs (see Chapters 2 and 3). It is easy to document, and every report on the natural history of *Limulus* features an account of spawning. Few of the early publications gave any data on spawning numbers—except those by Cook (1857), Fowler (1908), Shuster (1950, 1955), Baptist et al. (1957), Shuster and Botton (1985), and an annual survey along the shores of Delaware Bay begun in 1990 (Finn et al., 1991; Swan et al., 1992, 1999, 2000). The largest of all the major spawning populations is that centered on Delaware Bay. Later on, we discuss the Delaware Bay as a habitat.

Of all the environmental factors governing the survival of Limulidae, including temperature, salinity, and oxygen content of the water, the existence of beaches where successful spawning and egg incubation occur is probably the most critical. On Delaware Bay shores, although spawning occurs on less favorable beaches, Brady and Schrading (1997) selected four environmental conditions that define optimal conditions: (A) depth of sand over peat must be at least 20 cm (8 in), with 40.5 cm (16 in) or more being optimal, to avoid anaerobic conditions that could prevent egg development (Botton et al., 1994); (B) horseshoe crabs tend to nest in moist sand at an average depth of 9.4 cm (3.7 in) (Brockmann, 1990) that is 3 to 4 percent saturated (Penn and Brockmann, 1994); (C) beach slope is about 7 percent (Tim Jacobsen, Cumberland Community College, N.J., personal communication); and (D) median grain size of sediment from 0.6 to 0.8 mm (Botton et al., 1994). Although granular sediments form the most favorable beaches, those with particles as large as gravel or composed of fragments of bivalve shells may also be spawning sites. *L. polyphemus* has also been observed to spawn in masses of finely ground organic detritus and mud (a precursor to mud ball concretions? See Figure 8.2). The degree to which mud and detritus impede embryonic development and ultimate hatching has not been studied.

The suitability of beaches as spawning habitats has been examined in relation to natural processes and beach preservation projects and spawning activity. On a sandy beach subjected to considerable erosion, sediments were reduced and exposed peat of a former marsh (Botton et al., 1988). This was correlated with high levels of hydrogen sulfide and lower spawning activity, indicating the crabs tended to avoid such beaches. Jackson et al. (2002) evaluated the effects of the timing and frequency of storms, wave processes, and tidal range on the morphologic response of beaches, sediment activation, infiltration and exfiltration of water through the foreshore, and litter distribution on the surface, along with the impacts of human adjustments of beaches by bulkheads and beach nourishment. They recommended eight types of research: (1) biological sampling at times of episodic storm events to evaluate habitat modification; (2) identification of those spatial differences in the dimensions of the low tide terrace that control the dynamics and biologi-

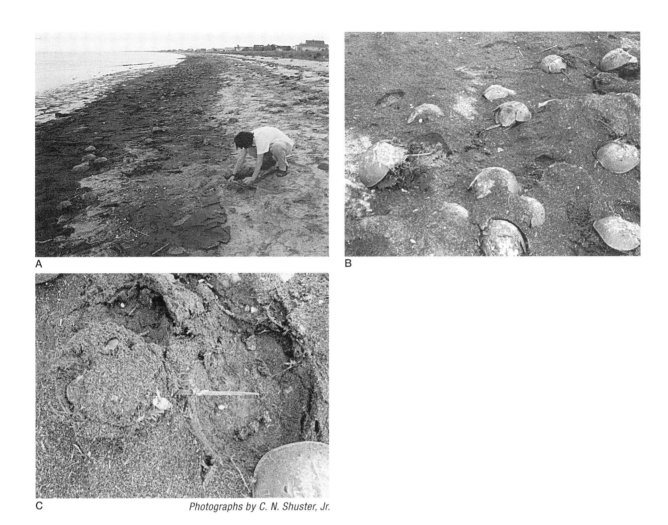

A

B

C

Photographs by C. N. Shuster, Jr.

FIGURE 8.2 A thick mixture of fine sediments and ground-up marsh grass coated the lower beach at low tide at Slaughter Beach, Del., in 1999, trapping many horseshoe crabs. (A) The crabs became mired in the muddy mat while making their way back to the Bay after spawning, as the water receded. (B) The telson of an overturned male extends over the depression in the mud where a female was removed. (C) As with many of the other crabs, the ventral area of this male was plastered so solidly with the mud/debris mix that it could not move its legs.

cal functions on the foreshore; (3) design of biological sampling plans according to geomorphic process regimes rather than static morphologic features on the beach; (4) standardization of sampling locations to facilitate comparison among different sites; (5) simultaneous studies of sediment activation by waves and fauna; (6) determination of the role of litter on faunal community structure and function; (7) detailed physical and biological studies of bulkheads to determine the significance of position on the intertidal profile; and (8) determination of the significance of nourishing a beach with fill material significantly different from native materials. Several of these suggested study areas are relevant to topics discussed in this and other chapters.

Smith et al. (2002) showed that potential food for migratory shorebirds (the quantity of horseshoe crab eggs in the surface sediments of a beach) was dependent upon the density of spawning females, beach morphology, and wave energy. The association between beach morphology and live eggs in surface sediment was strong, especially in late May. Perhaps horseshoe crabs prefer to spawn on narrow, bay-front beaches because of low wave energy. At peak periods of spawning activity, the density of spawning females was inversely related to foreshore width (distance from the spring high water line to the low tide terrace) on mid-latitude beaches within Delaware Bay where egg density was an order of magnitude higher on narrow beaches (less than 15 m wide = 3.38×10^5 m sq) compared to wider beaches (1.49×10^4 m sq). Other important findings were: (1) egg abundance and potential biomass for migratory shorebird forage are determined by density of spawning females, beach morphology, and wave energy; (2) prediction of egg abundance also depends on the age-related fecundity of the spawning females; (3) the timing of wave-generating winds, in relation to spawning, affects whether eggs rise to the surface sediment when the shorebirds need them; (4) egg-sampling methods must be robust to compensate for variation in beach morphology and applicable regardless of when the samples are taken because the distribution of eggs across the foreshore varies with beach morphology and widens as the spawning season progresses; and (5) protection is necessary for beaches critical to the conservation of shorebird foraging habitat, where beach morphology and wave energy are associated with the higher quantities of eggs retained in the surface sediments.

How spawning behavior, the habitats occupied, and environmental conditions have changed over geologic time is not known, except for possible influence of the moon, as explained by Robert Fripp (1992, The Impact Group, Toronto, Canada; personal communication). The moon has been receding from the earth through time, so gravitational forces that were strong in the Devonian were even stronger in the Silurian. Because the moon also once orbited the earth faster, the tides must have been more frequent, their amplitude greater, and the collision of water more forceful. Given that the tides influenced by the moon were much stronger when the spawning behavior of horseshoe crabs evolved, laying eggs at a lunar high tide would have led to an absolute advantage in the early Paleozoic. Thus, Fripp reasoned, what we see today is the legacy of a pattern of defensive breeding behavior originally induced by significantly greater lunar forces than occur today. Possibly such breeding behavior, even though there is the risk of stranding, led early chelicerates to develop mechanisms that took arachnids ashore, or, just as likely, an invasion resulting in predatorial advantage.

Nesting on low-wave-energy shorelines, as evidenced in the extant species, might have led to multiple spawning—the females could nest several times during a single high tide and spawn during several tides, usually but not always in the same area on the same beach (see Chapter 2). This behavior results in the crabs "not putting all of their eggs in one basket."

Sandy beaches have poor preservation and subsequent recognition potential in the geological record. Wave-protected shores should be relatively easy to spot within the geological record, due to the low-amplitude sedimentary structures that they produce. Bay beaches, on the other hand, may be difficult to detect, as they are often the first areas to be obliterated by severe erosion and by changes that accompany rising sea levels.

Behavior after Spawning

After spawning and feeding, many adult limuli probably retreat to the deeper waters of bays and sounds during the winter months. But not all venture out onto the continental shelf. Horseshoe crabs have not been collected in the deeper waters during any of the National Marine Fisheries Service resource surveys along the New England coast, east of and north of Cape Cod, although some have been found in shallower water. Trawl collections in 1985 by the fisheries resource agency of the Commonwealth of Massachusetts (cited by Botton and Ropes, 1987) obtained only thirty-four horseshoe crabs at sixteen of the ninety-four stations sampled at depths from 6 to 76 m (average depth, 28 m). At Cape Ann, Mass., Dexter (1947) found *Limulus* to be more abundant and ecologically significant within the tidal inlet of the Annisquam River than in Ipswich Bay. Similarly, Baptist et al. (1957) observed that the inshore migration of horseshoe crabs into Plum Island Sound, Mass., began early in March, peaked in June, and continued through August. The offshore migration was most evident during September.

Tagging recoveries indicated that there was a fairly discrete, relatively small population of *Limulus* within Ipswich Bay. Most of these recaptures in Massachusetts's waters were within a few miles of or even in the vicinity of the release points, indicating that although some individuals may roam, many tend to remain in or return to the area. Even though the range of most of the individuals is limited, this does not necessarily exclude the possibility that some interpopulation mingling might occur. In addition to abundance and the relatively limited range of individuals, other conditions probably are also factors in distribution—such as intermittent distribution of suitable habitats, steep continental slopes, and the configurations of the landmasses.

At any given period of time, discrete populations are identifiable on the basis of their adult dimensions and genetic makeup even though there has been a definite latitudinal gene drift (Saunders et al., 1986) in geologic time for *Limulus polyphemus*. Several morphometrically distinct populations of the American horseshoe crab exist within the overall range of the species along the coast of North America, from Maine to the Yucatán (Shuster, 1955, 1979; Sokoloff, 1978; Riska, 1981). This discreteness of the populations has been further reinforced by genetic studies (Saunders et al., 1986; Pierce et al., 2000).

Distances traveled in the Delaware Bay area, as much as 80.6 km (50 miles), are not uncommon (Shuster, 1997). In other areas, the distances obtained from earlier mark-recapture studies were much less over 3-year periods within Cape Cod Bay—up to 34 km (21 miles) (Shuster, 1950) and in the area of Plum Island Sound and Ipswich Bay, up to 30 km (19 miles) (Baptist et al., 1957).

Diet of Horseshoe Crabs

The intertidal and near-shore shoal areas are the nurseries for juvenile horseshoe crabs. They are also teeming with prey species. Because *Limulus* feeds on a wide spectrum of benthic organisms, from annelids to mollusks (see Chapter 6), availability of food within embayments may be a significant limiting factor only when the crabs are overly abundant and the local food resources are insufficient, as in Delaware Bay. This disparity between the numbers of predators and prey may be an important driving factor in the dispersal of *Limulus*.

Throughout geologic time, the relative importance of possible food items such as annelids, bivalves, other mollusks, and crustaceans has changed with the ongoing evolution of these groups. The near-shore environment was probably as densely populated back in time as it is now, but brachiopods were more numerous in the Paleozoic. Bivalves rose to importance, with the relative demise of the brachiopods toward the end of the Paleozoic. Fresh and brackish-water crustaceans radiated in the Lower Carboniferous, while groups such as annelids have left trace fossils of their existence in rocks of all ages.

One of the most important ways of placing biostratigraphic dates on the Carboniferous is the use of the rapidly evolving, colonizing, and changing nonmarine bivalve faunas (Eager et al., 1985). Shells are distinctive enough for correlations to be made between geographically disparate areas. Trace fossils made by bivalves *(Lockeia)* and annelid worms *(Cochlichnus)* are often found in close association with horseshoe crab trace fossils *(Kouphichnium)*. Interestingly, *Lockeia* are quite small (1–15 mm long)—a good match of smaller horseshoe crabs with smaller bivalves. Similar associations of species have been found in Devonian deposits in Erie County, Pa. (Babcock et al., 1998). If the Carboniferous xiphosurans fed on these kinds of organisms, then the diet of the extant species has not changed very much over time.

Evidence of Foraging

Although National Marine Fisheries Service resource surveys have found *Limulus* in many places on the continental shelf, large numbers exist only near estuaries with extensive spawning beaches (Botton and Ropes, 1987). This suggests that distribution on the shelf may be a density-dependent phenomenon, correlated with a search for food. Conversely, small populations are much less likely to be found on the continental shelf because adequate food is nearby in shallower waters. All studies considered, there is no doubt that *Limulus* is attracted to sites with abundant food supplies (see Smith, 1953). Lacking any other data, we can assume that the presence of large numbers of *Limulus,* widely distributed on the continental shelf, is primarily indicative of the foraging activity of crabs that originated in nearby breeding areas.

Abundance and Distribution

Abundance may be the key to the extent of distribution—the larger the population, the greater its range. While most of the animals from one population may remain in or return to their natal bay, some may roam. In large local populations, like the one centered on Delaware Bay, those animals out on the continental shelf may, in moving in the direction of the local bottom water currents during the flood stage of the tide, come ashore in any one of the several small bays on the New Jersey and Delmarva Peninsula coasts (B. L. Swan, 2002, Limuli Laboratories, personal communication). This behavior, in leading some horseshoe crabs into other embayments, is probably the main mechanism for the intermingling of genes among adjacent populations. Although such dispersal may not have an immediate, major impact, in geologic time it could have changed the genetic makeup of populations along the Atlantic coast and in the eastern portion of the Gulf of Mexico.

Distribution of the Extant Species

Each of the four extant species (see Chapter 7) has a relatively wide geographical range, but they are found on opposite sides of the planet —on the eastern shores of North America and Asia, including the Philippines and Indonesia. *Limulus* ranges from latitudes of about 21° N to 44.5° N, while the three Indo-Pacific species, together, cover about 40 degrees of latitude (5° S to 35° N) and more than 40 degrees of longitude. Unlike *Limulus*, these species extend across the equator, encom-

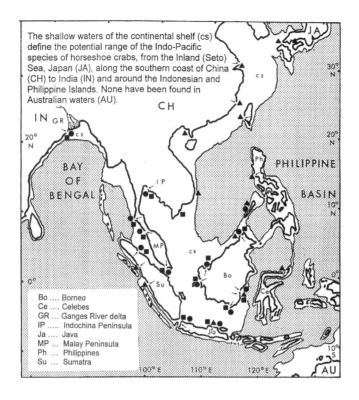

The shallow waters of the continental shelf (cs) define the potential range of the Indo-Pacific species of horseshoe crabs, from the Inland (Seto) Sea, Japan (JA), along the southern coast of China (CH) to India (IN) and around the Indonesian and Philippine Islands. None have been found in Australian waters (AU).

Bo .... Borneo
Ce .... Celebes
GR ... Ganges River delta
IP ..... Indochina Peninsula
Ja .... Java
MP ... Malay Peninsula
Ph ... Philippines
Su ... Sumatra

FIGURE 8.3 The continental shelf (cs) confines the distribution of the three Indo-Pacific species. One or more of the species were found near most research sites (indicated by small arrows): triangle = *Tachypleus tridentatus*, square = *T. gigas*, and circle = *Carcinoscorpius rotundicauda*. By permission from *Sekiguchi, 1988*.

passing an area from southwestern Japan to eastern India, from eastern Asia to the southernmost islands of Indonesia. Today, the continental shelf in each region is expansive. Although its dimensions, like those of other geological features, have changed over the millions of years the species have existed, it would have furnished an avenue of distribution and a place to forage.

In Indo-Pacific Waters

The three species have different ranges that overlap to a considerable degree (Sekiguchi and Nakamura, 1979; Sekiguchi and Nakamura, 1980; Sekiguchi, 1988a) (Figure 8.3). Like *Limulus,* these species also exist in discrete populations, recognizable by the ultimate sizes of the adults and other characteristics (Yamasaki, 1988). Of the three, *Tachypleus tridentatus* has the greatest latitudinal distribution, ranging from southwestern Japan and along the coast of China to the Philippines to the Java Sea; and, except for the west coast of Sumatra, it has not been found in waters of the Indian Ocean. *Tachypleus gigas* is found along with the other two species along the Indo–China coast and around the East Indies. *Carcinoscorpius rotundicauda* has been found in the Ganges

River delta area along the shore of the Bengal Sea. It has been found as far east along the Asiatic coast as the Gulf of Siam and is widely distributed among the islands of Indonesia as far as the coast of Borneo. *T. gigas* has a range mainly overlapping that of *C. rotundicauda,* from Indo-China and Indonesia to the western shores of the Bay of Bengal. Although *T. gigas* and *C. rotundicauda* are found together in some areas, they spawn in different habitats; the former is found on sandy beaches whereas the latter spawns on muddy banks (Sekiguchi, 1988b).

There are only a few data on the depths at which the Indo-Pacific species have been found. *T. gigas* has been found from the tide line to a depth of 36.6 m (20 fathoms) on the east coast of the Malay Peninsula and in Bengal Bay (Annandale, 1909). An adult, about 39 cm in length, was captured in a large surface tow net at night in December 1911 (Sewell, 1912). This was at a distance some 4 miles west of the entrance to the Hinze Basin on the south Burma coast where the water depth was about 10 fathoms.

Sekiguchi (1988b) summarized the ecology of the extant species of horseshoe crabs. Basically, behavior, morphology, and habitat may vary within and among the species. While much still is to be learned, it is clear that, in addition to basic similarities, there are some differences in habitats among the species as well as differences within species, related to their age and the seasons. Sekiguchi also noted that these differences are important in addressing questions about speciation and evolution. In reviewing what is known about the ecology of the Indo-Pacific species, we can see that their survival is tied, not surprisingly, to the shoreline. Indeed, the best example, for those interested in habitat preservation, is the situation of an endangered species, *Tachypleus tridentatus,* in Japan. Loss of spawning areas, due to construction of dykes, revetments, and polders, has been a major factor in its decline.

Further discussion on the Indo-Pacific species is beyond the immediate compass of this chapter. You can follow up on this subject in the many reports by Sekiguchi and his colleagues (referenced in practically every chapter of this book) and by Chatterji (1994), Chatterji et al. (1992), Chatterji and Paruleker (1992), Debnath (1991, 1992), Debnath and Choudhury (1988a, 1988b), Debnath et al. (1989), and Saha (1989).

The close similarity in morphology and behavior of the extant species indicates that their populations have either not been as isolated as we suspect or that their characteristics have not changed very much. Certainly sufficient geologic time has elapsed to expect genetic differentiation and isolation of pockets of breeding populations. Because many of the islands of Indonesia have populations of horseshoe crabs, is it possible that these are survivors of previously more continuous popu-

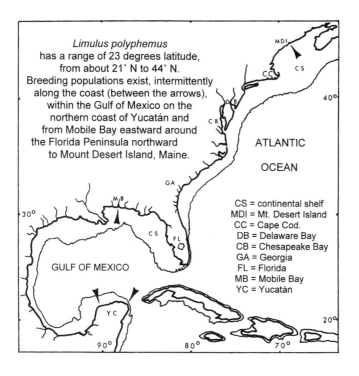

Limulus polyphemus has a range of 23 degrees latitude, from about 21° N to 44° N. Breeding populations exist, intermittently along the coast (between the arrows), within the Gulf of Mexico on the northern coast of Yucatán and from Mobile Bay eastward around the Florida Peninsula northward to Mount Desert Island, Maine.

ATLANTIC OCEAN

CS = continental shelf
MDI = Mt. Desert Island
CC = Cape Cod.
DB = Delaware Bay
CB = Chesapeake Bay
GA = Georgia
FL = Florida
MB = Mobile Bay
YC = Yucatán

GULF OF MEXICO

FIGURE 8.4 There is a large gap in the distribution of *Limulus polyphemus* in the Gulf of Mexico, from Yucatán to Mobile Bay.

lations that inhabited the shallow waters at a time of lower relative sea level?

Along the Western Atlantic Coast

The species featured in this book, *Limulus polyphemus,* is found in many of the bays along the western Atlantic coast of North America. Where it is most numerous, the species sometimes ranges far out onto the continental shelf (Figure 8.4). The Yucatán Peninsula and northern Maine are its boundaries, but it is not found everywhere within this range (Say, 1818; Ives, 1891). Ironically, less is known about the populations at the extremes of the distribution of *Limulus;* in addition to their natural history and ecology, their physiological and genetic characteristics are research opportunities.

The numbers of horseshoe crabs and their habitat along the Yucatán Peninsula have markedly decreased in recent decades. In recent years only a few thousand crabs have been found, mostly associated with mangrove areas and unperturbed seagrass zones (Gomez-Aguirre, 1979, 1987, 1993). *Limulus* habitats within the Gulf of Mexico are limited to the eastern portion, from around the Florida peninsula to Mobile Bay, Ala., and along the coast of Yucatán; they are probably en-

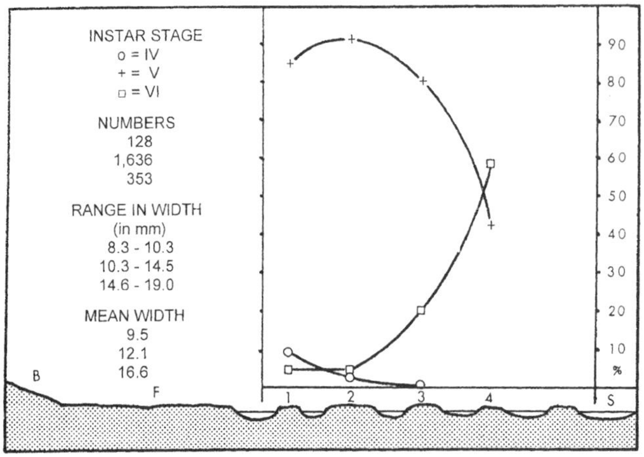

FIGURE 8.5 During the summer of 1951, first-year juvenile *Limulus* moved away from the beach at Cape May, N.J. (Shuster, 1955, 1979). By September 11, instar stages IV through VI were distributed unevenly over the first four intertidal flats (1–4) from shore. Legend: B (beach), F (intertidal flat), S (slough).

hanced by the broad shelves that extend some 100 miles off the coasts of Yucatán and western Florida.

Almost one million years ago, when southern Florida was composed of several large islands (Petuch, 1992), *Limulus* might have freely moved through straits into and out of a large inland sea between the Gulf of Mexico and the Atlantic Ocean. Such movement could explain why populations on either side of the Florida peninsula today have similar genetic signatures (Saunders et al., 1986). The crabs now in the coastal area from Miami to Cape Canaveral could represent a relict population, a survivor of an otherwise extinct fauna that made its way into the area from the Gulf and from the Keys. The localization of the crabs in the Cape Canaveral region may be a reflection of the small size of the population, perhaps sustained by an adequate food supply with no other pressures to cause dispersal. Across the Gulf, *Limulus* on the Yucatán Peninsula is far removed, geographically and genetically, from other populations (Tim King, 2003, U.S. Geological Survey; report to the Technical Committee, Atlantic States Marine Fisheries Commission Management Board).

Habitat and Behavior of the Juveniles

First-year juveniles are found in intertidal and shallow water areas adjacent to breeding areas (Shuster, 1955, 1979). At Cape May, N.J., on September 11, 1951, at low tide, they were distributed by size across several intertidal flats; the larger stages were more numerous farther from the shoreline (Figure 8.5).

Fisher (1975) noted the seasonal occurrence of adults but concentrated on determining the distribution and behavior of juvenile limuli

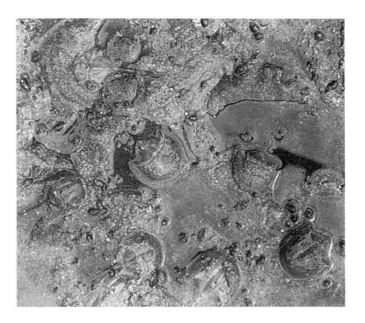

FIGURE 8.6 Can you find the seven juvenile limuli on an intertidal flat on Pleasant Bay, Mass. (early August 1949)? *(Photograph by Charles Wheeler)* These juveniles are within the size range of 3- to 4-year-olds (approximately 4 cm in prosomal width). The numerous dark ovoid objects are juveniles of the common mud snails *(Ilynassa obsoleta).*

more comparable in size to the fossilized specimens that he was studying. His detailed observations were made at a tidal marsh and intertidal flats complex at Duxbury, Mass., during numerous trips from April 1973 through January 1975. The occurrence of these Duxbury juveniles was typical of those at other New England sites—as at Pleasant Bay, Mass., in 1949 (see Figure 8.6), at Barnstable Harbor, Mass., and on intertidal mud flats on the Point Judith marsh, R.I., where juveniles were easily obtainable during the summer in the 1960s. Fisher documented differences in the habitats occupied by all of the age classes that occupied the study area. During low tide in the summer, he found instar stages 6 to 7 buried in the intertidal mud flat up to a maximum of 5 cm, with patchy distribution ranging from 0 to 12 individuals in a 4-m square. When stage 8 individuals became abundant after the July molt, they tended to be most abundant in the sandy intertidal. In late summer and fall, they moved progressively into subtidal areas. During fall and winter, these juveniles were still intertidal and burrowed deeper and deeper into the flats. This trend began in October when foraging had ceased. By November the individuals seemed to be less abundant but were found at depths from 15 to 20 cm. Subadults, instar stage 9 up to those of sexual maturity, moved into the intertidal flats during high tide but retreated mainly to sandy subtidal areas during the ebbing tide. Adults also moved with the tides but were more often likely to be stranded at low tide, whereupon they dug in. By middle and late sum-

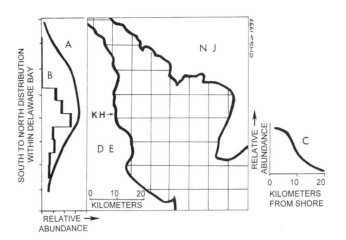

FIGURE 8.7 Relative distribution of juvenile and adult horseshoe crabs in the Delaware Bay area (KH = Kitts Hummock, Del.). The north-south abundance curves (A and B) are not based on the same scale. A = Adults spawned on beaches from Cape Henlopen north to Woodland Beach; the greater densities were midway up bay *(based on Swan et al., 2000)*. B = The higher concentrations of juveniles in trawl samples were from Little River to Mispillion River *(based on Michels, 1997)*. C = limuli on the continental shelf tend to be more numerous close to the coast *(based on Botton and Ropes, 1987)*.

mer, the adults mainly occupied subtidal areas. In the fall and winter, both adults and subadults were subtidal.

Young limuli with prosomal widths up to 160 mm were trawled from Delaware Bay, April though October, 1992 to 1995, from depths less than 9 m (30 ft deep). By comparison, small adult males in Delaware Bay may be 175-mm wide, averaging about 220 mm. Figure 8.7 (B and A, respectively) shows graphs of the approximate distribution of the juveniles and the abundance of horseshoe crabs spawning on Delaware beaches. A comparison of the two curves, the abundance of juveniles and the distribution of the spawning areas, suggests that the juveniles were already tending to move down bay toward the continental shelf.

Is Cape Cod a Barrier to Distribution?
The horseshoe crabs living in the waters of Cape Cod provide an interesting contrast to those in the Delaware Bay area. Neither the ultimate size of the individual adults north of the Cape (Shuster, 1955) nor the size of the populations in this region has reached those seen in the waters farther south. During the 1940s, Charles Wheeler (1949, fisheries biologist with the Commonwealth of Massachusetts; personal communication) observed that the distribution of juvenile and adult horseshoe crabs in the vicinity of Cape Cod was uneven and that they were abundant only in certain areas. Wheeler's observations, combined with the dredge results of Verrill and Smith (1873) and Sumner et al. (1911) and the results of tagging (Shuster, 1950), indicated that horseshoe crabs were present in many places in the vicinity of Cape Cod and the offshore islands of Martha's Vineyard and Nantucket.

This distribution suggests that there are at least two major populations in the Cape Cod area: one centered in Cape Cod Bay and the other south of the Cape. If any intermingling occurred between these two populations, it would have been limited to migrations through the Cape Cod Canal and around the eastern side of the Cape. It is possible, however, that the crabs in Buzzards Bay are separate from the population in the Nantucket Sound area.

What Is Special about Delaware Bay?

Today, the shores of Delaware Bay are renowned worldwide as the location of the largest spawning activity of horseshoe crabs anywhere. Why? At least several conditions are probably responsible:

(1) Because the Bay is located within the range of the largest-sized animals from Georgia to Long Island, it presumably provides an optimal thermal environment, or most nearly so, for spawning, growth, and survival.

(2) Delaware Bay is a relatively large estuary with 1,865 square kilometers (720 sq mi = 460,000 ac) of surface area and a shoreline totaling 206 km (128 mi: Delaware = 55, New Jersey = 73). Some 43 percent (surface area) of the Bay is in Delaware, with 74,000 ac (116 sq mi) less than 12 feet in depth, while New Jersey has 96,000 ac (149 sq mi). The Bay has extensive shallow areas less than 1 fathom (6 ft) in depth: 253 square km (98 sq mi = 63,000 ac) representing 13.5 percent of the total area of the Bay (Shuster, 1959). Within this area the salinity regime is favorable, from ocean seawater at the mouth to one-quarter the salinity up bay. Actually the area available to *Limulus* is much greater because a large number of the larger-sized juveniles and adults move out of the Bay onto a contiguous area of the continental shelf within a perimeter of at least 50 km (31 mi) from the mouth of the Bay.

(3) The Bay is essentially a nursery area—it has lengthy sandy beaches, totaling about 84 km (52 mi), many favorable for successful incubation of the eggs and extensive intertidal areas where the early stages of juveniles forage.

(4) Food resources are abundant in Delaware Bay and on the continental shelf. Invertebrate populations (Wigley and Theroux, 1981), especially of mollusk species such as the surf clam (Botton and Haskin, 1984), provide ample prey species on the continental shelf (Botton and Ropes, 1989). This large foraging area off the mouth of Delaware Bay is an important adjunct to the other resources of the Bay that are favorable to horseshoe crabs.

(5) In the context of geologic age, Delaware Bay as we know it is very young. If a coastal bay existed during the last great ice age (16,000–18,000 years B.P—before the present), when sea level was

some 100 m lower than at present, it must have been part of a former coastline, some 100 to 120 km to the southeast of present-day Cape Henlopen. Since then, sea level has risen and some of the once-coastal sediments have migrated landward and upward, in space and time. Along with this reshuffling of the sediments, the old proto–Delaware Bay migrated to the northeast, its basin continually undergoing modification by the change in sea level and the hydraulic regime of the old river system. The evolution of the estuary of the Delaware River, from the time of the last great ice age to the present, is described in a series of studies conducted by John C. Kraft (1971) and his students (including Weil, 1977; Belknap and Kraft, 1981; Chrzastowski, 1986; Knebel et al., 1988; Fletcher et al., 1990; and Hoyt et al., 1990) at the University of Delaware.

This evolution has been a more or less continuous process, with the sandy coastal areas constantly in a state of flux of erosion, transport, and deposition as the sea level rose and coastal sediments were moved landward, a process termed "marine transgression." During much of this history, the shoreline was fringed by tidal marshes. One can imagine that, over geologic time, alternation from tidal marshes to sandy shores and back again, with mixed conditions, was a natural process, much like today. Long before the Bay approximated its modern position, the increasing sandy shoreline acted as a barrier separating tidal marsh areas from the Bay. Given the existence of sandy beaches from the beginning of the proto–Delaware Bay, it is very probable that *Limulus* was able to occupy the embayment as soon as the annual temperature regime was sufficiently warm to permit growth and reproduction, perhaps as long ago as 5,000 years B.P. The present asymmetrical shape of the Bay, with the prominent Maurice River Cove (see Figure 14.1) may have been, at least in part, due to the ancient Maurice River and West Creek drainages.

Similar comments could be made about many other Atlantic coast estuaries, but the characteristics of the nearest major estuary, Chesapeake Bay, stand in sharp contrast to those of Delaware Bay, especially in the lack of long, continuous sandy beaches and the much smaller horseshoe crab population (colorplate 14). Perhaps the answer lies in the stage of geomorphic evolution of an estuary—whether it is in a dominant sandy beach front stage or that of a marshland shoreline.

On the Continental Shelf

Because earlier surveys collected *Limulus* on the shelf (Verrill and Smith, 1873; Sumner et al., 1911), it was not surprising when marine resource surveys by the National Marine Fisheries Service (NMFS) found large numbers of *Limulus,* mostly adults and large-sized juveniles, at many sites south of Cape Cod (Shuster, 1979, citing Ropes, personal communication; Ropes et al., 1982). The bulk of the horseshoe crab population on the western Atlantic continental shelf was located between northern New Jersey and southern Virginia (Botton and Ropes, 1987) and the greatest reported distance from shore was a tagged horseshoe crab found some 161 km (100 mi) off Ocean City, Md. (Benjie Lynn Swan, personal communication). Evidence for onshore-offshore migrations comes mainly from subjective evaluations bolstered by the recovery of a limited number of tagged animals that have moved up to 100 mi from their release points.

An analysis of the NMFS data by Botton and Ropes (1987) proved that large numbers of horseshoe crabs are far-ranging and unevenly distributed on the continental shelf. They examined the data compiled during the years 1976–1983 by NMFS bottom trawls conducted from the Scotia Shelf south to Cape Fear, N.C., and by clam surveys using hydraulic dredges on the Middle Atlantic region north to Georges Bank. They divided the data obtained from the Middle Atlantic Bight into two collections: inshore areas (defined as stations of 9–27 m in depth) and offshore, those deeper than 27 m. In this area, they estimated a minimum inshore population ranging from 1.8 to 3.1 million individuals (0.5 to 1.4 million offshore).

Southward, from southern Virginia to North Carolina, horseshoe crabs were taken at the edge of the continental shelf, at depths up to 290 m. Especially notable was the horseshoe crab photographed with a submersible camera at the depth of 1,097 m at 32°38′ N, 76°33′ W, approximately 10 mi east of Charleston, S.C. (David Bunting, Duke University Marine Laboratory, cited by Botton and Ropes, 1987). How did it get there—did it fall down from the edge of the shelf? Or did it die and currents swept the carcass there? If it were alive, could it have gotten back up? How? By crawling? Swimming probably was not an option because *Limulus* sinks as soon as it stops moving its appendages. If a few horseshoe crabs regularly fall off the shelf and get buried in sediments slumping off the slope into the abyss, they could become fossils. What would be the scientific determination of their habitat?

Horseshoe crabs are numerous along the coast of New Jersey near the entrance to Delaware Bay, but this varies seasonally. During surf clam surveys within 5.5 km (3 mi) of the shoreline and employing a hydraulic dredge, Botton and Haskin (1984) documented a latitudinal gradient of horseshoe crabs along the New Jersey coast from Cape May to Beach Haven Inlet. The crabs decreased in abundance from Delaware Bay northward in the spring and summer. Another gradient, in a perpendicular direction (onshore-offshore; see Figure 8.7c), occurred along the northern New Jersey coast, where few crabs were found beyond 1.8 km of the shore (Botton and Ropes, 1987). The transition between areas of high and low concentration took place between Great Egg Harbor Inlet (Ocean City) and Absecon Inlet (Atlantic City). Horseshoe crabs were more abundant inshore in the late spring and early summer than in the late summer and fall.

Recently a pilot study established that only a few considerations were needed, beyond trawl design and operation, to conduct a trawl survey to monitor horseshoe crab populations. This study, carried out in the fall of 2001, was on the continental shelf in the vicinity of Delaware Bay (Hata and Berkson, 2002). It confirmed that distance from shore, bottom topography, and activity of the crabs at night were prime considerations. Crab catchability by trawl was greater: (1) within 3 nautical miles of the shoreline than between 3 to 12 nautical miles offshore, (2) in sloughs, and (3) at night. Based on the nighttime sampling, the numbers of horseshoe crabs in the study area were conservatively estimated at between 5.9 and 16.9 million. We wonder if the crabs were in the sloughs because hydraulic conditions concentrated food organisms there and also produced currents conducive to movement into and out of the Bay; sloughs are also the lowest point crabs sink to after scuttling or swimming.

Errant Distribution of *Limulus*

Wolff (1977) reasoned that the occurrence of horseshoe crabs in European waters was probably mostly due to trans-Atlantic transport by fishermen, with an occasional escapee from an aquarium. Reports of *Limulus* in waters along Israel and western Africa (Mikkelsen, 1988) were probably also due to transplanted animals. Those on the coast of Israel may have been specimens from the Gulf of Mexico that formed the basis of an experiment in breeding at a mariculture laboratory 7 km south of Elat (Kropach, 1979). Horseshoe crabs along the coast of Mauritania and around Agadir, Morocco, may have arrived with oceanographic and fisheries-support activities from America.

Other aberrant distributions have also been reported. One unwelcome transport occurred during the settlement of San Francisco, in the days of the 1848–1849 gold rush. Upon the arrival of a ship from the East Coast, it was discovered that the sand it was carrying as ballast was so full of *Limulus* eggs and larvae that it was unfit for use in masonry, so it was jettisoned. Perhaps this report was based on the observation that, a year or two previously, a vessel took on a load of sand from a Delaware Bay beach and, within 2 or 3 days, so many *Limulus* larvae appeared in it that the whole load was thrown overboard (Cook, 1857). Anyway, there were no further reports of *Limulus* on the West Coast.

Examples of the Distribution of Fossil Species

Today, the fossil remains of horseshoe crabs are often found on dry land in areas remote from the sea. Plate tectonic movements, uplift of marine-deposited sediments to form dry land, and major changes in relative sea level have all created areas that are in obvious contrast to the coastal distributions of the extant species.

That said, all of the sites where fossil horseshoe crabs have been collected are linked by the commonality that the exoskeletons of xiphosurans are relatively unmineralized, unlike the often-compared trilobites. As such, very specific sediment chemistry conditions must exist for horseshoe crabs to be preserved as fossils. Because these conditions are most often associated with near-shore, restricted-salinity waters or deep-water anoxic sediments, these are the environmental habitats in which the fossil associations are best known. It seems that slow sediment deposition in lagoons, or high sedimentation in brackish water and lakes, provided the preservation window most likely to aid in their fossilization. Thus, the fossil record of the early xiphosurans may be biased toward those environments that were most likely to have had the conditions conducive to the preservation of their unmineralized integuments.

Now those land formations preserving former coastal shallow water habitats are where the collector can break open rocks and find their remains. The existing environmental conditions must have been well within the capacities of those early horseshoe crabs to cope or changed so slowly that the crabs could move from less to more favorable climes. Although the fossil record and distribution of xiphosurans are patchy, we know that possible ancestors of the modern limulid line existed as far back as the Lower Carboniferous (some 340 million years ago) in the United States (*Paleolimulus longispinus* at Bear Gulch Creek, Mont. [Schram, 1979]) and Europe (*Xaniopyramis linseyi* in England [Siveter and Selden, 1987] and *Rolfeia fouldenensis* in Scotland [Waterston, 1985]). Also, if their spawning habitats and breeding cycles differed, as they do in the extant species (see Chapter 5), that may explain how the closely related Carboniferous horseshoe crabs coexisted without interbreeding and homogenization.

The origin of the extant species is not known, but they or their antecedent species might have migrated from their original center of abundance when the Mesozoic shallow seas where Europe now stands became land (Ives, 1891). One ancestral type moved westward, giving rise to *Limulus,* while the type(s) giving rise to three species went eastward. That may have happened, but there is also the possibility that the lineage leading to *Limulus* may have been established much earlier in its part of the world, while *Mesolimulus* or related species gave rise to the Indo-Pacific horseshoe crabs. Further morphological study and comparison of these Mesozoic fossils is required to test this particular hypothesis.

If diurnal tides of amplitudes greater than 1 ft are as important for the other extant species as for *Limulus*, then this may have important implications for using fossil xiphosurans as gauges of tidal conditions, as in the Carboniferous. Would the distribution of fossil horseshoe crabs reinforce the studies on lunar cyclicity in the Carboniferous by Broadhurst (1964) that were based on tidal rhythms (couplets of sand and mud laid down one per tide)?

Death in Hypersaline Lagoons

Fossil-Lagerstätten, a term derived from German mining tradition relating to rocks containing constituents of economic interest (Seilacher, 1990), are those localities that preserve the soft and unmineralized components such as teeth, bones, and shells. Among such localities, the Solnhofen limestones (Plattenkalk) are famous for their fine-grained surface texture, which was the basis for their use in early lithography. Solnhofen limestones are also famous for their fossilized animals (Barthel et al., 1990). The most significant of these is the first example of a feathered creature, *Archaeopteryx,* which is intermediate between dinosaurs and birds (Hecht, 1985; Viohl, 1985; Chatterjee, 1997; Shipman, 1998), but it was *Mesolimulus* that most attracted us. The deposition of a limey mud during the Mesozoic that led to the formation of the thick limestone strata was not a life habitat for most creatures—far from it; the mud was a death trap. These limestones are on large isolated beds in a wide area in Bavaria, Germany, some 30 by 70 km in extent (Figure 8.8).

During the Mesozoic, extensive coral reefs and islands shielded a shallow, warm-water coastal sea from the buffeting of waves from the ancient Tethys Sea. A number of deeper basins or lagoons, surrounded by algal-sponge reefs, were within this shallow sea where the limestone was ultimately formed (Barthel et al., 1990). The size, depth, and position of these lagoons created a physical pocket within which water stagnated, heavy brine from evaporation in the shallow sea settled, and lime dissolved in the seawater precipitated out. During the summer months, excessive blooms of phytoplankton such as the coccolithophorids further poisoned the water mass in the back-reef basins. Few organisms other than bacteria and algae could have lived in the lagoons. The prevailing water and sediment conditions within the basins

FIGURE 8.8 Two views of a limestone quarry in the vicinity of Eichstätt, Germany. Sheds adjacent to the work areas, at each level in the quarry, protect the workers and their products from the weather after the men have removed slabs from the open sites. *Photographs by C. N. Shuster, Jr., 1987.*

were conducive to the preservation of organic remains that sank down through them.

But some animals did venture into this hostile environment and apparently remained alive, at least long enough to produce a recognizable, consistent trail (Figure 8.9). Among these were a variety of sizes of the horseshoe crab *Mesolimulus,* the smallest of which were probably juveniles that either wandered into the lagoons or were swept in with

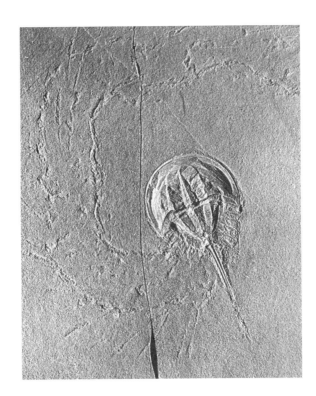

FIGURE 8.9 This is an internal mold of one of the most illustrated of the Solnhofen limestone horseshoe crab fossils (prosomal width, 92 mm). The final struggles of an exhausted *Mesolimulus walchi* crab that died at the end of a long, tortuous, and spiraled trail are revealed by the thrusts of the pusher legs and the telson. *Photograph by permission of Theo Kress, Solnhofen Aktien Verein (given to C. N. Shuster, Jr., by K. W. Barthel, 1974).*

ocean water during an occasional storm. Their trails in the limy mud and impressions of their bodies provide an intimate story of their death. Several of the trace fossils recorded in the Solnhofen limestone were circuitous, disoriented trails that were accompanied by much thrashing about by *Mesolimulus,* particularly with its telson (Barthel, 1974). This activity has been interpreted as consistent with individuals in their death throes due to murky waters, lack of oxygen, hypersalinity, lack of oxygen, or presence of hydrogen sulfide in the lagoons. One of the best-fossilized trails extends over a meter in length, with the fossilized small horseshoe crab at the end (see Figure 8.9). Elsewhere, similar Plattenkalks (layered limestone) of different ages have also yielded exceptional fossils, including those of horseshoe crabs (see, for example, Bear Gulch, Mont. [Schram, 1979] and the Permian of Kansas [Dunbar, 1923]).

Tropical Swamps and River Deltas

Some 300 million years ago, an array of aquatic animals were distributed within a salinity gradient from freshwater to brackish to marine within a river-influenced estuary (Baird, 1997b). Today this area, in northeastern Illinois, near the southern tip of Lake Michigan, is historically important in the coal industry. It is also important for its fossil biota, named after the Mazon Creek, a tributary to the Illinois River. The paleohabitats of the area and the plants and animals that are preserved there in rocks are described in a recent book edited by Shabica and Hay (1997). The sites most conducive to burial and preservation of Mazon Creek organisms were areas of migrating muddy tidal channels and mud bars bordering a swamp. In these regions, where seaward deposition of sediments was active, sediments collected along the edges of muddy bar fingers, ensuring a rapid burial of organic remains. This area once lay near the equator, within a supercontinent called Pangaea. Lush vegetation within an ancient equatorial band across Pangaea was the origin of many of the coalfields in America, Europe, and Asia where fossils of horseshoe crabs have been found. The Mazon Creek fauna is presumed to have existed in a tropical area during a period of extensive rainfall and a rising sea level that produced coastlines that migrated landward over the swampy areas. At the same time several river deltas extended out into the estuary. Consequently, the coastline had river deltas interspersed with tidal streams that penetrated into peat swamps (Baird, 1997a, 1997b, 1997c).

The two most prevalent xiphosurans within the different habitats of the Mazon Creek estuarine embayment were *Paleolimulus sp.* in the marine sector, while *Euproops danae* inhabited brackish to nonmarine sectors (Baird, 1997b). Compared to other species in the Mazon Creek biota, the fossils of xiphosurans are rare. For a description of the paleoenvironmental conditions of the Mazonian estuary, see the discussion on *Paleolimulus* in Chapter 7. *Euproops* was one of a group of species that inhabited freshwater habitats. Presumably these were also habitats with slow-flowing, quiet water in low-lying swampy areas. Few comparable habitats exist today outside the river deltas and mangrove swamps in Indo-Pacific coastal areas (Sekiguchi, 1988b) and the mangroves of southern Florida and Yucatán (Daiber, 1960) where horseshoe crabs have been recorded. Fisher (1975, 1979) presented evidence that *Euproops* inhabited the land, crawling among the vegetation, while Anderson (1994) claimed that it would be unable to breathe or eat on land. There remains, however, the possibility that *E. danae* clambered and survived among the vegetated surface of a very humid tropical swamp. This conjecture is based on the survival of *Limulus* in an artificially moist environment in which a hundred or so live adult limuli were kept alive, out of water, for 3 to 4 months, stored in a pit for later use as eel bait. The pit was lined with moistened leaves in a shaded wood lot in the vicinity of Sandy Hook, N.J., during the early 1950s. If large specimens of *Limulus* can survive in a quiescent state when their book gills remain moist, what would be the chances of survival of tiny creatures only one-twentieth of their dimension? Or would they dry out more rapidly if they were not constantly bathed by a moist environment?

Intraspecific Variation in the Fossil Record

When considering the fossil record of xiphosurans—or of any other group of organisms—we face the challenge of intraspecific variation. Being able to distinguish extant populations from each other on morphometric grounds also has important implications in the taxonomy of fossil species (the taxonomic determination or raising of additional species from within a former species). Fossilization effectively strips away much of the information that a zoologist would use to classify an animal and to define different species. Paleontologists use the term "taphonomy" (events occurring to the organic remains during and after burial) to encompass all the processes that occur during fossilization, including chemical and physical changes to the remains.

The first steps in taphonomy involve the death and decay of the animal. Decay often removes the details of the soft parts; hence the respiratory, reproductive, and digestive structures are not typically preserved. Furthermore, the remaining material may suffer predation, scavenging, or physical damage (as in wave-dominated settings) before burial removes the carcass to the next stage. Upon burial, increasing temperature and pressure regimes further modify the remains. Burial can physically compact the remains, and temperature tends to destroy any remaining biomolecules. While we can easily subject living populations to DNA analysis, the same is not true of fossils. Contrary to the picture presented by the book and movie *Jurassic Park,* DNA molecules are rarely fossilized due to the temperature, pressure, and time span involved. Thus, although we wish we could, we cannot use DNA to help sort out the fossilized species of horseshoe crabs. So, the sooner the animal is buried and preserved, the more information can be retained. Such quick burial occurred most often at Konservat-Lagerstätte—and these sites supply almost all of the horseshoe crab fossils discovered so far.

What Can We Conclude about the Distribution of Xiphosura?

In terms of their geographic distribution and number of species, the recent species of horseshoe crabs may have attained the broadest level of environmental compatibility ever reached by any horseshoe crab species. If so, they are the most environmentally successful of all horseshoe crab species, extinct and extant. There were more species at another time in their geologic history, during the Carboniferous period, but one line lived primarily in freshwater and low-salinity habitats. Interestingly, these members of the Bellinurina were all relatively small compared to *Limulus,* and smaller-than-expected limuli are invariably found in lower-salinity habitats. Have form and function been important in the abundance and distribution of the species? Why have they survived so long? We may not be able to answer such questions, but we have suggested some possibilities in this chapter and in Chapter 7.

Survival and Extinction

The end of the Permian period of geologic time (ca. 250 million years ago) was marked by a major extinction event that severely affected both marine and terrestrial plants, vertebrates, and invertebrates. This extinction event was orders of magnitude larger than the one that saw the final demise of the dinosaurs at the end of the Cretaceous period. Both trilobites and eurypterids were very much in decline by the end of the Carboniferous, with only a few families apiece. Competition from better-adapted crustaceans for the same food resources had severely curtailed their species diversity. However, horseshoe crabs pulled through this crisis and indeed also survived the Late Cretaceous extinction event. Perhaps their ability to withstand a multitude of adverse environmental conditions provided an adaptational advantage during these times.

When thinking about *Limulus polyphemus* as an occupant of geologic time, we have wondered what the chances would be of discovering a fossil in an area where this species is now abundant. Might one have been preserved in some ancient marsh peat bank or in the muddy sediments of an intertidal flat? Or, like *Limulus coffini,* found in a hard, calcareous concretion? A potential mud ball environment was noted at Slaughter Beach, Del., in 1999. There, spawning adults had become embedded in a muddy material, including fragments of marsh grass on the beach (see Figure 8.2).

Interpretation of the geographical distribution of extinct species is constrained by the limited fossil record and the lack of a detailed account of changes in the juxtaposition of landmasses and coastal waters during the millions of years over which horseshoe crabs have evolved. Few fossil sites have been studied extensively, however, as the two (Mazon Creek and the Solnhofen) reviewed in this chapter. Overall, the environmental factors that limit the distribution and abundance of horseshoe crabs today offer clues to the distribution patterns of horseshoe crabs throughout geologic time.

The horseshoe crabs were evolving, as were the bodies of water bordering the shifting landmasses. Profound changes have occurred in the shape, size, depth, and distribution of water bodies, due to the drifting of continents and to the rise and fall of the oceans in response to glacial events. Coastal habitats have changed markedly as mountain chains have eroded. Relative shifts in land and sea levels have also modified coastal habitats. Based on the ecology and distribution of the extant species of horseshoe crabs and the occurrence of fossils, migrations and distribution probably occurred in the relatively shallow coastal waters. If ancient horseshoe crabs moved from the shores of one continental mass to another, they must have followed "bridge-forming" shelves in shallow water, as *Xaniopyramus* in a Carboniferous stratum of England (Siveter and Selden, 1987). Where and when the juxtaposition of landmasses, their embayments, and associated shallow seas occurred is important to understanding the distribution of the species. A sequence of maps of geologic eras showing fossil locations and the drifting of landmasses helps to illustrate the overall distributions of plants and animals. But the small scale of the maps usually found in books on fossils and paleontology generally do not show the extent of the coastal embayments and shallow seas in which extinct species once lived, nor do they fully illustrate any possible connections between the seas that could have served as routes of distribution. In describing the locale of specific fossil species, we have used only those charts and habitat scenes that help to describe specific sites.

There are two places where relatively large numbers of fossil horseshoe crabs of one species can be found (Ward, 1992; Mikulic, 1997): the Solnhofen limestone of Bavaria, Germany *(Mesolimulus walchi)* and the Mazon Creek concretions in Illinois in the United States *(Euproops danae)*. The Essex fauna of the Mazon Creek marine biota is unique. In fact, the fauna (scientifically designed as the Braidwood type) accounts for all other Carboniferous siderite concretion fossil sites in the United States, England, and the mainland of Europe (Anderson, 1994).

Presumably all species could have been similarly distributed. But tracking horseshoe crabs in geologic time is problematic because of limited records; some species are based on only one or two fossils. Then, too, habitats differed from freshwater to brackish shallow coastal embayments to marine waters, and from tropical to temperate waters. Even more disconcerting, in our attempts to create a semblance of a family tree, are the huge geologic time and distribution gaps in the fossil record. Connecting lines between the branches of the tree is invariably altered with each new fossil find. We suspect that the rarity of fossils may be balanced out somewhat. While every Limulidae fossil adds a disproportionately large amount of information in distribution studies, it is just another flag to pin on the global map.

Tagging experiments on *Limulus* provide an intriguing perspective. If the maximum distance that an individual is likely to travel in a short period of time is 21 mi, could this distance be used as a guide for fossil occurrences? Could it provide a radius of distance from the sea, perhaps inland into brackish and freshwater environments? Such a distance could weed out organisms with less tolerance to water salinity variations and perhaps place them at sites with preservation potential with terrestrial arthropod biota such as arachnids, scorpions, and millipedes.

How useful is information on the physical environment? Geologic and geographical dimensions are both parts of the available record. The four extant species of horseshoe crabs are widely spread along the east coasts of North America and Asia, and the fossil species have been found in those continents and in Europe and Australia. Because each of the extant species has a large geographical distribution, the fossil species also may have been widespread, even though the present record does not firmly support such a conclusion. One exception is the wide distribution of species of *Euproops* (Størmer, 1955). Therein lies a paradox. Despite the overall lengthy geologic time span that xiphosurans have existed and their wide distributions, only four living species and a few dozen fossil species of horseshoe crabs are known to have existed, each within a specific geographic range. Perhaps there were so few species because they were well adapted to environmental changes.

What Governs the Distribution of Limulidae?

What can we infer about the occurrences of fossilized horseshoe crabs in comparison with the living species? Do their distributions reveal any similarities, particularly in their environments? Can we assume that the distributions of extinct and extant species have been governed by the same or similar environmental factors? The best short answer may be that xiphosurids have always lived in embayments—in the shallow waters behind barrier islands or in the estuaries and deltaic streams associated with river mouths. If so, possibly from the beginning, their homes and life cycles were affected by the tides. In this, as with practically every other characteristic of these animals, there have been exceptions. One early branch of the Xiphosura—Bellinurina—was apparently at home in freshwater streams and swamps, but even these may have been tidal waters or tidally influenced (Anderson, 1996; Anderson and Selden, 1997). High tides could have backed up the river flow, producing minor cyclic variations in the water level of these areas. Perhaps even these small variations were sufficient to make this branch of Xiphosura feel at home. Members of this group persisted at least into the Permian period. Then the mass extinction at the end of the Permian (ca. 250 million years ago) may have marked the finale for this particular branch of the Xiphosura, as it did for many other species.

We assume that horseshoe crabs have always been bottom dwellers, exposed to muddy and sandy-subtidal and intertidal areas and beaches. There they existed within wide ranges of water temperatures, oxygen levels, water depths, and salinities. Today, when populations of these estuarine creatures are large, individual crabs may wander out of estuaries into the higher-salinity waters on continental shelves. Maybe some of the ancient species also did so. After hatching, the juveniles of the extant species move farther and farther away from the beaches as foragers, later returning as adults to spawn. We do not believe that their return to the same beaches is obligatory. Spawning and foraging now drive and may always have driven Limulidae migrations. But, assuming that horseshoe crabs follow predominant bottom currents shoreward, this characteristic could impose randomness in their ultimate destination, depending on where and when they started. Although their visual and chemical senses probably keep them more or less oriented to their immediate surroundings, the somewhat random, broader orientation to currents takes them into various environments where they may or may not survive. This randomness is so broad, with reorientation during migration, that many (if not most) of the crabs reach suitable habitats. These migrations, coupled with the abundance of the crabs, are the chief determiners of the extent of the distribution of any population of horseshoe crabs.

Not every place where extant horseshoe crabs might be found has been searched. Extensive reconnaissances, however, have mapped out the major geographic regions for *Limulus* (Shuster, 1955, 1979) and the three Indo-Pacific species (Sekiguchi and Nakamura, 1979; Sekiguchi, 1988a). From these and other studies, we can see two main constraints to the distributions of horseshoe crabs—those that restricted the geographic range of the species and those that affected local distributions. The problem is more difficult when you consider the distribution of fossil horseshoe crabs, due to the dimension of deep time. In essence, the patterns of distribution of the extant species are frozen moments in time, compared to what happened earlier. For every geologic age, there have been different geographical distributions, particularly as a result of shifting tectonic plates (continental drift), but also resulting from compression and expansion of the climatic belts and sea levels. For these reasons, the fossils can and frequently do turn up almost anywhere on the globe. The rarity of horseshoe crab fossils presents problems in pinpointing migration and colonization of different areas through time. Morphological conservatism is also a problem; as explained in Chapter 6, horseshoe crabs have changed little, making tracking difficult. In a group like the trilobites, faunas can be tracked as they developed, colonized, expanded, and contracted due to well-defined lineages on the basis of marked morphological changes through time.

Information on the juxtaposition of landmasses and coastal waters is fairly well developed for most geologic periods, based on a combination of paleontological and sedimentological studies. We do not have the level of resolution (map size), however, that is required to track migrations and species distributions of horseshoe crabs. A further problem is that the fossil record of xiphosurans is limited and the number of taxa (species, genera, and higher taxonomic classifications) in the group is relatively small compared to those of other groups of animals.

CHAPTER 9

COPING WITH ENVIRONMENTAL CHANGES: PHYSIOLOGICAL CHALLENGES

David W. Towle and Raymond P. Henry

Perhaps the most dramatic environmental challenge, and the *Limulus* behavior most obvious to human observers, is the congregating of horseshoe crabs on intertidal sediments during the breeding season (see Chapter 1). Although they spend most of their lives offshore, in a habitat that is characterized by relative stability with regard to the major physical parameters of temperature, salinity, and oxygen content, horseshoe crabs make a yearly migration into inshore waters and estuaries. This migration, which is timed with the annual breeding cycle, takes advantage of the fact that estuaries are rich in nutrients and low in predatory and competing species. Therefore, when the eggs hatch, they are in a favorable environment for survival and growth. While estuaries may be favorable with respect to biological factors, they are considered hostile environments with respect to physical factors.

Estuaries, by definition, are characterized by low and fluctuating salinity, which reflects the total concentration of dissolved salts, mainly sodium chloride (NaCl) but including magnesium chloride ($MgCl_2$), magnesium sulfate ($MgSO_4$), calcium chloride ($CaCl_2$), potassium chloride (KCl), and other minor salts. Mixing fresh water from rivers with ocean water creates a gradient along the length of the estuary, from approximately 35 parts of salts per thousand of water (ppt) at the mouth to near 0 ppt at the head. Superimposed on this gradient are changes due to the daily tidal cycle, seasonal rainfall and evaporation patterns, and catastrophic events such as hurricanes. In addition to salinity fluctuations, because inshore and estuarine waters are relatively shallow, temperature fluctuations are also common. And finally, the very organic decay that makes the waters so rich in nutrients also depletes the waters of oxygen, a problem that can be exacerbated by high temperatures in summer.

When *Limulus* migrates into estuarine waters to mate in the late spring or early summer, water temperatures are rising and rainfall is at its highest. The combination of these conditions presents a severe challenge to the horseshoe crab's physiological adaptations. Reduced salinity alone can lead to water uptake and tissue swelling, as well as to the loss of salts from the hemolymph (the circulatory fluid or blood) and tissues. The physical exertion of migration and breeding places heavy demands on the animals' ability to obtain oxygen; the fact that this occurs during high temperatures and in areas of potentially low oxygen places further environmental stress on the animals. The final stage of the process is perhaps the most stressful: *Limulus* emerges from the aquatic environment to lay its eggs in the sand. The resulting air exposure can result in desiccation, elevated body temperature, reduced gas exchange, and an inability to excrete metabolic waste products.

What physiological mechanisms have evolved in horseshoe crabs that enable them to cope with these environmental challenges? What organs and tissues are involved with these processes and how does their structure relate to their function? How may physiological processes in *Limulus* interact and counterbalance each other? In this chapter, we describe highlights of the ecological physiology of horseshoe crabs and at the same time identify areas in which our knowledge is incomplete.

How Does *Limulus* Cope with Changing Salinity?

Horseshoe crabs have been found living in waters as dilute as 7-ppt salinity (McMannus, 1969), but in the laboratory they have been acclimated to as low as 2 ppt (Warren and Pierce, 1982). In salinities above 23 ppt (690 mOsmol l^{-1}), horseshoe crabs allow the osmotically active molecules in their hemolymph to fluctuate with the surrounding environment. That is, *Limulus* behaves as an osmotic conformer (see Exhibit 9.1). Its hemolymph osmolality, while slightly above that of the environment (~50 mOsmol l^{-1}), changes in direct proportion to changes in external salinity. In salinities below 23 ppt, *Limulus* becomes an osmotic regulator, maintaining its hemolymph osmolality significantly above that of the external medium (Figure 9.1). However, *Limulus* is considered to be a weak regulator, keeping its hemolymph osmolality at a maximum of 200 mOsmol l^{-1} above ambient, while strong regulators

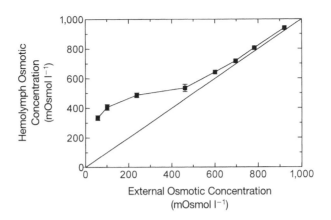

FIGURE 9.1 Osmoregulatory behavior of *Limulus polyphemus,* exhibiting hyperregulation of blood osmolytes in external salinities below about 500 mOsm kg^{-1} H$_2$O (17.5 ppt salinity). The diagonal line represents osmotic equivalence between hemolymph and the environment. *Redrawn from Warren and Pierce, 1982.*

such as the blue crab *Callinectes sapidus* can maintain an osmotic difference of up to 600 mOsmol l^{-1} (Robertson, 1970; Mangum et al., 1976; Towle et al., 1982; Warren and Pierce, 1982). Below about 2 ppt in the laboratory, osmotic regulation collapses in *Limulus* and the animal dies.

What Are the Important Components of the Osmotic Response of *Limulus* to Salinity Change?

The major osmotically active components of *Limulus* hemolymph are sodium ions (Na$^+$) and chloride ions (Cl$^-$), contributing over 90 percent of the total concentration of osmolytes (Robertson, 1970). Thus the regulation of Na$^+$ and Cl$^-$ is at the heart of the ability of horseshoe crabs to tolerate estuarine salinities. How might these animals maintain blood levels of Na$^+$ and Cl$^-$ that are higher than ambient concentrations? At least three mechanisms are possible: reducing water permeability, increasing the active uptake of Na$^+$ and Cl$^-$, or increasing the retention of Na$^+$ and Cl$^-$.

Reduction in water permeability appears to be an unlikely strategy. Water permeability of adult *Limulus* carapace and gills measured *in vitro* is very high (Dunson, 1984) and does not change with salinity. This pattern is similar to that seen in other estuarine arthropods (primarily decapod crustaceans), in which even the strongest osmoregulators exhibit high and salinity-insensitive water permeability (for example, Cameron, 1978). In juvenile *Limulus,* water exchange rates decline as salinity is decreased from 32 ppt to values below 17 ppt (Hannan and Evans, 1973). However, such a decline may simply be the result of the reduced heartbeat rate and consequent decrease in gill perfusion observed in animals treated in this way (Mangum et al., 1976). Water exchange rates in embryonic stages of *Limulus* are high and do not respond to salinity manipulations at all (Laughlin, 1981).

Because horseshoe crabs of all developmental stages face changes in salinity, high water permeability presents an important physiological challenge and seems not to be subject to significant physiological control. Thus the observed "ballooning" of *Limulus* gills when animals are diseased or exposed to sudden and extreme salinity reductions is easily explained (Robertson, 1970; Mangum et al., 1976; Bang, 1979; Dunson, 1984). Controlling water permeability thus seems not to offer a resolution to the challenges of salinity change, at least over the time scale necessary to account for the observed physiological response in *Limulus*. Active ion uptake by the gills provides a second alternative.

How Are *Limulus* Gills Specialized to Facilitate Ion Transport?

The gills of many euryhaline and freshwater animals provide a well-documented route of active ion entry from dilute salinities into the blood (Towle, 1990; Taylor and Taylor, 1992; Perry, 1997). The book gills of *Limulus* offer an equivalent possibility. Composed of 150 to 200 leaflets on each of five pairs of gill plates, providing about 11,000 cm² of surface area in a 28-cm-wide adult (Shuster, 1982), the book gills are actively irrigated by seawater when they are moved in a wavelike motion produced by sets of muscles that are controlled by branches of the ventral nerves (Figure 9.2) (Fourtner et al., 1971).

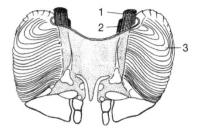

FIGURE 9.2 Book gill of the horseshoe crab, *Limulus polyphemus.* Diagrammatic representation of the anatomy of a book gill (branchia): the promotor muscle (1) and the remotor muscle (2) work together to move, hence, ventilate, the gill's lamellar surfaces; the lamellae of a book gill (3) look like a stack of thin plates. See also colorplate 15. *By permission from Meglitsch (1967).*

Covered externally by cuticle, each gill leaflet consists of two epithelial cell layers held apart by pillar cells and enclosing a blood space (Figure 9.3). The more darkly pigmented central region of each leaflet contains one epithelial cell layer (the ventral layer) that closely resembles ion-transporting epithelia from gills of other arthropods. These cells are substantially thicker than the epithelial cells in the dorsal layer or in the periphery of the ventral layer, and they have an extensive intracellular tubular system that appears to originate from infoldings of the basal membrane (Figure 9.4) (Henry et al., 1996). The cell layer in the central ventral epithelium of *Limulus* gills is about 6 μm thick, a value that is typical of arthropod ion-transporting cells in general (for example, Henry, 1994). Cell thickness in the peripheral region is less than 1 μm, consistent with respiratory gas exchange epithelia in which the water-to-blood distance is minimized.

The cells in the central region of the gill lamellae contain high levels of two transport-related enzymes, the sodium pump (also called $Na^+ + K^+$-ATPase) (Towle et al., 1982) and carbonic anhydrase (CA) (Henry et al., 1996). Thus the book gills of *Limulus* are equipped with the subcellular machinery thought to characterize ion-transporting cells (see Exhibit 9.2). No direct measurements of ion fluxes have been

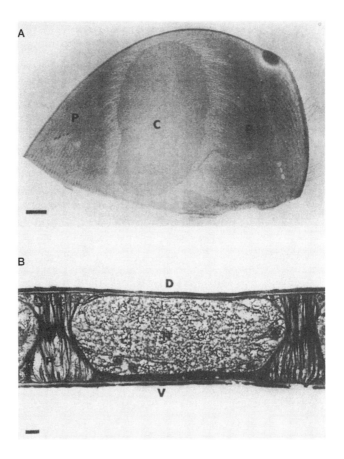

FIGURE 9.3 Single lamella of a book gill of *L. polyphemus*. (A) Whole mount showing central, C, and peripheral, P, regions (scale bar = 2 mm). (B) Longitudinal section through the central portion of a gill lamella, showing the thin dorsal, D, and thick ventral, V, epithelial cell layers, pillar cell network, P, and the hemolymph space, H (scale bar = 10 micrometers). *By permission from Henry et al. (1996).*

made with *Limulus* gills. However, measurements of transepithelial potentials (the voltage gradient across the gill epithelium) indicate that Na+ and Cl− are both maintained out of electrochemical equilibrium in animals subjected to salinities below 15 ppt, suggesting that the high hemolymph concentrations of these two ions are maintained via active transport (Towle et al., 1982).

The carbonic anhydrase activity of the central gill region is salinity-sensitive, increasing by as much as 65 percent when horseshoe crabs are transferred from 30 to 10 ppt. Salinity-sensitive carbonic anhydrase activity is considered to be a marker for ion-transporting tissue, especially in euryhaline arthropods (Henry, 1984; Henry 1988). Its presence in the central region of *Limulus* gills further supports the idea that this epithelium is the site of ion transport. Carbonic anhydrase activity in the gills of *Limulus* is, however, significantly lower than that in the gills of other arthropod species. This difference may represent part of the physiological basis for why *Limulus* is at best a weak osmoregulator.

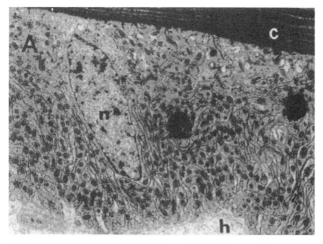

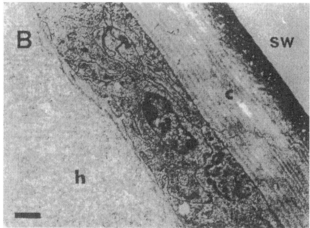

FIGURE 9.4 Ultrastructure of presumptive ion-transporting cells of the ventral epithelium in the central region, A, of a gill lamella from *L. polyphemus* compared with the presumptive respiratory cells of the peripheral region, B. The peripheral epithelium lacks the extensive basolateral membrane infolding and numerous mitochondria of the central ventral epithelium. The basolateral membrane is in contact with the hemolymph space, h. The apical membrane of the epithelial cells faces the cuticle, c, that is bathed by the environmental seawater, sw. A nucleus, n, appears in each view (scale bar = 1 micrometer). *By permission from Henry et al. (1996).*

What Is the Mechanism of Ion Uptake by *Limulus* Gills?

In nearly all epithelial cells, the sodium pump is restricted to the basolateral membrane (rather than the apical), where it actively exchanges intracellular sodium ions for extracellular potassium or ammonium ions. Due to the nearly universal presence of basolateral potassium channels, potassium ions are recycled back into the hemolymph across that membrane. Basolateral export of sodium ions, however, produces a net accumulation of sodium in the hemolymph, the result being the net active uptake of sodium ions across the epithelium (Towle, 1990). Although this model has not been investigated directly in the *Limulus* gill, the presence of substantial sodium pump and carbonic anhydrase activities supports a role of the gill in ion uptake.

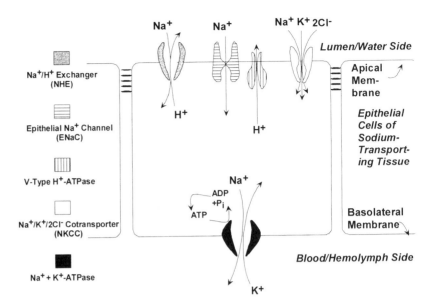

FIGURE 9.5 Model of a sodium-transporting epithelium, showing basolateral $Na^+ + K^+$-ATPase and candidate apical transporters. Note the polarity of transporter distribution with respect to apical and basolateral membranes, affording a directionality of the Na^+ transport.

The mechanism by which sodium ions enter the *Limulus* gill epithelial cell across the apical membrane from the external medium is not known. Candidate apical sodium transporters in gills of other euryhaline arthropods include sodium/hydrogen exchangers, epithelial sodium channels coupled to vacuolar-type ATPase, and the sodium/potassium/2 chloride cotransporter (Exhibit 9.2 and Figure 9.5) (Towle, 1997). Preliminary evidence supports the existence of a sodium/hydrogen exchanger in the *Limulus* gill (Towle and Litteral, 1999), but none of the remaining candidate transporters has been explored in *Limulus*. The presence of branchial carbonic anhydrase activity suggests that some type of sodium/hydrogen or chloride/bicarbonate exchange is taking place. It is believed that branchial cytoplasmic carbonic anhydrase catalyzes the hydration of respiratory carbon dioxide to hydrogen ions and bicarbonate, thus providing the necessary counterions for sodium and chloride uptake (Henry, 1984). Whether hydrogen ions and bicarbonate are directly involved, and what specific role they play in active ion uptake, are questions that have not been answered.

Whatever the mechanism, the ion uptake processes in the *Limulus* gill respond rapidly to salinity change, counteracting ion losses in low salinity within 12 to 24 hours after salinity reduction (Figure 9.6). Whether ion retention processes are also at work in this situation is the subject of the following section.

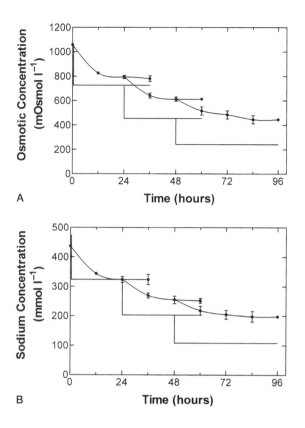

A

B

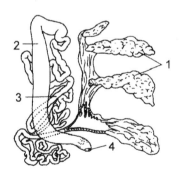

FIGURE 9.6 Adjustments in hemolymph (blood) osmotic concentration (A) and hemolymph sodium concentration (B) of *L. polyphemus* (lines with solid circles) following changes in seawater salinity (the step-wise change in the medium = solid lines without symbols). The salinity shifts were 35 to 24 ppt (0 hour), 24 to 15 ppt (at 24 hours), and 25 to 8 ppt (at 48 hours) *Redrawn from Towle et al. (1982).*

How Are Sodium and Chloride Ions Retained in the Hemolymph of *Limulus?*

Active ion retention by means of a kidney-like organ is not a common mechanism of ion regulation in gill-bearing arthropods. For example, the antennal glands of most true crabs excrete urine that is osmotically and ionically equivalent to hemolymph, especially with regard to monovalent ions, regardless of acclimation salinity (Cameron and Batterton, 1978; Wheatly, 1985). The major difference in urinary output in low salinity is an increase in urine volume, removing excess water from the extracellular fluid but adding a significant route of salt loss in these animals.

Surprisingly, the excretory organ in *Limulus* appears to serve an ion-conserving function rather well. This rudimentary "kidney" is the reddish coxal gland, a multilobed structure buried in the musculature of the second through fifth walking legs (Figure 9.7). The urinary opening, or nephropore, is located on the posterior surface of the fifth walking leg (colorplate 16). At high salinity, the coxal gland produces urine

FIGURE 9.7 Diagrammatic representation of the coxal gland anatomy of *L. polyphemus*. 1, lobes of coxal gland; 2, nephric duct; 3, convoluted tubule; and 4, nephropore. *By permission from Meglitsch (1967).*

that is osmotically and ionically equivalent to hemolymph; but in low salinity (9 to 15 ppt), the urine is less concentrated than hemolymph by as much as 81 mOsmol l^{-1}, with ionic concentrations also lower (Na$^+$ by 55 mmol and Cl$^-$ by 20 mmol l^{-1} (Mangum et al., 1976; Towle et al., 1982). Thus, the coxal gland appears to serve as an organ of active ion reabsorption, contributing an important component to the suite of physiological adaptations in horseshoe crabs living in reduced salinities.

High levels of sodium pump and carbonic anhydrase activity have been found in coxal gland homogenates, indicating specialization for transepithelial ion transport. Moreover, the levels of activity of both enzymes increase as the animals are subjected to stepped decreases in salinity (Towle et al., 1982). Carbonic anhydrase levels in the coxal gland are lower than in the gill but showed a tripling of activity 2 weeks after transfer from 32 to 10 ppt (Henry et al., 1996).

How Is the *Limulus* Coxal Gland Organized to Achieve Ultrafiltration, Ion Reabsorption, *and* Nitrogen Excretion?

The microscopic structure of the *Limulus* coxal gland clearly indicates functionality in both ultrafiltration and ion reabsorption. The inner cortex of each lobe contains specialized cells called podocytes; these cells have foot processes and slit diaphragms that are identifying characteristics of ultrafiltering renal organs (Figure 9.8) (Briggs and Moss, 1997). Hemolymph proteins such as hemocyanin appear to be retained on one side of these foot processes while a transparent urinary filtrate accumulates on the other. The primary urine is then apparently subjected to reabsorptive activities of epithelial cells in the medulla of each lobe. These medullar cells possess apical microvilli, basolateral membrane infoldings, and numerous mitochondria—all distinctive properties of ion-transporting cells (Figure 9.9) (Briggs and Moss, 1997). The ultrastructural evidence clearly suggests that ion reabsorption takes place in these medullary cells, producing the final urine that exhibits a urine-to-blood ratio of substantially less than 1.0 for Na$^+$ and Cl$^-$ (Towle et al., 1982).

Although no direct measurements have been reported of ion transport by *Limulus* coxal gland, the ultrastructure, enzymatic profile, and urine composition data all point to an important role played by the coxal gland in regulating blood ions. In sharp contrast to the antennal glands of crustaceans, the coxal glands of horseshoe crabs thus appear to offer a second level of ion regulation, namely active ion retention, in addition to active uptake at the gill. The outcome of such a dual system is seen in the horseshoe crab's ability to maintain a hyperosmotic blood in low salinities, even in the face of very high water permeability.

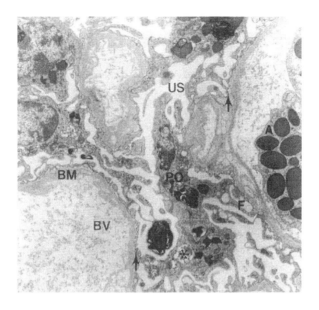

FIGURE 9.8 Fine structure of the ultrafiltration mechanism in the inner cortex of the coxal gland of *L. polyphemus*. Blood vessels, BV, lined by an irregular basement membrane, BM, are separated from the urinary space, US, by foot processes, F, of specialized cells known as podocytes, PO. The foot processes form filtration slits with slit diaphragms (arrow) that permit small molecules to pass into the urinary space but prevent the passage of large molecules such as proteins (flocculent material in the blood vessel). *By permission from Briggs and Moss (1997).*

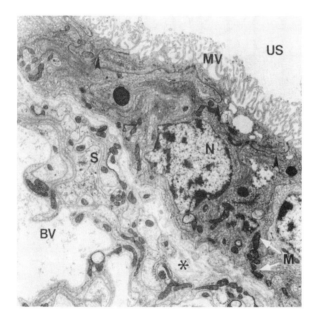

FIGURE 9.9 Fine structure of medullary tubules of *L. polyphemus* coxal gland, likely involved with reabsorption of sodium and chloride from the urinary filtrate produced in the inner cortex. Blood vessels, BV, are separated from the urinary space, US, by support cells, S, and an epithelial cell layer. These epithelial cells are similar to ion-transporting cells in the tissues of other animals and are characterized by branching, microvillous-like projections, MV, extending into the urinary space, interdigitating membranes between adjacent cells with separate junctions (black arrow points), a central nucleus, N, and an extensively invaginated basal membrane with numerous adjacent mitochondria, M (shown by white arrows). *By permission from Briggs and Moss (1997).*

A second important function of the coxal gland is excretion of nitrogenous wastes. In horseshoe crabs acclimated to 32 to 33 ppt salinity, the coxal gland produces urine that contains 0.65 mmol l^{-1} ammonia, compared with blood concentrations of 0.09 mmol l^{-1} (Mangum et al., 1976). In more dilute salinities, urinary ammonia concentrations are lower but the ratio of urine-to-blood concentrations remains well above 1.0. In contrast, the gills of most aquatic arthropods are generally considered to be the main route of ammonia excretion. Whether the gill of *Limulus* excretes ammonia is not known and thus the relative contributions of coxal gland and gill cannot be estimated. It is also unclear whether urea or uric acid represents a significant component of nitrogenous waste excretion in *Limulus.* It has been reported that *Limulus* muscle contains 0.4 to 1.1 mmol l^{-1} urea but, to our knowledge, no data for urea in blood or urine are available (Robertson, 1970).

What Intracellular Adjustments Are Made in the Face of Changing Salinities?

Because *Limulus* is a weak osmotic and ionic regulator, its hemolymph experiences significant dilution when the animal moves into low-salinity waters. For example, when *Limulus* moves from 33 to 7 ppt (about 1,000 to 200 mOsmol l^{-1}), hemolymph osmolality decreases by about 600 mOsmol l^{-1} (from about 1,000 to 400 mOsmol l^{-1}). As a result, cells within its tissues experience a wide range of osmotic environments as the animal moves from one salinity to another. The major physiological consequence is that, upon exposure to low salinity, hemolymph and therefore cells gain water and cells swell due to increased volume. In contrast, on exposure to high salinity, hemolymph and cells lose water, and cells tend to shrink due to decreased volume (see Exhibit 9.1).

Water and cell volume changes are usually measured by changes in the total wet weight of whole animals or isolated tissues. Intact horseshoe crabs gain as much as 20 percent of their total body weight through the uptake of osmotically obligated water during the transition to low salinity, and lose about 6 percent when transferred to hypersaline (60 ppt) conditions (Robertson, 1970). Furthermore, isolated tissues such as the heart gain 40 percent of their total weight when transferred from 940 to 400 mOsmol l^{-1} (Warren and Pierce, 1982).

To avoid massive water loss or gain, cells must accumulate or excrete osmotically active molecules as the cellular milieu dictates. The cells of many euryhaline marine arthropods contain a pool of nonprotein free amino acids that constitutes between 40 and 60 percent of the total intracellular osmolality (Henry, 1995). It is that pool of intracellular osmolytes that is regulated to adjust cell volume in response to salinity changes. The intracellular pool of free amino acids in *Limulus,* however, is small and does not change significantly with acclimation salinity (Warren and Pierce, 1982). Rather, the primary intracellular organic osmolyte in horseshoe crabs is glycine betaine, a quaternary ammonium compound (Figure 9.10).

The physiological challenge of moving from high to low salinity, with the consequent dilution of blood osmolality and resultant cell swelling, is apparently met in *Limulus* by two cellular mechanisms working in tandem. When isolated hearts from *Limulus* are transferred from 930 to 400 mOsmol l⁻¹, a rapid loss of monovalent ions (Na⁺ and Cl⁻) occurs during the first 24 hours (Warren and Pierce, 1982). This is followed by a slower depletion of glycine betaine from about 650 to 250 μmoles gm dry weight⁻¹ between 48 hours and 7 days post-transfer. During that time, intracellular concentrations of Na⁺ and Cl⁻ increase, indicating that cell volume readjustment via inorganic ion loss can be tolerated by the tissue only briefly, and that long-term intracellular volume control must still be accomplished by regulation of the pool of organic osmolytes. The cellular depletion of Na⁺ and Cl⁻ appears to be mediated by the sodium pump and is stimulated by the release of the biogenic amine octopamine from the cardiac ganglion (Edwards and Pierce, 1986). The metabolic fate of the expelled glycine betaine is not known, although it does appear to leave the muscle cells in an intact form.

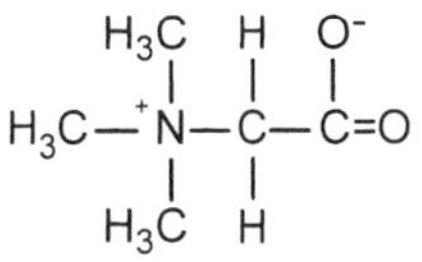

When *Limulus* moves from low to high salinity, the hemolymph becomes more concentrated, causing cells to lose water and shrink in volume (Figure 9.11). Cellular volume is readjusted through an increase in the glycine betaine pool, leading to rehydration of tissues and restoration of cell volume. Levels of glycine betaine in heart muscle cells increase from 550 to 800 μmoles gm dry weight⁻¹ when horseshoe crabs are transferred from 32 to 64 ppt salinity (Dragolovich and Pierce, 1992). The mechanism of glycine betaine accumulation within heart cells experiencing a hyperosmotic stress depends on the uptake of choline by mitochondria and conversion of choline to glycine betaine by a two-step enzymatic process (Dragolovich and Pierce, 1994). Whether synthesis of the relevant enzymes and/or transport of proteins is induced upon exposure to high salinity is not known.

FIGURE 9.10 Chemical structure of glycine betaine, the major organic osmolyte in cells of *L. polyphemus*.

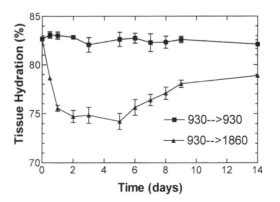

FIGURE 9.11 Volume regulation, expressed as loss and regain of water content, in heart tissue after transfer of horseshoe crabs *(L. polyphemus)* from 930 mOsm seawater to 1860 mOsm. *Redrawn from Dragolovich and Pierce (1992).*

How Does *Limulus* Respond to Changes in Oxygen Availability?

The horseshoe crab is not only euryhaline but also euryoxic, tolerating very low environmental oxygen concentrations. Indeed, individual horseshoe crabs have been maintained successfully in oxygen concentrations less than 5 percent of normoxic conditions over periods of 36 to 60 hours. The heart rate slowly declines to 25 percent of the basal rate after 24 hours of exposure to such oxygen-depleted conditions (Falkowski, 1974).

A much more rapid response to oxygen depletion is observed in the respiratory movements of the book gills. These rhythmic movements of the gills and gill plates are induced by motor output from branches of the ventral nerves (Fourtner et al., 1971). These movements not only irrigate the external surfaces of the gill leaflets with ambient water, but they may also serve as a secondary blood-pumping mechanism, assisting the flow of blood to the heart as the leaflets empty and fill with blood (Freadman and Watson, 1989). The rate of respiratory gill movement is reported to be proportional to the logarithm of the ambient oxygen concentration (Figure 9.12; Page, 1973), ceasing completely in oxygen-depleted water. Upon restoration of oxygen to anoxic water, respiratory movements resume within 5 seconds (Waterman and Travis, 1953). It should be noted that one study found no correlation between respiratory movement rate and ambient oxygen supply (Mangum and Ricci, 1989).

The signaling pathway for the regulation of gill respiratory movements is thought to originate at two anatomical sites: on the surface of the gills themselves (Page, 1973; Crabtree and Page, 1974) and on the cuticle between coxal segments of the walking legs (Thompson and Page, 1975). Neurophysiological measurements have identified three

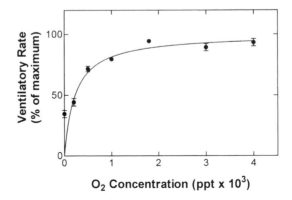

FIGURE 9.12 Effect of environmental oxygen levels on ventilatory rate of book gills in *L. polyphemus. Redrawn from Page (1973).*

types of receptors in the gill leaflet: (1) receptors stimulated by oxygen, (2) receptors inhibited by high oxygen, and (3) tactile receptors responsive only in the presence of oxygen (Crabtree and Page, 1974). In the intercoxal cuticle, only the first type of receptor has been identified (Thompson and Page, 1975). The "branchial warts" on the endopodites of *Limulus* gill leaflets contain goblet-shaped chemosensory appendages 15 to 25 μm in diameter (Griffin and Fahrenbach, 1977). However, these are not likely to contain the oxygen-sensing receptors because endopodites may be removed with no effect on oxygen sensitivity of respiratory movements (Page, 1973).

Sensory input from gill and intercoxal receptors does not require processing outside of the ventral nerves, nor is it necessary to maintain an intact blood supply. Instead, central pattern generation in the ventral nerves is modulated by sensory input from the receptors (Wyse and Page, 1976). The outcome of this signaling pathway is cessation of gill movement in oxygen-depleted waters and immediate resumption of gill movement when oxygen is restored. Shutting down respiratory movements in low-oxygen environments will conserve energy reserves, for example, when individuals are stranded by low tides or encounter oxygen-depleted water during migrations in stratified estuaries. However, it may also lead to important changes in acid-base balance, considered later in this chapter.

Coping with Limited Oxygen Supplies

How do horseshoe crabs cope with a reduction in oxygen availability, other than shutting down gill respiratory movements? Several physiological strategies seem to exist, all based on properties of specific proteins that are important in controlling oxygen delivery and metabolic activity. In the latter case, one can monitor the metabolic status of a tissue by examining the abundance of the energy-providing molecule adenosine triphosphate (ATP) and its immediate precursor, phosphoarginine. In isolated *Limulus* hearts subjected to hypoxia (less than 15 torr), phosphoarginine levels decline more rapidly than ATP levels, the ATP still showing 75 percent of normoxic levels after 45 minutes (Dykens et al., 1996). Upon reoxygenation, ATP and phosphoarginine levels begin to recover within 15 minutes, although normal levels are still not reached by 2 hours. The enzyme in *Limulus* tissues that interconverts ATP and phosphoarginine, arginine kinase, seems less susceptible to reoxygenation injury caused by oxygen-derived free radicals compared with other similar enzymes (Dykens et al., 1996). Thus the recovery from transient hypoxia is thought to be more effective in the horseshoe crab than in other less euryoxic species.

During exposure to oxygen-depleted water, *Limulus* produces lactate, the most common end product of anaerobic glycolysis and the molecule familiar to athletes as the cause of muscle pain after anaerobic exercise. Twelve hours after a reduction in ambient oxygen levels from 155 torr to 50 torr, blood lactate reached levels of 6.5 mmol l^{-1}, more than eightfold higher than normoxic levels (Towle et al., 1982). Upon restoration of oxygen, the accumulated lactate is rapidly metabolized to carbon dioxide and is also incorporated into free amino acids, glucose, and glycogen (Gaede et al., 1986). Concomitant with increased lactate in hypoxic conditions is a substantially reduced blood pH, falling from a normoxic pH of 7.42 to 7.11, approximately doubling the concentration of hydrogen ions (Towle et al., 1982). This decline in blood pH is a result of metabolic acidosis, which is the buildup in the blood of acid end products of metabolism at a relatively constant CO_2 concentration, and is a significant factor in determining how the oxygen delivery system in *Limulus* adjusts to oxygen depletion.

Dissolved oxygen passing over the gills during active gill ventilation diffuses across the cuticle and the thin epithelial layer of each leaflet, where some portion of the oxygen binds to the protein hemocyanin. This copper-containing protein serves as an extracellular oxygen carrier in the blood (see Exhibit 9.3). Measurements of the oxygen affinity of *Limulus* hemocyanin indicate that the ionic environment exerts a strong influence on its oxygen-carrying properties. For example, removing chloride ions from purified hemocyanin produces an increase in oxygen affinity (Sullivan et al., 1974). Thus, as animals encounter lower salinities in an estuary, the oxygen affinity of the blood actually increases. In well-oxygenated waters, Mangum et al. (1976) predicted that hemocyanin would be fully oxygenated and would deliver an insubstantial portion of its bound oxygen to the tissues. However, in moderately oxygenated waters more closely resembling those encountered by animals entering the estuary, the importance of hemocyanin in oxygen delivery would become greater. Indeed, the higher oxygen affinity caused by declining blood chloride would help to ensure adequate extraction of oxygen from the environment.

How Does *Limulus* Cope with Air Exposure?

The dramatic emergence of horseshoe crabs into air during the mating season leads to an important shift in the animal's physiology. The quiescent gills limit the capacity to expel respiratory carbon dioxide, leading to a buildup of carbon dioxide in the blood and a consequent increase in proton concentration according to the following reactions:

$$CO_2 + H_2O \rightleftharpoons H_2CO_3 \rightleftharpoons HCO_3^- + H^+$$

COPING WITH ENVIRONMENTAL CHANGES

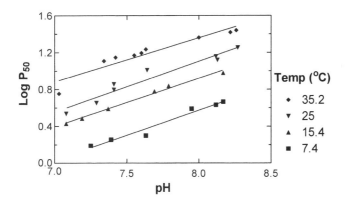

FIGURE 9.13 The effect of temperature and pH on oxygen affinity, expressed as log P_{50}, of *L. polyphemus* hemocyanin. *Redrawn from Burnett et al. (1988).*

Such respiratory acidosis, observed in many intertidal organisms subjected to air exposure, is thought to be compensated in mollusks and crustaceans by dissolution of calcium carbonate from shell or exoskeleton, the calcium ions contributing to the strong ion difference (Burnett, 1988). However, the *Limulus* exoskeleton is not calcified and thus cannot provide an immediate source of calcium for compensation.

A clue to this physiological dilemma rests in the properties of *Limulus* hemocyanin, which interestingly exhibits a reverse Bohr shift (Mangum et al., 1975). That is, unlike the majority of species that use hemocyanin as an oxygen carrier, the hemocyanin of *Limulus* exhibits increased oxygen affinity as pH decreases (Figure 9.13). Thus, during hypoxia (metabolic acidosis) or air exposure (respiratory acidosis), the reduced pH elicits an increased oxygen affinity in hemocyanin.

The outcome is that the horseshoe crab experiencing air exposure is capable of maintaining oxygen uptake at about 36 percent of the immersed level, with nearly 90 percent of the oxygen transported by hemocyanin (Mangum et al., 1975; Burnett, 1988). At least as importantly, the reverse Bohr shift also compensates partially for the reduced oxygen availability in hypoxia. Thus this peculiar property of *Limulus* hemocyanin enables the species to tolerate air exposure and hypoxia, two important environmental challenges faced by intertidal organisms, more effectively than many of its neighbors.

Response to Temperature Change

Horseshoe crabs in the western Atlantic regions may encounter water of widely varying temperatures. During the mating season, when animals emerge onto beaches, body temperatures may reach 40°C

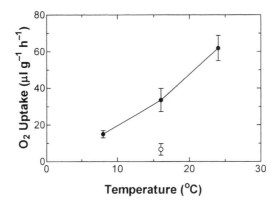

FIGURE 9.14 Effect of temperature on oxygen uptake by horseshoe crabs *(L. polyphemus):* comparison of active animals (solid circles) with buried animals (open circle). *Redrawn from Johansen and Petersen (1975).*

(Fraenkel, 1960). Not surprisingly, the rate of oxygen uptake by intact animals is strongly influenced by temperature, exhibiting a $Q_{10} = 1.84$ between 15 and 25°C (Figure 9.14) (Johansen and Petersen, 1975). That is, the rate of oxygen uptake nearly doubles with a 10°C rise in temperature over this range.

As ambient temperature increases, the affinity of *Limulus* hemocyanin for oxygen decreases (see Figure 9.13) (Burnett et al., 1988). This effect of temperature, however, is counterbalanced by a decline in blood pH that accompanies increased temperatures. In moderately oxygenated waters, for example, blood pH changes from 7.60 at 4°C to 7.29 at 24 to 25°C, representing a doubling of hydrogen ion concentration (Mangum and Ricci, 1989). The pH decrease produces an increased oxygen affinity, as just noted. This reverse Bohr effect thus leads to a stabilization of hemocyanin function with varying temperature in *Limulus,* an example of the physiological phenomenon known as enantiostasis, in which two opposing physiological forces tend to offset each other (Mangum et al., 1976; Mangum and Towle, 1977).

◼   ◼   ◼

The annual migration of horseshoe crabs to their spawning sites presents a series of physiological challenges, the responses to which we are just beginning to understand. Particularly in the area of molecular physiology, which attempts to join gene expression and protein structure with organismal function, little is known about the fundamental adaptive processes in *Limulus.* Additional studies on the interaction of those processes may reveal unexpected relationships.

Multicellular organisms that have the advantage of an extracellular fluid such as blood or hemolymph may use two distinct mechanisms to combat the problems caused when the osmotic composition of the environment changes. If the concentration of dissolved substances increases in the environment (for example, the salinity increases), the organism will tend to lose water by osmosis, resulting in a higher concentration of dissolved materials within the organism. The decrease in volume that accompanies water loss may be directly damaging to the organism's function, but the increase in solutes may be even more damaging to the catalytic activity of enzymes within cells of the organism. Conversely, a reduction in salinity tends to cause cell swelling and dilution of solutes; if the swelling proceeds without control, cells may become damaged and/or have their functions compromised.

Organisms from bacteria to oaks and humans have evolved mechanisms to cope with such environmental changes. Cells are able to regulate the levels of certain organic osmolytes, organic compounds freely dissolved in the cytoplasm that contribute to the total osmotic concentration of the intracellular fluid. These molecules are very important to both the osmotic balance of the cell and its overall general physiology and, as a result, they can accumulate to high concentrations within the cell, accounting for up to 60 percent of the total intracellular osmolarity.

Organic compounds are used as osmotic effectors in favor of salts (for example, sodium chloride or potassium chloride) because they do not perturb the cell's metabolic machinery. High intracellular salt concentrations have been shown to disrupt the normal function of important metabolic enzymes (Yancey et al., 1982), but equivalent concentrations of organic molecules such as amino acids and trimethyamine oxide (TMAO) preserve normal enzyme activity. As a result, these molecules may be accumulated to very high (or very low) concentrations, balancing the osmotic composition of the external medium and thus controlling cell swelling and shrinkage. Such cell volume regulation occurs widely in nature.

Animals endowed with an extracellular fluid such as hemolymph can use a second mechanism to combat environmental osmotic changes: the composition of the fluid can be regulated, thus reducing the need for cells to operate mechanisms of cell volume regulation. In most animal species, the most important extracellular osmolyte is sodium chloride (NaCl), although sharks notably use urea as well. If we can describe how animals regulate the concentration of extracellular NaCl, then we will understand (at least in part) how they may tolerate fluctuations in their osmotic environment. Kidneys and other renal organs are often important in this process; many aquatic animals use their gills as well.

EXHIBIT 9.2

EPITHELIAL CELL STRUCTURE AND FUNCTION

Cells that line a tubular or lamellar structure usually form a continuous epithelial layer, providing a permeability boundary between the inner compartment of the structure and its environment. This epithelium is composed of cells that have a distinct polarity of structure and function. The surface membrane of the cell facing the inner compartment, the so-called apical membrane, is often thrown up into folds and projections, increasing the effective surface area for the passage of materials across that membrane. The membrane facing the environment of the tube or lamella, which is most often the blood or hemolymph, is also often folded, but by invagination into the interior of the cell. This basolateral membrane is the site of one of the most commonly found transport proteins in animal cells, the sodium pump (designated more specifically as the sodium-plus-potassium-dependent adenosine triphosphatase, or $Na^+ + K^+ - ATPase$).

The sodium pump uses the energy from the breakdown of adenosine triphosphate (ATP), the energy "currency" of cells, to pump sodium ions out of cells in exchange for potassium ions. The resulting chemical gradient of sodium ions provides the driving force for many other transport systems in epithelia, including several that may participate in the movement of sodium ions from the external aqueous environment across gill epithelia into the blood. These candidate transporters, residing in the apical membrane of epithelia that facilitate uptake of NaCl, include the sodium-hydrogen ion exchanger (NHE), the epithelial sodium channel (ENaC), and the sodium-potassium-chloride co-transporter (NKCC) (Towle, 1997). An important partner in achieving ion transport across gill epithelia is the protein carbonic anhydrase, an enzyme that catalyzes the formation of hydrogen ions (H^+) and bicarbonate ions (HCO_3^-) from carbon dioxide and water. The resulting hydrogen ions may participate in Na^+/H^+ exchange facilitated by the NHE (Henry, 1996).

HEMOCYANIN, THE OXYGEN-CARRYING PROTEIN

EXHIBIT 9.3

Hemocyanin gives the hemolymph of horseshoe crabs its characteristic bluish color, a result of the copper bound within the protein molecule. *Limulus* hemocyanin is a very large protein, not contained within blood cells, giving the appearance of hexagonally packed cylinders in electron micrographs. The oxygen affinity of hemocyanin is estimated by measuring the change in color (from colorless to blue) when it takes up oxygen. By varying the amount of oxygen provided to the protein in a gas-tight vessel, the degree of hemocyanin oxygenation can be determined as a function of oxygen availability. The level of oxygen giving 50 percent oxygen saturation of the protein is known as the "P_{50}", a value that depends on temperature, pH, and the ionic environment of the protein.

GLOSSARY OF TERMS

EXHIBIT 9.4

anaerobic glycolysis: The metabolic pathway leading from glucose to lactate (or, in yeasts, ethanol) that occurs in the absence of oxygen.

carbonic anhydrase: The enzyme that catalyzes the addition of water to carbon dioxide to form carbonic acid.

central pattern generation: Rhythmic output of groups of nerve cells.

choline: An organic chemical that serves as the precursor to glycine betaine.

Cl^-/HCO_3^- exchanger: A membrane-bound protein that transports chloride ions in exchange for bicarbonate.

counterion: An ion, such as bicarbonate, that serves as an exchange substance during transfer of a second ion, such as chloride, across a cell membrane.

euryhaline: Describing the capability of a limited number of species to tolerate a broad range of environmental salt concentrations.

ganglion: A group of nerve cells involved in coordinating organ functions.

glycine betaine: The organic chemical that serves as an important osmotic effector within cells of the horseshoe crab.

hyperosmotic: Describing the condition in which one fluid contains a higher concentration of osmotically active molecules (such as sodium chloride) than another fluid. For example, the hemolymph of the horseshoe crab is hyperosmotic to the environment in dilute salinities.

hypoxia: The presence of significantly reduced oxygen levels in the surrounding medium.

ion reabsorption: The process in the coxal gland by which inorganic ions (mainly sodium and chloride) are transported from the primary urinary filtrate back into the hemolymph.

mmol l^{-1}: Millimoles per liter, a measurement of the concentration of any dissolved molecules in a fluid.

monovalent: Having a single positive or negative charge, compared with divalent.

mOsmol l^{-1}: Milliosmoles per liter, a measurement of the concentration of osmotically active molecules in a fluid.

Na$^+$/H$^+$ exchanger: A membrane protein that transports sodium ions into animal cells in exchange for hydrogen ions exiting.

Na$^+$ + K$^+$ − ATPase: Sodium-plus-potassium-dependent adenosine triphosphatase, the sodium pump protein that transports sodium ions out of animal cells in exchange for potassium (or ammonium) ions, driven by the energy from the hydrolysis of ATP.

osmolality: The concentration of osmotically active molecules in a solution.

ppt salinity: A measurement (parts per thousand) of the concentration of dissolved salts in seawater or brackish water.

ultrafiltration: The process in the coxal gland by which small molecules and ions are removed from the hemolymph to produce the primary urinary filtrate.

vacuolar-type H$^+$ − ATPase: A complex protein originally identified in plant vacuoles that transports hydrogen ions across a cellular membrane, using energy from the hydrolysis of ATP.

DISEASES AND SYMBIONTS: VULNERABILITY DESPITE TOUGH SHELLS

Louis Leibovitz and Gregory A. Lewbart

This chapter is the result of a fortuitous, long-term opportunity to observe morbidity and mortalities in *Limulus polyphemus* at the Marine Biological Laboratory (MBL) in Woods Hole, Massachusetts. With the realization and recognition that all laboratory animals, including invertebrates, deserve veterinary care and diagnostics, an effort was made to perform routine medical examinations of the MBL's *Limulus* population (Exhibit 10.1). During a 9-year period, 1981 to 1989, we examined about 25,000 horseshoe crabs during almost daily rounds of inspections of the laboratory tanks. These included newly harvested, healthy specimens as well as moribund and dead animals that we found in the central holding aquaria and in the tanks maintained by individual research workers. Of these, 1,904 adults were collected for necropsy studies.

Tissues were fixed and saved for histological studies by light and electron microscopy; bacteriological cultures were taken in many cases. Almost all of these animals were originally obtained from Cape Cod waters during the spring, summer, and fall. Occasionally wild moribund and dead animals were submitted for examination at the laboratory from geographical areas of the United States other than Cape Cod. Some of these included specimens collected during ongoing epizootics of disease in adults in natural marine environments. We made only a few observations on laboratory cultured embryos, larvae, and juveniles.

Prologue

Despite the early recognition of some of the problems that horseshoe crabs have encountered, a paucity of information is available on the natural and spontaneous epizootic diseases of this important animal. For example, although massive mortalities of horseshoe crabs occur each year on beaches, these events have been accepted as natural occurrences during the breeding season. They have been attributed to predation, exhaustion, inability to return to the water, hyperthermia, and dehydration that occur during the breeding season. Even less is known about the causes of mortality in free-swimming, wild, migrating marine populations, including larvae and juveniles.

Historically, interest in *Limulus* began growing toward the end of the nineteenth century, when the animal was recognized as an important comparative experimental laboratory animal. The animal was large, readily available, and easy to maintain in laboratory aquaria. Studies of the animal's blood cells (amebocytes), its circulatory system, and immune responses to bacterial infections were published first (Dekhuysen, 1901; Loeb, 1902). Since this early work, the horseshoe crab has continued as an important experimental animal, called a "living fossil" (Gould, 1989). The study of the *Limulus* amebocyte lysate (LAL) provided a further impetus to horseshoe crab research (Bang, 1953, 1956; Levin and Bang, 1964; see also Chapter 12). You can find a summary of biomedical and physiological research in which horseshoe crabs were used in the published proceedings of two separate symposia (Cohen et al., 1979; Bonaventura et al., 1982).

Although horseshoe crabs have been observed worldwide as living and fossil animals since ancient times, there are relatively few reports of the causes and description of their natural diseases (Bang, 1979; Leibovitz, 1986; Leibovitz and Lewbart, 1987) or in review articles (Shuster, 1982). Surprisingly, the study of diseases of horseshoe crabs has been a static research area despite the fact that they have been the basis of important biomedical research since the turn of the last century. Much of the research has pursued comparative implications and applications of the modern horseshoe crab for man and higher animals. Unfortunately, such research, while pioneering, does not reveal much about the evolution of diseases in *Limulus*. Such research was limited because it was focused on modern microorganisms in a modern horseshoe crab, which is not a direct descendent of the primeval animal.

While mortality has been frequently observed in captive populations, descriptions of the causes of such mortality, or the recognition of illnesses, have been rare. Loss of all ages of horseshoe crabs in the laboratory was generally considered normal and accepted by the researchers and those that maintained the aquaria because the animals were easily replaced. This was the situation when we began our studies at the Marine Biological Laboratory—little was known about the stresses placed on horseshoe crabs under natural and research conditions (see Exhibit 10.2 on how to study diseases). Now, however, we are able to present new evidence of specific disease entities. In this chapter, we describe the etiologic agents associated with disease and the related anatomical and pathological conditions for each age group and environment (wild, captive, and cultured). Unless otherwise stated, the observations pertain only to adult crabs. Because the terminology is often technical, definitions of many of the terms are given at the end of this chapter in Exhibit 10.3.

Our studies allow us to integrate the new with the old findings of the evolutionary history of this animal and its associated flora and fauna. As a result, we now have a better understanding of the animal that has survived today as well as of those species of horseshoe crabs that have become extinct. We have also begun describing the dynamic, progressive changes that occurred in the symbionts, commensals, parasites, microorganisms, and diseases that developed during their long relationships with the crab. So, this chapter builds a retrospective record of what transpired over the eons and places it in contemporary science. While there are many gaps in our information, a real beginning has been made and more research will follow. One recently published work describes a reference range of biochemical and immunological parameters for *Limulus polyphemus* (Smith et al., 2002).

What Are the Main Causes of Illnesses?

Limulus is host to many pathogens, several of which have not been described. Thus, our studies at the Marine Biological Laboratory opened a door onto a new research vista. We now invite you to follow with us as we describe the etiologic agents associated with the diseases and classify them on the basis of anatomical/pathological descriptions for each age group and environment (wild, captive, and cultured). We have divided the discussion about these diseases into three groups: physical abuse, stresses from biological entities, and environmental conditions such as nutrition and water.

Physical Abuse

Traumatic injuries were the most common condition we saw in newly harvested wild adults. Long linear fractures of the hard dorsal prosomal surface of the exoskeleton occur when animals are dropped onto a ship's metal deck or are dragged over the rough surface of the ocean bottom in the nets of collecting vessels. Often the viscera are visible through the open wounds (colorplate 17). Perforating, stab-like wounds of the softer ventral gills and at the bases of the legs occur when the serrated spike-like telson of one animal punctures the tissues of an adjacent animal when they are compressed together or, during capture, when a crab is struck by a tine on a dredge. External hemorrhage from the wounds follows, frequently extensive enough to prove fatal. Unless the hemorrhaging is massive, it may not be noticed when the animals are harvested. Improved methods in handling are needed when these animals are harvested to avoid such wounds and to allow time for healing.

We observed wound repair and healing (see colorplate 17) in both wild and laboratory animals that conformed to healing described by Bursey (1977). The process of healing is initiated by a primitive "inflammatory" response, namely, coagulation of hemolymph that promotes stasis and allows time for healing by reducing hemorrhage and providing a barrier to invading microorganisms. Repair is initiated by the migration of amebocytes into the wound area to protect and restore the damaged tissue. The same process was found to occur in fossilized trilobites millions of years ago (Isberg, 1917; Ludvigsen, 1977) and, perhaps, also occurred in the horseshoe crab species of the time.

Can Organisms Invade the Crab?

There are a surprising number of single-celled and multicellular organisms—fungal, algal, and animal—that attach to horseshoe crabs and then damage their hosts. Some of these are invasive and cause infections; others are ectoparasitic. In each case there are specific effects, with each "invader" presenting different problems for horseshoe crabs (see Exhibit 10.4 on identifying diseases of *Limulus*).

What Is the Nature of a Green Algal Infection? The most common infection of wild and captive adult horseshoe crabs is a green algal (chlorophycophytal) infection of the dorsal exoskeletal surface and the associated organ structures on the dorsum of the animal. Fresh growth of the invading algal cells imparts a greenish hue to the outer and inner layers of the dorsal exoskeleton. Older infections produce a mottled green and dark gray color on the shell due to the mixture of necrotic and degenerating layers of shell with the invading green algal cells (Figure 10.1). Occasionally, where algal growth extends outward at the lateral margins of the shell, elevated, localized, brush-like, colonial collections

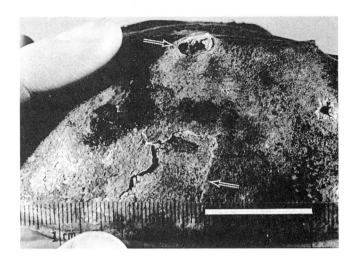

FIGURE 10.1 Advanced green algal infection in an adult *Limulus.* Arrows point (above) to the empty infected ocular orbit and (below) to the area of the flaking, mottled, fused layers of the exoskeleton of the prosoma, which are green, brown, and gray (ruler in increments of 1 mm).

of the green algal filaments can be seen projecting upward from the shell.

Morphological and culture studies of this tissue-invading organism, at the light and electron microscopic level (colorplate 18 and Figure 10.2) demonstrate that the unit of infection is the green chlorophyll- and starch-containing algal cell. These cells are surrounded by cellulose walls at various stages of their life cycle. The sexual developmental stages of the organism conform to that of the eukaryotic family Ulvacea (Braten and Lovlie, 1968). Laboratory-cultured green algal cells grew into young germlings (zygotes) with extended rhizoidal processes (Braten, 1975) (Figure 10.3). These same processes were probably responsible for the observed tissue invasion of the shell, and they corresponded to those observed in histological tissue sections invading the spaces between the deeper cells of the chitinous layers of the outer wall (Figure 10.4). Many developmental algal stages were observed extending deep into the host tissues. This resulted in the pathological fusion of old shell to the newer shell layers being formed inwardly. Surprisingly, even with the presence of bacteria, we did not observe an inflammatory response. The end result, as the disease progressed, was the inability of the animal to cast off the fused shell layers and molt normally. Molting is a prerequisite for growth, and if the infection continues, the disease can be fatal.

The progress of the disease can be monitored by the extent of algal growth and the dorsal lesions of shell necrosis, which are grossly evident as contrasting mottled areas of green and grayish necrotic tissues. The old necrotic shell areas are irregularly raised above the surface, brittle,

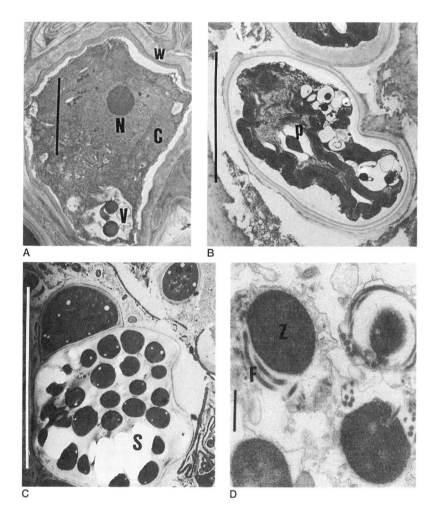

FIGURE 10.2 Photographs of individual developmental stages of green algae in exoskeletal tissues of *Limulus polyphemus* (all scale bars are approximately 1 micron). (A) Young vegetative stage with outer cellulose wall, W; chloroplast, C; vacuole containing starch granules, V; and nucleus, N. (B) A more mature vegetative stage with features similar to 10.2A. Note the pyrenoid body, P, and multiple vacuoles with starch granules. (C) A cystic sporangium, S, containing many immature spores. (D) Asexual quadriflagellated zoospores, Z, with flagella, F.

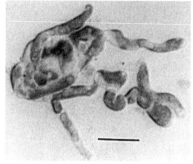

FIGURE 10.3 Photomicrograph of a 48-hour laboratory culture of an *Ulva* sp. germling with rhizopodial extensions into the growth media (scale bar = 260 microns).

and friable at the margins (see Figure 10.1). In sharp contrast, the deeper exposed, newly formed layers of the exoskeleton are pliable and uniform, and they adhere to the smooth grayish-green sheen of new shell. As the disease progresses, lesions become deeper and loss of symmetry and structure are more pronounced. On histological sections, secondary bacterial and mycotic infections are apparent on the surface and follow the algal growth into the deeper layers of the dead shell tissues. We also observed mottled necrotic fused layers of old and newly formed shell at other locations on the exoskeleton, especially at the base of the telson attachment.

In addition to the shell, the softer (less chitinized) accessory surface structures (the paired lateral eyes, the ocelli, and the arthrodial ligament)

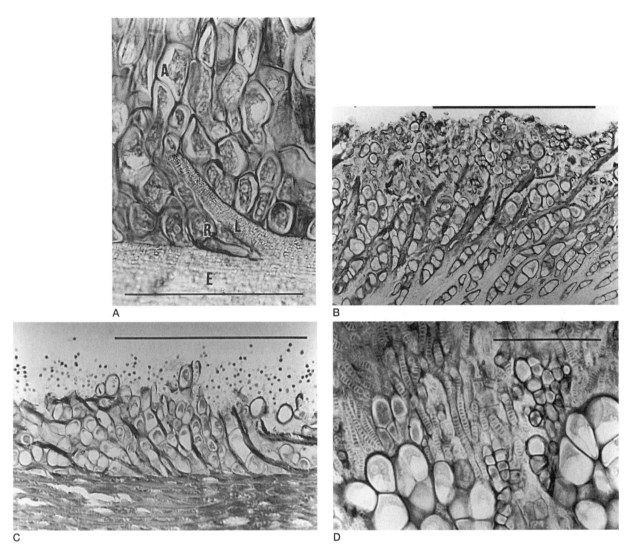

FIGURE 10.4 Photomicrographs of stained histologic sections of various stages of green algal infections of the exoskeleton of an adult *Limulus*. (A) Early stage infection. Note the thick surface layer of developing young vegetative stages of invading green algal cells, A; the laminated layer of the chitinous exoskeleton, E; a lamina, L, of the exoskeletal tissue being forced upward and outward by the invading rhizopodial extension of the germling vegetative cell, R, into the deeper tissues of the exoskeleton (scale bar = 50 microns).

(B) Midstage infection. Note the extensive invasion of the exoskeletal tissues by the asexual vegetative stages of the green algae; the great number of displaced chitinous laminae toward the surface between the rhizopodia, the lysis of chitinous tissues surrounding the invading organisms, and the loss of tissue organization into distinct layers (scale bar = 200 microns). (C) Late-stage asexual infection with the discharge of infective zoospores from the exoskeletal surface. Note the remnants of displaced fragments

of chitinous laminae as vertical and diagonal dark bands, the lysed tissue between the algal thalli, and the holes in the remaining exoskeleton at the base (scale bar = 180 microns). (D) Late-stage sexual infection characterized by the formation of gametes and sexual zoospores in the remains of the exoskeleton. Note the absence of asexual stages, chitinous laminae, and any organized tissue structures. All normal tissues have been displaced by sexual stages of the green algae (scale bar = 70 microns).

A

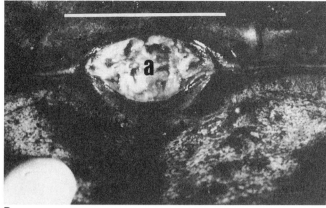

B

FIGURE 10.5 Dorsal views of the arthrodial membrane over the heart between the prosoma and the opisthosoma. (A) The normal, smooth dorsal surface of the intact, uninfected arthrodial membrane, a, between the prosomal and opisthosomal segments of the exoskeleton. Many white, soft-bodied, elongate, adult, triclad turbellarid worms are on the dorsal surface (scale bar = 4 cm). (B) The irregular, eroded surface of an arthrodial membrane and adjacent exoskeleton infected with green algae, a (scale bar = 4 cm). (C) Microscopic stained cross-section of a green-algal–infected arthrodial ligament. Note the layer of green algal organisms, G, on the dorsal infected surface of the arthrodial membrane, AM, above the chamber of the heart, H. A perforating, green-algal–infected wound, P, was probably caused by a hypodermic needle during the process of bleeding. Note that the perforating wound extends from the dorsal surface through the arthrodial membrane and into the heart chamber (scale bar = 1 mm).

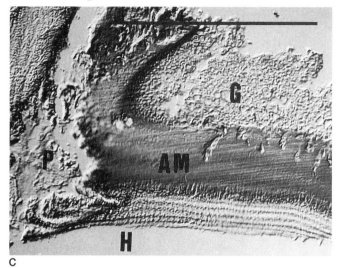

C

are especially vulnerable to destruction. Progressive algal invasion from the surface of the eyes inward results in destruction of the corneas and the deeper mosaic-like sensory units of vision (ommatidia) of the lateral compound eyes as well as the ocelli. We observed the complete destruction of the eye, with only empty eye sockets remaining (color-plate 19).

The arthrodial ligament (or hinge), critically located on the medial dorsal surface and joining the prosoma and the opisthosoma above the heart, is especially vulnerable to green algal invasion (Figure 10.5A). The disease can produce progressive erosion from the external surface to complete perforation of the ligament into the cardiac cavity, resulting in external hemorrhage and death (Figure 10.5B). Experimental

DISEASES AND SYMBIONTS

animals that were bled by contaminated hypodermic syringes, introduced through infected arthrodial ligaments, were especially predisposed to more advanced tissue destruction (Figure 10.5C), hemorrhage, and death. This conclusion is supported by the studies of Rudloe (1983), in which animals, bled through the arthrodial ligament, had a mortality rate about 10 percent higher than that of the control group.

Because this disease is confined to the dorsal surface of the animal and the growth of the pathogen is light dependent, direct sunlight presumably promotes the pathogenicity of the organism and darkness retards infection. *Limulus* is predisposed to this disease when the animals are in direct sunlight on the marine surface, and when they are returning to the beaches during breeding migration. Infection is less likely when *Limulus* returns to deeper waters and darkness during benthic migration after breeding.

How Do the Cyanobacteria Infect? Blue–green algae (Cyanophyta) are omnipresent in the marine environment and a common cause of disease in *Limulus. Oscillatoria* is an important genus of blue–green algae found in many aquatic environments (Vasconcelos and Pereira, 2001). The genus *Oscillatoria* can cause disease in a wide range of animals, including domestic mammals and humans (Kerr et al., 1987; Belov et al., 1999). In mammals, the organisms produce a toxin that, when ingested, can be fatal. An in vitro study showed that lipopolysaccharides from *Oscillatoria tenuis* and *O. brevis* caused gelation of *Limulus* amoebocyte lysate (Keleti and Sykora, 1982).

Acute Oscillatoria spp. Infection of the Gills. During spring and summer months, we observed an epizootic gill disease characterized by swollen gills, sudden death, and high mortality in both wild and captive laboratory populations of horseshoe crabs. The earliest sign of the disease is the rapid, dense, cotton-like growth of delicate filaments on the gill lamellar surfaces (colorplate 20). Microscopically, these blue-green filaments (trichromes) are composed individually of long chains of typical abutting blue-green cyanobacterial cells, ensheathed in a mucilaginous outer layer (colorplate 21). Microscopic examination of direct wet smears of the trichromes demonstrated the distinctive gliding motility and the tapered rounded differentiated polar end of the filament (hormogonium). The multicellular chain of cells that forms the unbranched filament are wider than high and are separated by cross-walls containing pigment granules and intracellular networks of thylakoid structures. Histological and electron microscopic examination of filaments and infected gill tissues conformed morphologically to a species of *Oscillatoria,* a cyanobacteria. Reproduction is by cell division and by fragmentation of the trichromes.

As the disease progresses, the cyanobacteria adhere and grow rapidly on and into the gill surface (colorplate 22), penetrating through the thin chitinous walls into the vascular sinuses, with a resulting marked vasculitis and loss of osmoregulatory control. The normal architecture of the gills is greatly distorted by the enlarged vascular sinuses. Fractured fragments of the trichromes and bacteria can be demonstrated in the vascular sinuses of the gill and from blood samples taken from the heart of the horseshoe crab during this acute phase. During the course of the disease, the gill's vascular sinuses become further distended with infiltrating fluids, gas, necrotic tissue, bacteria, and fragments of the invading organism. Gill swelling reaches its outward limits (colorplate 23) with the rupture of its ballooned lamellae and release of blood, trapped gas, putrid odors, and necrotic tissue. Bang (1979) photographed and described the swollen gill lesions in detail, but did not ascribe an etiologic agent to the disease. Shuster (1982) reported dilated gill leaflets on numerous adults during mass mortalities in Delaware Bay.

During the acute phase of the disease, the joint-capsules of the appendages and the entire ventral vascular sinus become greatly enlarged and distorted, displacing the ventral body wall downward and outward, giving the animal a ballooned, rounded appearance. It is likely that the invading cyanobacteria are responsible for these deformities due to the loss of osmoregulation through specific toxic effects (Snyder and Mangum, 1982).

Finally, death and rapid postmortem decomposition follows. The cavity that contained the gills becomes putrid, blackened, and empty, and devoid of recognizable gill tissues within a very short period of time (colorplate 24). The entire course of the disease ranges from 24 to 72 hours.

These sudden epizootics apparently occur when water quality, available nutrients, and optimal temperatures favor the rapid growth and proliferation of the *Oscillatoria.* These conditions can exist in captivity or during unusual environmental events (Paerl and Bebout, 1988). It is also likely that the movement and structure of the gills selectively retain the organisms, allowing for invasion of the gill tissues. Perhaps the mucilaginous surface of the cyanobacteria allows for the rapid attachment and growth of the organism into the gills. Bang (1979) accurately referred to this lesion, in a comparative sense, as an "external pneumonia." In contrast to the crab's "fortified" dorsal exoskeleton that apparently resists invasion, the ventral surface, especially the gills with their thin one-celled membranes, are extremely vulnerable.

The pathogenesis of this disease may offer an explanation as to how the horseshoe crab became anatomically and physiologically predisposed to it. Cyanobacteria may have been one of the earliest gram-negative endotoxin-producing organisms. The production of *Limulus* amebocyte lysate (LAL) was possibly an evolutionary development by the animal to a common pathogen in its environment. Likewise, modern gram-negative bacteria, which probably evolved from the cyanobacteria, may retain some of their endotoxin-producing abilities. These modern bacteria appear to lack the specific pathogenicity that the cyanobacteria have for the horseshoe crab. While man and modern bacteria did not exist during the evolution of horseshoe crabs, man remains sensitive to endotoxin. However, modern bacteria comprise many specific pathogens for man and higher animals that are far less pathogenic for the horseshoe crab.

Chronic Beggiatoa Gill Infection. This disease is chronic and most common during fall and winter months in captive horseshoe crabs maintained in tanks or aquaria for long periods of time. During spring and summer months, there is a short-term residence and rapid turnover of horseshoe crabs in the central holding tanks of the laboratory and thus very low incidence of *Beggiatoa*. A decrease in water temperature, food availability, and overall water quality during the winter holding period may predispose *Limulus* to *Beggiatoa* infection by favoring the growth of this organism. The horseshoe crabs are less active, and they display inanition, visceral atrophy (as determined by necropsy), and chronic gill fouling.

The gills are thickened, darker than normal, and covered with a layer of colorless filamentous organisms (colorplate 25) composed of long chains of prokaryotic cells. There are morphological similarities between this organism and *Oscillatoria* spp. Historically, these organisms were referred to as colorless *Oscillatoria*. Recent research, however, has identified morphologic and biochemical differences between these two groups of organisms, revealing a distinct taxonomic group of gliding sulfur-oxidizing bacteria (*Beggiatoa* spp.) containing sulfur granules. These organisms are associated with marine sediments and organic debris similar to the accumulations in holding tanks.

Beggiatoa is not highly invasive in gill tissues, nor does it produce the profound tissue and osmoregulatory changes attributed to *Oscillatoria*. *Beggiatoa* lesions are characterized in the early stages by the non-invasive growth of the organism on the thickened walls of the gill lamellae covering slightly distended vascular sinuses. During the late stages of infection, the gills become filled with a few intact amebocytes (blood cells) and organized clots of fragmented blood cells. While the gill surface becomes covered with bacteria, bacterial invasion of the gill sinuses does not occur until very late in the disease. The gills are not characteristically ballooned outward or broken as in *Oscillatoria* infection.

While mortality is limited, it is progressively additive, especially at the end of the longer holding period of fall and winter. This offers an explanation for the chronic response of the horseshoe crab to the less pathogenic organism over a longer time period with a more limited but extended mortality. The role of toxins in the course of both diseases is unknown, although the acute gill lesions and ballooning of the gills are strongly suggestive of the actions of toxins. Cyanobacteria have been implicated in the intoxication of many animal species, including man.

To prevent this disease, caretakers need to change tank water more frequently, place fewer crabs in each tank to avoid overcrowding, and reduce the amount of time crabs spend in the tanks. The role of malnutrition as a predisposing cause of the disease should be investigated. In the fall and winter when food is less available, poorly nourished crabs have a lowered immunity and greater susceptibility to infection.

Colonial Cyanobacterial Diseases. A chronic noninvasive disease of the gill surface that occurs throughout the year is characterized by the growth of dense colonies of small, rounded cyanobacterial cells that occlude the respiratory surface of gill filaments. When we examined the surfaces of the gill lamellae microscopically, the overall colonial growth was gold in color (colorplate 26). The individual organisms in the colony are round and surrounded by a clear mucilaginous layer that separates one cyanobacterial cell from another (Figure 10.6). Electron microscopic examination revealed characteristic morphological structures of cyanobacteria, suggestive of the order *Chroococcales*. More studies are needed to further characterize this organism.

Histopathologically, the degenerating gill lamellar walls contain many tiny holes and the gill vascular sinuses, containing coagulated hemolymph, are surrounded by a thickened wall. On the surface of the lamellae, older growths of the organism form a blackened, degenerating, and less differentiated layer of organisms with an overlay of bacteria and minute gas bubbles. These dense layers of cyanobacteria and bacteria frequently cover the entire respiratory surface of the gill epithelium.

The estimated annual mortality (based on necropsied animals) ranges from 1 to 10 percent. There is no seasonal incidence or epizootic of this disease, nor any observed associated predispositions. The role of toxins in the pathogenesis of this disease is unknown. There is a need to

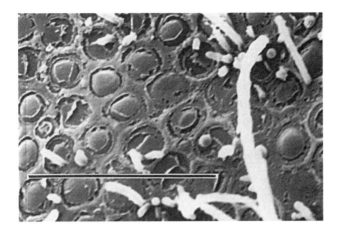

FIGURE 10.6 Scanning electron micrograph of the surface of an adult *Limulus* filament covered with a growth of colonial cyanobacteria (*Chroococcales* sp.). The organisms have a general spherical shape and are embedded in a continuous mucilaginous layer (scale bar = 50 microns).

identify the species of cyanobacteria involved and to study the pathogenesis of this disease in order to develop preventive and control measures.

What Are the Impacts from General Bacterial Infections? Leo Loeb and his students were pioneers in comparative studies on experimental infections of the horseshoe crab. His investigations extended from 1902 to 1938 and stimulated the work of many other scientists. The comparative biomedical implications of the experimental response of the horseshoe crab to bacterial infections in higher animals, including man, became apparent through his work. Central to these studies was the characterization of horseshoe crab blood (hemolymph) and its cellular elements. Additional work by the Loeb group identified the immune response of the horseshoe crab hemolymph to bacterial agents.

Originally (early 1900s), only a single type of granulocytic amebocyte was thought to present in the blood. Later, another less-abundant nonphagocytic cell (cyanocyte) was found in small numbers. This cell was considered to be responsible for the synthesis and release of hemocyanin, the oxygen-carrying pigment, and was not thought to be involved in the response to experimental bacterial infection (Fahrenbach, 1970).

The results of more recent research establish that bacterial endotoxin released into a crab's vascular system initiates the coagulation of blood (Bang, 1956; Levin and Bang, 1964). If sufficient endotoxin is released into the blood, death results due to massive intravascular clotting (Bang, 1956). Both humoral (Johannsen et al., 1973; Furman and Pistole, 1976) and cellular elements (amebocytes) are involved in the defense against bacterial infections and endotoxin. The amebocyte granules contain an enzyme-like substance (called LAL) that inhibits or neutralizes the action of bacterial endotoxin from a wide variety of gram-negative bacteria (Nachum et al., 1978). LAL, produced from the blood of horseshoe crabs, is employed commercially for a highly sensitive test to detect the presence of endotoxin (lipopolysaccharides, pyrogens) (Sullivan and Watson, 1975; see Chapter 13).

A great number of gram-negative bacterial genera have been employed in the experimental studies mentioned above, including *Vibrio, Escherichia, Shigella, Salmonella, Klebsiella, Serratia, Flavobacteria, Pseudomonas,* and *Pasteurella.* The endotoxins that are produced by these bacteria were also studied. Some gram-positive bacteria have also been studied, including *Staphylococcus, Streptococcus, Listeria,* and *Micrococcus* spp.

Pistole (1978) concluded that there are two separate agglutinins in *Limulus* serum: one reactive with gram-positive bacteria and the other reactive with a variety of gram-negative bacteria, with each species of bacteria varying in the degree of reactivity it initiates. In addition, lectins in the serum play a specific role in the immune response of *Limulus* blood.

Our studies provided an opportunity to compare the experimental, pathological, and immunological response of *Limulus polyphemus* to modern bacteria as well as the crab's response to "ancient natural disease agents" like the archaebacteria (premodern bacteria). Cyanobacteria belong to the archaebacteria group and predate modern bacteria (Baumann et al., 1983). Both gram-negative groups (Cyanobacteria and modern gram-negative bacteria, for example, *Vibrio, Salmonella,* and *Pseudomonas*) have retained their endotoxin and disease-producing characteristics, further supporting their evolutionary relationship. Our work suggests that the protective substances produced by *Limulus polyphemus,* in addition to LAL, have a general rather than a specific protective effect. The horseshoe crab's vascular response to many species of bacteria is similar, and experimental and natural infections cannot be differentiated.

Brandin and Pistole (1985) reported that it was not unusual to isolate bacteria from the blood of horseshoe crabs that had been held in laboratory tanks, even for short periods of time. Apparently, *Limulus* has the ability to tolerate low levels of these organisms due to its immune system (Nachum et al., 1978). If massive experimental doses of bacteria are introduced into the vascular compartment, the result is lethal due to the classic overwhelming response that endotoxins elicit. This suggests that the pathogenic bacteria found in the wild require a smaller dose to cause disease than the random, less pathogenic bacteria found in the captive environment. As already noted, injuries and the captive environment predispose *Limulus* to serious bacterial infection (Smith, 1964).

A wide range of bacterial species produce infections that are especially common in laboratory-cultured larval and juvenile horseshoe crabs. These immature forms are extremely sensitive to infection by microorganisms, indicating age-related immunity in *Limulus*. Chemotherapy and strict sanitation are required to control such infections (French, 1979). French also noted that juveniles raised without sand eventually became covered with long, filamentous epiphytes.

High doses of many species of gram-negative bacteria, when injected experimentally into the systemic circulation (Nachum et al., 1978), produce the same pathologic response observed in spontaneous infections. Additionally, low doses of bacteria, in either experimental or natural infections, are well tolerated and cleared without pathology by adult horseshoe crabs.

Modern bacteria appear to be most pathogenic when introduced intravascularly in high concentrations, or when they are secondary invaders following serious traumatic injury. We isolated a number of bacterial genera, including *Vibrio, Escherichia, Flexibacteria, Shigella,* and *Salmonella,* from normal and diseased horseshoe crabs during necropsy studies. While bacteria do not account for mass mortality (epizootics) in the horseshoe crab, these microorganisms can have a significant impact on individual horseshoe crabs, especially those maintained in captivity or suffering from trauma.

Do Fungal (Mycotic) Infections Also Occur? We have not observed epizootic mycotic infections in wild or captive adult horseshoe crabs. Branchial (gill) mycoses have been documented in laboratory-maintained adult horseshoe crabs. In juvenile laboratory-reared horseshoe crabs, mycotic infections (Figure 10.7) are enzootic and part of the disease spectrum found in this population. The following mycotic organisms have been isolated from juvenile horseshoe crabs (L. Ajello, 1987, personal communication): *Penicillium* sp., *Scopulariopsis brumpti,* and *Exophiala psciphilus.* Although the number of juveniles we examined was limited, the incidence of this disease suggests that juveniles are more susceptible to mycotic infection than adults.

What about "Fellow Travelers on the Way"?
When Allee (1923) described horseshoe crabs of New England waters as "walking museums," he highlighted a well-known phenomenon; but the significance of many of these travelers on *Limulus* did not become

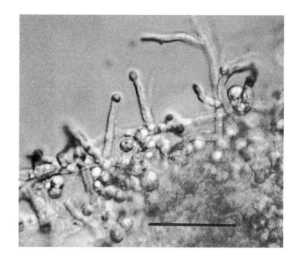

FIGURE 10.7 Photomicrograph of mold growth on the surface of a juvenile *Limulus* exoskeleton. Note the many branched filaments (hyphae), fruiting bodies, and spores (scale bar = 250 microns).

known until much later. Viewed broadly, such attachment to surfaces, objects, or organisms, whether by microscopic or macroscopic substances or organisms, is commonly referred to as *fouling* in the aquatic environment. The same term is generally applied to extraneous materials attached to the exoskeletal surfaces of the horseshoe crab, a condition common in both wild and captive adult horseshoe crabs, but uncommon in juveniles. Because this may result in surface encrustations, clogging, choking, impediment to movement of the appendages, and so on, the amount of fouling is significant. It is also usual to refer to organisms "traveling" on *Limulus* as symbionts—those species living on the crabs in one way or another, superficial or specific. The location and degree of fouling (symbiotic) organisms on the exoskeletal surfaces of the horseshoe crab are not random and are usually defined by the evolutionary preferences and abundance of the fouling organisms in the environment. When the horseshoe crab molts, it sheds its exoskeleton, liberating itself of symbionts that it has acquired. When the new shell is formed, new symbionts attach.

From the standpoint of evolution, the development of symbiotic relationships represents a marker for tracing the ancient history and migration of the horseshoe crab. The horseshoe crab became the host and the fouling organism the symbiont. This biotic relationship must have been an important influence on the survival of each species (horseshoe crab and symbiont) from the origin of trilobites to modern times. Its evolution was expressed and differentiated as changes in morphology, life cycle, migration patterns, and defense strategies as well as in host/symbiont adaptation. Species that could not make needed changes for survival probably became extinct, terminating their symbiotic relation-ships. The symbiotic relationships of the extant species of horseshoe crabs in different geographical areas need to be studied and their flora and fauna compared. This work is currently in progress (Kawakatsu et al., 1988).

We have integrated a considerable amount of experimental and natural knowledge of the horseshoe crab *(Limulus polyphemus)* and have arrived at what we believe is a unique biomedical observation: we have been examining a large phenomenon, an archeozoic relationship, both in time and evolution. When we experimentally challenge the modern horseshoe crab in the laboratory, the frame of reference cannot be the archeozoic animal. We can, however, assemble the new information for a greater understanding of the duality of these animals, as they were and as they are today, for the first time. Here we describe the changes in symbiotic relationships (mutualism, commensalism, and parasitism) of a single species, *Limulus polyphemus.* Both the symbionts and the horse-shoe crab have made progressive transitions during their evolution that can be recognized in their modern forms. Perhaps future studies will compare the results of this work with studies of other surviving species in order to reconstruct some of the evolutionary gaps.

Symbiosis. The horseshoe crab has probably acted as a transport host (phoresis) for a dynamic variety of organisms for millions of years. While the act of transport may be passive, it has important biotic impli-cations, as a starting point for exceptionally primitive mutualistic and free-living organisms. Among these, bryozoans, barnacles, and the com-mon blue mussel *(Mytilus edulis)* frequently encrust the shells of adult *Limulus.* Because the mussel abundantly sets on the shell and then grows rapidly to a large size, it can blanket a crab with a heavy mass. When at-tached to the bases of the legs (Shuster, 1982) the mussels can immobi-lize the crab. When attached to the book gills (Botton, 1981), they may interfere with respiration. Attachment of the larvae to overturned crabs may account for some of those found around the bases of the append-ages. They may have also reached the leg bases and the gills by detach-ing, moving, and reattaching several times (Seed, 1976). Larval blue mussels may reach the gills via the "respiratory" flow of water over the gills (see Figure 7.5).

Then there are oddball associations, as when a sea star *(Asterias* sp.), oyster drills *(Urosalpinx cinerea* and *Eupleura caudate;* MacKenzie, 1979), or a whelk *(Busycon* sp.) "rides" on a horseshoe crab. Perhaps the whelk, which has a "toothed" tongue (radula), is positioning itself to rasp a hole through the crab's carapace just like a southern predaceous snail,

Melongena, would (Perry, 1940). Or imagine the astonishing sight of a horseshoe crab hobbling along with a quahog (the hard clam, *Mercenaria mercenaria*) attached to the tip of a leg. The "ball and chain" effect impedes the movements of the crab. Such attachments are probably accidental and happen when a crab is either digging in the mud or just passing over a clam flat.

Thus, mollusks (for example, slipper shells, mussels, oysters, clams, oyster drills, and others), bryozoa, barnacles, coelenterates, nematodes, and annelids have all been provided the solid, protective, and mobile environment of the horseshoe crab carapace. Feeding, fertilization, and reproduction of these species all occur within this habitat. Settlement of symbiont larval stages occurs and larvae are transported, dispersed, and introduced to new locations during horseshoe crab migration. Symbiotic organisms may have physically interfered with the vital functions of the horseshoe crab (movement of appendages, respiration, and feeding). The degree of the harmful effect is determined by the density and location of the fouling agent. Once the primitive horseshoe crab evolved migratory patterns from the estuarine to the oceanic benthos, estuarine symbionts probably alternated with oceanic ones or became adapted to both conditions.

Correspondingly, when wild adult modern horseshoe crabs are harvested and placed in holding tanks, the importance of macroscopic fouling organisms decreases with time and subsequent molts. In captivity, microscopic fouling becomes more important as accumulated surface contaminants of the horseshoe crab. It is possible that migration of the horseshoe crab may have been initiated to escape unfavorable populations of mutualistic organisms, either macroscopic or microscopic.

Commensals and Parasites Most of the microscopic organisms that coat the exoskeleton of healthy horseshoe crabs are commensals (ectocommensals). This includes many species of bacteria, molds, algae, protozoa, and nematodes that feed on the shell surface. From an archeozoic aspect, the commensal status of these organisms could have easily progressed to a more dependent relationship over long periods of time, especially during the development of a migratory life cycle in xiphosurans. In the event of trauma or disease, these organisms (opportunists) can become secondary invaders of damaged tissues and further contribute to ongoing disease processes. Etiologic agents of disease (pathogens) of the horseshoe crab and immune mechanisms of the horseshoe crab may have evolved for the first time in this manner. True pathogens of the horseshoe crab, such as ancient cyanobacteria, could have been responsible for initiating an acute, protective, general, circulatory response

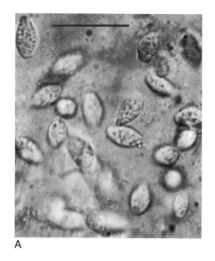

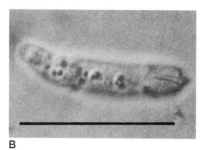

FIGURE 10.8 Photomicrographs of two protozoan parasites of juvenile and adult *Limulus*. (A) Flagellated protozoans (*Hexamita* sp.) on the surface of the gill leaflets of a juvenile *Limulus* (scale bar = 25 microns). (B) A Phytomastigophoran protozoan organism found in the hemolymph of an adult *Limulus* (scale bar = 25 microns).

A   B

for the initial survival of the horseshoe crab to cyanotoxins. It is likely that, as new pathogens arose, new mechanisms of defense against them developed.

Protozoa. Among the ciliated protozoa, *Pananophrys* spp. are the most common tissue invaders of embryos and juvenile horseshoe crabs. Sarcomastigophorans, including *Hexamita* spp. (Figure 10.8A), members of the family Paramoebidae, and a species of Phytomastigophoran (Figure 10.8B) previously observed in molluskan larval cultures (Leibovitz et al., 1984; Leibovitz and Capo, 1982), were also found colonized in tissues and invading the hemolymph of juveniles or debilitated adult horseshoe crabs. There appears to be an age-related immunity after long periods of exposure to these protozoa.

Triclad Turbellarid Worms. The continuous evolutionary association of symbiotic turbellarid worms (flatworms) with the horseshoe crab, from free-living, rock-dwelling, aquatic, or marine organisms to commensals to parasites, is well documented. Since the first descriptions, they have been classified as both commensals and parasites, yet they retain a strong anatomical resemblance to the free-living forms (Ryder, 1882; Wheeler, 1894; Verrill, 1893, 1895; Groff and Leibovitz, 1982; Shuster, 1982). All of the living species described from horseshoe crabs are classified as triclad turbellarid worms, suggesting that they share a common ancestry. Triclad species have been described from all the living horseshoe crabs, so there may be a continuous parallel association in each geographical range for each species of horseshoe crab. There are four (possibly five) different species of triclads (all belonging to the family Bdellouridae) described on the Atlantic horseshoe crab, *Limulus*

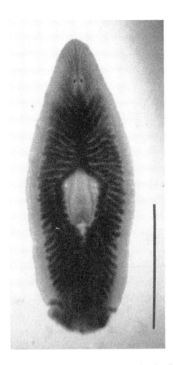

FIGURE 10.9 Photomicrograph of a fixed, stained, and compressed adult triclad turbellarid worm *(Bdelloura candida)* after removal from a gill of an adult *Limulus*. Note the paired dot-like eyes and the darkly stained digestive system. Many diverticulae extend from the anterior and two posterior lateral main trunks of the system. A central mouth and the large plicate, muscular pharynx lead into the digestive tract. The muscular posterior end of the worm functions as a sucker (scale bar = 5 mm).

polyphemus. Of these, we commonly observed two species, *Bdelloura candida* (Figures 10.5A and 10.9) and *Syncoelidium pellucidum.* Three other species of triclads (*Bdelloura parasitica, B. propinqua,* and *B. wheeleri*) have also been reported (Wheeler, 1894) from *L. polyphemus,* but we did not find them in our study. *Bdelloura candida* has been redescribed by Kawakatsu and Sekiguchi (1989).

Syncoelidium pellucidum is a small commensal living on the edges of the gill lamellae of modern horseshoe crabs. Neither the worm nor its egg capsules produce lesions on the gills. Both *Bdelloura* and *Syncoelidium* are members of the family Bdellouridae. Judging from the phylogeny of extant horseshoe crabs and the results of taxonomic and karyological studies, the origins of these triclads are very old and have no direct relationship with those of Southeast Asian triclads (Kawakatsu et al., 1988). In sharp contrast to *Syncoelidium pellucidum, Bdelloura candida* is a true parasite and a very important pathogen of the modern adult Atlantic horseshoe crab.

Both species may be found on the same animal; however, *Bdelloura* is much larger and more numerous, and it occupies the central, most vascular portion of the gill lamellae. The adults of each species are capable of leaving the body of the horseshoe crab (Figure 10.5A) and escaping into the marine environment. Once free, they are able to relocate a host. It is this ability to leave the host and revert to the free-living form that has led many observers to classify them as commensals. Both species of triclads, when adults, are positioned flat between the gill lamellae for long periods of time and derive much of their nourishment from the liberated hemolymph via wounds created during egg capsule (cocoon) deposition (colorplate 27). Studies of the digestive contents of triclads indicate both host tissues and other food items are present in the digestive tract (Ryder, 1882).

Bdelloura and *Syncoelidium* cocoons (colorplate 28) differ from one another in size, shape, number of embryos, and location and attachment on the gill lamellae. Both are covered with the same epiphytes and both have egg case parasites that destroy their larvae.

The most common of the epibiota associated with the triclads is the loricated ciliated protozoan, *Folliculina* spp. (colorplate 29A). It is usually attached to the margins of triclad egg capsules (Andrews and Nelson, 1942). Nematodes (*Chromodarida* spp.) can be seen within the interior of some of the embryonating cocoons, actively feeding on the triclad larvae (colorplate 29B). All of the developmental stages, in both sexes, of an immature parasitic copepod (*Niticra bdellourae*) have been observed in the developing cocoons (colorplate 29C; Liddell, 1912). As many as forty immature copepods have been counted in a single cocoon. The life history of the mature copepods is unknown.

DISEASES AND SYMBIONTS

The cocoons of *Syncoelidium* are smaller, thinner, with longer stalks, and usually contain two developing embryos (colorplate 28B). They are usually distributed uniformly along the outer margins of the gill lamellae. The cocoons of *Bdelloura candida* are much larger, wider, with short, stout stalks, and they contain multiple embryos (usually six to eight) (colorplate 28A). They are usually in groups on the central surface of the gill lamellae. The broad, cemented base of attachment of the stalk is firmly attached to the single-celled layer of chitinous epithelium. The entire process of curing (hardening and tanning) of the egg case on the gill lamellar surface is distinctive. The cocoon becomes brittle, compressed, and cemented by special secretions to the chitinous epithelial layer of the gill lamella. The cocoon color changes from a light tan to a dark brown or black color (Figure 10.10A). The cemented stalk attachment (Figure 10.10B) and the body of the cocoon fracture from the lamellar surface, leaving multiple surface wounds from which the endolymph escapes (see colorplate 27).

The adults and embryonated larvae feed actively on the surface endolymph. Some larvae migrate into the vascular sinuses of the lamellae (Figure 10.11) and take up residence. The movement of the gills and triclads further enlarges and merges adjacent holes into much larger coalescing surface defects of the gill lamellae with increased blood loss (colorplate 27). Coagulation of the hemolymph at the margins of the defects occurs with time, and these oval and round lesions become organized and delineated by colored borders of coagulated hemolymph and infiltrating amebocytes (colorplate 30). A formalin bath treatment has shown promise in safely reducing the number of adult and juvenile triclad worms on the gills of *Limulus* (Landy and Leibovitz, 1983).

Sluys (1983) described a new species of triclad *(Ectoplana undulata)* from both the Malaysian (Japanese) and Indonesian horseshoe crabs (*Tachypleus tridentatus* and *Tachypleus gigas*). In addition, the triclad *Ectoplana limuli* was described earlier on *Tachypleus tridentatus* (Ijima and Kaburaki, 1916). Triclad cocoons and adult *Ectoplana undulata* have also been found on the gills of *Carcinoscorpius rotundicauda* (Sluys, 1983; Kawatatsu and Sekiguchi, 1988). Sluys collected crabs in the Pacific Rim area and placed them in preservative for shipment to Amsterdam, The Netherlands. Compared to living specimens, the triclads and their respective cocoons on the preserved crabs probably were of a lesser study value. Ball (1977) suggested it would be interesting to study, from evolutionary, parasitic, and pathologic standpoints, the relationship of the *C. rotundicauda* triclads to each other and their pathogenicity to their respective hosts, and then compare this data to observations of triclads found on *Limulus polyphemus*. Several such comparative evolutionary studies of triclads and their cocoons have been completed (Ka-

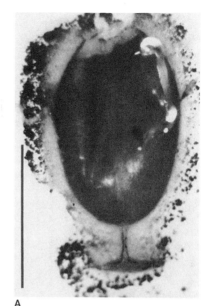

A

B

FIGURE 10.10 Cocoons of the triclad *Bdelloura candida*. (A) Photomicrograph of a collapsed, empty cocoon cemented to the surface of the gill lamella. Note the cement-like material surrounding the partially tanned cocoon that is compressed against the gill surface and the ruptured cap of the cocoon from which the triclad larvae have escaped (scale bar = 2 mm). (B) In this photomicrograph of a tanned, detached pedicle of a cocoon and attached gill tissue, the diameter of the pedicle is about one-third that of the tissue (scale bar = 0.5 mm).

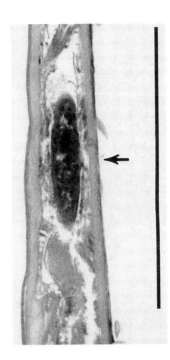

FIGURE 10.11 Histopathological cross-section of a gill lamella containing a larval triclad (arrow) feeding on hemolymph in the lamellar vascular sinus. The eyes in the anterior end of the larva are visible. The point of entry of the larvae into the vascular sinus is not in this section; it probably occurred through an open ulcer where a cocoon was attached (scale bar = 1.0 mm).

wakatsu et al. 1988; Kawakatsu et al., 1989; Kawakatsu and Sekiguchi, 1988).

Trematodes. Microphallus limuli (Waterman, 1950; Stunkard, 1950, 1951, 1953, 1968) is an important digenetic trematode parasite that uses the Atlantic horseshoe crab as its second intermediate host. The encysted metacercaria are found in great numbers throughout the bodies of juvenile horseshoe crabs, which feed on the marine snail *Hydrobia minuata*. They are especially common in the lateral eyes, brain, connective tissues, muscle, and viscera (colorplate 31). The snail serves as the first intermediate host infested with the sporocyst and cercarial stages of the parasite. The herring gull *(Larus argentatus)* is the definitive host for the adult trematode, whose eggs and miracidia infect the snail *Hydrobia minuata*. Adult horseshoe crabs are infected, but they carry a smaller parasitic burden, which indicates that the marine snail is less important in their diet. Apparently, infestations attained during juvenile stages are retained to maturity.

The importance of this parasite for the horseshoe crab is unknown. It has been postulated that the parasite affects the behavior, physiology, and development of the horseshoe crab by interfering with the orientation of the animal and the use of the telson. As a result of this impairment, the juvenile horseshoe crab is more easily captured and eaten by gulls, with a resulting increase in the parasite population (Penn, 1992, personal communication).

Nematodes. A few species of normally free-living nematodes (for example, *Monystera* spp. (Figure 10.12) and *Grathponema* spp.) invade and tunnel into shell tissues of the modern horseshoe crab without apparent harm to the animal, suggesting a mechanism (immunity) to invasion. The horseshoe crab does not encapsulate invading organisms with its own tissues; the parasite may produce a protective wall about itself (for example, metacercaria). The walling off of invaders (encapsulation) by the host is characteristic of higher animals.

What Are Some of the Nutritional and Water Quality Problems?

Very little is known about the nutritional requirements of juvenile or adult horseshoe crabs throughout their life cycle. It is known that they consume invertebrates, especially mollusks, polychaetes, and nematodes (see Chapter 5). How the availability of required foods influences the health, migration, and survival of the various developmental stages of the animal is unknown. Certainly, in its long evolutionary history and universal distribution, the geographic availability of food must have been a selective and limiting factor, in terms of the horseshoe crab's survival and migratory patterns in a continually changing habitat.

DISEASES AND SYMBIONTS

The state of health and nutrition of a living horseshoe crab is difficult to evaluate because its rigid exoskeleton limits observation of internal changes. Most laboratory workers judge the state of health of the animal by its mobility and response to handling rather than changes of physiognomy. While some scientists feed the animals in captivity, many do not because of fouling of the water after feeding. It is not known how long a horseshoe crab can do without food, but some animals in captivity are not fed for many months (see Exhibit 10.5 for information on maintaining *Limulus* in captivity). While the exterior (exoskeleton) remains the same, the tissues within may be completely atrophied (wasted away) without any detectable change in the movement of the animal. Histologic sections of such animals reveal extreme cachexia and progressive Zenker's muscle degeneration (colorplate 32) similar to conditions described in vertebrates. The latter is characterized by hyaline degeneration of muscular tissues, the fibers losing their striations, becoming sinuous, and later undergoing fragmentation and absorption. The etiology is attributed to diet- or disease-related malnutrition.

The importance of water quality as an environmental variable in the evolutionary survival of this euryhaline animal must be recognized. As seen previously in ballooning of the gills, osmoregulation is a critical part of respiration. Toxic changes, whether via biotoxins (cyanobacteria) or other toxic pollutants, can be fatal. Water temperatures may be another factor in survival. In our MBL study, *Limulus* showed an unusual tolerance for temperature extremes and sudden changes in temperature. In captivity the adult animal can tolerate water temperatures below 0 degrees C and above 35 degrees C for long periods of time. The influence of temperature on migration and physiology is unknown.

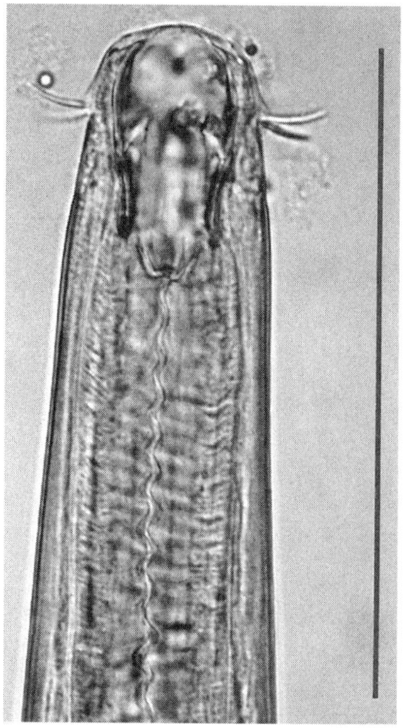

FIGURE 10.12 Photomicrograph of the anterior end of a nematode (*Monystera* sp.) removed from the exoskeleton of an adult *Limulus*. Note the bristles on the head of the worm surrounding the oral cavity and muscular esophagus (scale bar = 0.5 mm).

What Have We Learned? What Next?

Why doesn't the horseshoe crab's shell keep out disease? Until recently, there have been many misconceptions of the formidable protection afforded the horseshoe crab by its exoskeleton. As with all living organisms, disease agents adapt to the defenses of the host.

How Can Specific Pathogens Cause Disease on Different Parts of the Horseshoe Crab?

There are diseases of the dorsum of the crab (for example, green algal infection) in which sunlight or host age are contributing factors. There are diseases of the ventral surface in which the gill is predisposed to lethal agents (for example, cyanobacteria) in selected environments. During its long evolution, the migratory patterns of the animal probably helped it avoid some of these diseases due to changes in the environment (for example, lower to higher salinity waters). If avoidance was not possible, methods of defense and adaptation, including immunity, were required for survival.

In the case of the green algal infection, the pathogen does not require intracellular invasion, because it relies on lytic action. This lytic action dissolves intercellular tissues of the host and probably supplies nutrients for the green alga. Dense tissues like the exoskeleton dissolve slowly, while more pliable structures like the arthrodial ligament and eyes dissolve relatively quickly.

The duration of pathogenic action is important and limited in the case of the green algae by exposure to sunlight. In the case of cyanobacteria infections, the host's migration routes that avoid exposure are a means of defense against the pathogen. The host defense against cyanobacteria also requires resistance to the toxic effects of these bacteria.

Where Is the Source of These Pathogens?

While we have described a number of "new" diseases and the pathologic response of the horseshoe crab, the etiologic agents of these diseases are not new (*Limulus* has been around for several hundred million years). All of these agents and their related forms either existed before the horseshoe crab, or were acquired during the archeozoic evolution of the animal. Remember that the modern horseshoe crab and the disease agents described are modern representatives of previously existing forms that are genetically related. While these findings are not the same as those that would have been found in primeval times, they offer us the best evolutionary insight to date. This is especially true of the new discoveries of limited generalized immunologic response of the horseshoe crab, both to primeval and modern bacteria. Medical science is now aware of what systems predated the sophisticated immune response of higher animals. The horseshoe crab's response to modern microorganisms, which are related but not true pathogens of this animal, result in a graded protection via the animal's partial recognition of the modern pathogen's phenotype.

What about Viral Diseases and Cancer?

We did not find viral infections and true tumors (neoplasms) in *Limulus,* but these diseases probably do exist in this animal. We found one instance of a reported neoplasm of the horseshoe crab in the literature (Hanstrom, 1926). The tumor was described as a chitinous foreign body resembling a dermoid and could easily have been scar tissue from a wound. Again, the absence of viral-induced neoplasia from this ancient animal in its long history is remarkable; viruses are a common cause of tumors in other animals. Because viruses require specialized cell receptors to induce infection, and tumors are induced in complex manners in vertebrates, this is another important area for researchers to explore. Hypothetically, *Limulus* should be at risk to such infections. The question as to why they have not been detected remains. The answer may be quite simple—we just haven't looked hard enough.

What about Inherited Diseases?

We made no attempt to evaluate genetic diseases of the horseshoe crab in this study. A number of anatomical variations of the exoskeleton, especially those of the telson, have been reported (Shuster, 1982), but whether these are genetic in origin is unknown. Chapter 5 addresses malformed embryos and adults. This area of horseshoe crab medicine and pathology is wide open for study.

What Disciplines Are Applied to the Field of Aquatic Animal Medicine?

We have attempted to integrate and unify studies in a wide range of individual disciplines dealing with the horseshoe crab (for example, biology, microbiology, evolution, ophthalmology, neurology, and pathology). We have also emphasized the need to consider many nonbiomedical disciplines such as geology, climatology, and paleontology. The fact that this species has survived for eons while retaining its essential form and function is proof of its adaptability.

EXHIBIT 10.1

DR. LOUIS LEIBOVITZ

Dr. Leibovitz discovered much of what is known about the diseases of *Limulus*. This chapter is an apt testimony to his research achievement for much of it is based on unpublished information. For many years he was the medical consultant on the ills of the horseshoe crab encountered by researchers at the Marine Biological Laboratory in Woods Hole, Mass. A quiet man, Dr. Leibovitz (Figure 10.13) found fame among other horseshoe crab scientists in 1988, when he participated in a small international symposium organized by James J. Finn at Somers Point, N. J. Appropriately, his talk was entitled "Diseases of the Horseshoe Crab." Louis was a cadre member of the "*Limulus* Book Team." When he realized that, dying of cancer, he could not finish this chapter, he recruited a former student, Gregory A. Lewbart, to finish its preparation.

Louis Leibovitz was born in Philadelphia, the son of Harry and Lily (Cohen) Leibovitz. He graduated from Pennsylvania State College with a Bachelor of Science degree in veterinary science. He later received his doctor of veterinary medicine from the University of Pennsylvania. He published many papers on aquatic animal disease and was professor emeritus at Cornell University Veterinary School. In his later years he was the director of the Laboratory for Marine Animal Health at the Marine Biological Laboratory in Woods Hole. He enjoyed his retirement years in Falmouth, Massachusetts.

A World War II veteran, he served in Europe. He married Anne T. Twer, who died just a few days before him in 1998. She was born in Philadelphia and graduated from the Moore School of Design. She worked as a textile designer in New York City for several years. They had three sons: Daniel Leibovitz of Hilliard, Ohio; Henry Leibovitz of North Kingston, Rhode Island; and the late Mark Leibovitz.

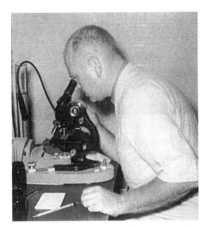

FIGURE 10.13 Dr. Louis Leibovitz examining microscopic details of parasites of *Limulus*.

EXHIBIT 10.2

HOW TO STUDY A DISEASE

The medical science concerned with the study and understanding of disease is termed pathology. To quote *Stedman's Medical Dictionary* (Williams and Wilkins, 1988), "Pathology is the medical science, and specialty practice, concerned with all aspects of disease, but with special reference to the essential nature, causes, and development of abnormal conditions, as well as the structural and functional changes that result from the disease processes."

The investigation of a recognized disease process requires an organized and systematic plan. The following approach would be employed to study a sick horseshoe crab:

1. Obtain an accurate history of the case.

   Is the animal wild, caught, or captive? How is it housed? How has it been fed? What are the water quality parameters? Is it part of a group of animals? If so, are other crabs sick, too? Has the crab been treated with any medications? Have similar problems occurred in the past with this particular crab or population?

2. Perform a thorough physical examination.

   The clinician should first take a step back and view the animal as a complete creature. Is the animal alert? Does it appear responsive? What is its general appearance? Next, each organ system (exoskeleton, appendages, gills, eyes, cardiovascular system) should be evaluated if possible and abnormalities noted in the developing medical record.

3. Antemortem sample collection.

   If the goal is to save the animal in question, then necropsy (autopsy) is not an immediate option. A number of useful samples can be collected from a live crab without harming or endangering it. Lesions can be biopsied and fresh samples examined under the microscope. Similar small tissue biopsies can be placed in 10 percent neutral buffered formalin (dilute formaldehyde) and preserved for microscopic examination using a technique called histopathology (this is one of the most widely used diagnostic procedures in both human and veterinary medicine). Suspect areas on the crab's body can also be swabbed for microbiological culture and sensitivity (the sensitivity indicates the drugs to which a particular microorganism might be susceptible). Finally, a blood sample can be easily obtained by placing a hypodermic needle directly into the heart beneath the arthrodial ligament at the top of the crab. The removal of one or two milliliters of blood will not harm the crab, and the blood can be analyzed in the laboratory for infection and other abnormalities.

4. Necropsy.

   This valuable technique is usually employed when the animal has already died or is near death and must be euthanized (humanely killed). A necropsy may also be in order if there is a "herd health" or population problem. In this case, sacrificing one or even several moribund individuals helps the clinician develop a more complete clinical picture. The major advantage of the necropsy is that all of the animal's organs are readily available for examination and sample collection.

EXHIBIT 10.3

PATHOLOGICAL AND BIOLOGICAL TERMINOLOGY

arthrodial Pertaining to the broad hinge (articulation = joint) between the forepart and middle part of the dorsal exoskeleton of *Limulus.*

cachexia A profound and marked state of constitutional disorder; general ill health and malnutrition.

commensalism A symbiotic relationship in which one species derives benefit and the other is not harmed.

disease Generally a departure from a state of health, an illness; specifically, a definite morbid process that affects all or part of a body and has a set of characteristic symptoms.

endemic A disease that has a low incidence but is continually present in a given population.

endotoxin A toxic substance, often a lipopolysaccharide (a starch complex united with a fatty compound), that is released from a bacterial cell when it disintegrates.

enucleated Removed whole and clean, as a tumor from its envelope.

enzootic Occurring endemically among animals.

epizootic Occurring as an epidemic disease among animals, rapidly spreading and widely diffused.

eukaryotic Pertaining to a visibly apparent nucleus.

gram-negative bacteria Bacteria that lose the stain or are decolorized by alcohol in Gram's method of staining (gram-positive bacteria retain the stain).

humoral Pertaining to the body fluids.

inanition The physical condition that results from complete lack of food.

lectin A substance not known to be an antibody but that combines specifically with an antigen and produces phenomena resembling immunological reactions.

lesion Any pathological or traumatic discontinuity of tissue or loss of the function of a part.

moribund In a dying state.

mutualism A symbiotic relationship in which both species in the relationship benefit.

mycoses Diseases caused by a fungus.

necropsy A postmortem examination; an autopsy.

necrosis Death of a circumscribed portion of tissue.

parasitism A symbiotic relationship in which one species benefits at the other's expense.

pathology The branch of medicine that treats the essential nature of disease, especially the structural and functional changes caused by disease.

prokaryotic Pertaining to an early stage in mitosis.

pyrogen An agent or compound that causes a rise in temperature.

quarantine Originally used to define the forty-day period of detention for human emigrants coming from an area where an infectious disease or diseases prevailed. This term is now more generally applied to the isolation of new animals being added to a collection or population.

symbiosis The living together of two or more species, as in commensalism, mutualism, parasitism, phoresis, and so on.

systemic Pertaining to or affecting the body as a whole.

thylakoid Lamellae. Sheet-like membranes that occur within the chloroplast. Each consists of a pair of membranes with a narrow space in between. Some 3,000 per chloroplast, they position the chlorophyll molecules to receive the maximum amount of light.

vasculitis Inflammation of a vessel.

A KEY TO THE IDENTIFICATION OF DISEASES OF *LIMULUS*

EXHIBIT 10.4

Carapace punctured, fractured, or abraded = physical injury (1); otherwise, no surface wounds = biological agent (2).

1. Physical Injury

There are three cases of physical injury in this key (1.1, 1.2, and 1.3).

    1.1. Puncture wounds
        A. Small wounds, which occur mainly during harvesting/capture when the telson of one crab punctures a soft part of the exoskeleton (around appendages and the hinge area), or
        B. Large wounds, resulting in major exoskeletal trauma, as when the crab is pierced by the tines of dredges.

1.2. Fractured carapace
   A. Carapace damaged by battering, as when the crab is dropped on a boat deck during harvesting or hit with a blunt instrument (in some localities, horseshoe crabs are killed by striking them with a mallet or ax), or
   B. Sharp cleavage of shell, generally due to axing.
1.3. Abraded carapace
   A. Due to mating activity, such as the mating scars and imprints from the male prosoma and claspers, respectively, on the opisthosoma of the female (the male claspers also show abrasion), or
   B. General abrasion, resulting from wear and tear from burrowing, and so on. Erosion of the lustrous outer protective layer of the shell may give disease organisms an easier entrance into the lower shell layers.

2. Biological Agents

There are two cases of injury due to biological agents (2.1 and 2.2).

2.1. Microscopic (not visible to the naked eye)
   A. Green alga, or
   B. Cyanobacteria (blue-green algae), or
   C. Gram-positive and gram-negative bacteria, or
   D. Protozoan organisms.
2.2. Macroscopic (clearly visible by the naked eye)
   A. Flatworms (turbellarians or trematodes), or
   B. Nematodes, or
   C. General fouling organisms (annelids, bryozoans, mollusks, and crustaceans).

EXHIBIT 10.5

MAINTAINING *LIMULUS* IN CAPTIVITY

The most important component to maintaining healthy animals in captivity is a sound and complete approach to animal husbandry and environmental management. Once a disease is diagnosed in an animal or population of animals, an uphill battle begins, usually with the employment of chemotherapeutants (drugs) and other measures. There are currently no compounds licensed by the FDA for use in treating horseshoe crabs for *any* disease, thus drug selection and dosing regimen usually consist of a shotgun approach based on empirical data and clinical experience. If the following checklist is followed closely, diseases of captive horseshoe crabs should be kept to a minimum.

1. Quarantine. All new animals should be quarantined for a minimum of 30 days before introduction into an aquatic system with existing animals. During this period of isolation, the new crabs should be sampled and tested for the presence of any diseases they could introduce to the colony.

2. Water quality. Good, clean water is paramount to the successful maintenance of any captive aquatic animal. While the horseshoe crab is probably less sensitive to poor water quality than many other aquatic invertebrates, suboptimal water can be a chronic stress, leading to a depressed immune system and secondary disease problems. A fresh supply of flow-through seawater or well-filtered artificial seawater should provide the appropriate level of water quality. The following parameters should be tightly monitored and controlled:

Parameter	Ideal Value
Salinity	30–35 g/L
Temperature	Varies with season
Total ammonia	Less than 0.3 mg/L
Nitrite	Less than 0.1 mg/L
Nitrate	Less than 40 mg/L
pH	8.0–8.5
Dissolved oxygen	4.0–8.0 mg/L

3. Nutrition. It appears that a mixed diet of mollusks, particularly bivalves such as native mussels and clams, will provide adequate nutrition for horseshoe crabs in captivity. (See also Kropch, 1979; Brown and Clapper, 1981.)

CHAPTER 11

A BLUE BLOOD: THE CIRCULATORY SYSTEM

Carl N. Shuster, Jr.

The color of the fluid in the circulatory system of a horseshoe crab is gray-white to pale yellow because the fluid carries so little oxygen. The compound that binds oxygen, comprised mainly of protein and copper, is dissolved in the crab's blood, in contrast to the iron-protein complex in the hemoglobin in human red cells. When a horseshoe crab bleeds, the copper-protein (hemocyanin: from the Greek terms *hema-*, blood, and *cyanea,* dark blue) absorbs oxygen from the air, turning the fluid dark blue. Many other invertebrates, including mollusks and crustaceans, also have a hemocyanin blood; but the blue color of air-exposed blood is much more evident in a horseshoe crab due to its large size and the greater volume of hemolymph (colorplate 33).

This chapter briefly describes the circulatory system of *Limulus.* It provides the foundation for more profound discussions—on immunity (Chapter 12) and on *Limulus* amebocyte lysate (Chapter 13).

Hemolymph

By virtue of its access to all parts of the body, blood serves several important functions in higher animals and has comparable functions in *Limulus.* It transports nutrients and oxygen to the tissues and carries carbon dioxide and other wastes from the tissues to the respiratory organs and kidneys for their removal from the body. It distributes hormones to all tissues of the body and plays a key role in the resistance against attack by pathogens and parasites.

Although the fluid in this system is most often referred to as blood, it is actually a mixture of blood and lymph; hence, technically it should be referred to as hemolymph. The proportions of the two main constituents of the hemolymph, blood cells (amebocytes) and the fluid plasma, vary in juvenile horseshoe crabs, whereas adults maintain a higher cell-to-fluid ratio (Table 11.1). Differences in the number of cells, as measured by the ratio of clotted cells to sera, is probably due to the frequency of molting among the juveniles, which have more dilute sera just after molting. Because adults do not molt, the quantity of their cells probably reflects a more normal or maximum level. In adults that have been bled, the volume of cells may not return to normal levels for 6 months (see Chapter 13). The juveniles may have 50 percent fewer blood cells than the adults (Yeager and Tauber, 1935). The amount of hemolymph in the circulatory system of *Limulus* has not been accurately determined because not all of it can be drained from the sinuses. A composite of measurements on representative female *Limulus* from Delaware Bay indicates that such adults, averaging a prosomal width of 28 cm and a total body volume of 3,200 ml, have blood volumes of at least 300 ml (Shuster, 1982).

Fluid Portion of the Hemolymph

The respiratory functions of the blood of *Limulus* are conducted by hemocyanin, the most abundant protein in the blood. Hemocyanin is extracellular—in other words, free in solution—unlike the respiratory protein of mammalian blood, hemoglobin, which is contained within the red blood cells, the erythrocytes. Hemocyanin is present at high concentration in *Limulus* plasma, typically 50 g per liter of blood. It is hemocyanin that gives the blood of *Limulus* its striking blue color (see colorplate 33) because it uses copper in the oxygen-binding domain of the protein—in contrast to hemoglobin, which is red and uses iron. The hemocyanins of different invertebrates are typically organized as large aggregates of smaller protein subunits. *Limulus* takes this situation

TABLE 11.1 Pooled sera of five size groupings of *Limulus* collected from Barnstable Harbor, Cape Cod, Mass., during August 1949: groups A–C were juveniles and D and E were adults (Shuster, 1955).

Group	No. of Specimens	Sex (M&F)	Width in mm	Clot in mg	Serum in mg	Clot/ Serum
A	45	—	35–48	5	460	0.01
B	10	M	94–125	15	590	0.02
C	10	F	95–114	16	598	0.03
D	3	M	160–175	254	875	0.29
E	3	F	222–269	520	1310	0.40

to an extreme with its hemocyanin organized as a huge protein containing forty-eight molecules of protein subunits with molecular masses of 65,000 to 70,000 daltons. The total molecular mass of the hemocyanin protein free in the blood is in excess of 3 million daltons (Johnson and Yphantis, 1978). The volume of blood is estimated to be 10–33 percent of the total volume of the animal.

Respiratory Pigment

The hemocyanin of *Limulus* is a large oligomeric protein containing forty-eight polypeptide subunits (Brouwer et al., 1982). It is produced in inconspicuous rounded cells (cyanoblasts) that apparently occur sporadically in the circulatory spaces of tissues shortly after an animal molts (Fahrenbach, 1999). These cells develop into cyanocytes containing hemocyanin molecules that spontaneously assemble, end to end, to form hexagonally packed columns that bond into hexagonal side-by-side arrays. Ultimately these coalesce, creating a giant cell totally filled with the hemocyanin crystals. When the cell membrane ruptures, the crystalline hemocyanin is released into the bloodstream, where it disperses into small granules in the serum, constituting 90 percent to 95 percent of the plasma protein. This hemocyanin is the only protein in the serum of *Limulus* after complete removal of the clot, and the clotting process does not remove the hemocyanin from the serum (Cole and Allison, 1940). The copper/nitrogen ratio was constant at 0.0023 (measured in mM/l) but the amount of hemocyanin varied considerably—from 4.6 percent to 9.8 percent—depending on the amount of water in the serum. The copper content of the hemolymph of *Limulus* is 0.17 percent to 0.18 percent (Roche, 1930). The protein nature of *Limulus* hemolymph is reflected in the relative amounts of carbon (53.4 percent), hydrogen (6.9 percent), nitrogen (16.9 percent), and sulfur (1.10 percent); the relative amount of the copper constituent of the protein was 0.173 percent (Redfield, 1934).

Only One Cell in the Hemolymph

Most higher animals, including most arthropods, have several different types of cells in the blood in their respiratory and immune systems. In contrast, *Limulus* has but a single cell type in the general circulation—the granular amebocyte, also called granulocyte (Fahrenbach, 1999). The adhesive and secretory functions of these cells are central to the function of the immune system in horseshoe crabs (for more detail, see Chapter 12). Because juvenile crabs, 65 to 80 mm wide, have only 50 percent of the mean cell count of adults (Yeager and Tauber, 1935), the fewer blood cells in juveniles may be due to the frequency of molting and the rate at which the cells develop.

An Open System

In contrast to the closed systems of mammals (heart, arteries, capillaries, and veins), horseshoe crabs have an open circulatory system that lacks capillaries. Blood flows from the arteries into body spaces (sinuses) before collecting in a long ventral channel leading to the veins that enter the book gills. Oxygenated blood then flows into a sinus surrounding the heart and on into the heart. The long tubular heart pumps blood into arteries that branch many times throughout the body. John I. Stagner revealed the extensive nature of the arterial branching and an apparent lack of large venous sinuses through a corrosive cast of the circulatory system of a 5-cm-wide juvenile (Redmond et al., 1982).

Development of the Circulatory System

Because development of the embryo advances most quickly along its ventral axis, segmentation of the body and formation of the limbs occur before the circulatory system starts to form (see Figure 5.2). Actually, formation of the vascular system and the body wall complete the enclosure of the remnants of the original egg yolk. The heart begins as a portion of the middle layer of embryonic cells (the mesoderm) that extends over the egg yolk toward the median dorsal line of the developing embryo (Kingsley, 1885, 1893). Then, while the walls of the heart are being formed by the edges of this advancing tissue, a series of segmentally arranged openings, the ostia, form in the walls. Formation of the heart begins posteriorly and then gradually extends forward. At first the walls of the heart have no definite arrangement of cells, but a few blood cells are already in the heart chamber. Even in the larvae, the heart wall is mostly a single cell layer. Although the heart appears to be floating within the pericardium, cords of cells attach the heart to the wall of the pericardial sac. This sac is essentially one of the blood sinuses, only specialized.

A pair of arteries soon forms after an invagination of the outer layer of cells of the embryo when the ectoderm produces the anterior part of the alimentary canal (the stomodaeum). These blood vessels grow downward from the anterior end of the heart, on either side of the stomodaeum, to rest as two tubes on the upper surface of the central nerve; they are the aortic (sternal) arches. By the time the embryonic heart resembles the adult heart in its segmentation, its anterior portion is already enclosed within the pericardial sac. Further development of the anterior arteries results in the fusion of their sides as they enclose the nerve cord and form beneath it, leading to the creation of the neural artery that is so characteristic of the adult horseshoe crab. But the nerve cord is not bathed in blood. It is surrounded by a sheath of cells, the neurilemma, which is part of the arterial wall.

The heart of the *Limulus* embryo commences to beat some 6 days before there are any nerve cells on the heart. Carlson and Meek (1908) observed that the heart rhythm begins on the twenty-second day after the eggs are laid. At this stage the heart tube is a single-layered mass of multinucleated protoplasm. When nerves appear on the twenty-eighth day, the median nerve cord appears first, on the dorsal aspect of the middle third of the heart, but the lateral nerves are not present as yet. Transverse striations on the heart muscle are very distinct by the thirty-third day. During early embryonic life, the automatic beat of the heart is in the muscle. This automacity of the heart muscle transfers to the cellular elements of the nerves when the nerves to the heart begin to form. An early stage in the circulation of the blood is clearly shown by the flow of the blood cells, as viewed by magnification through the thin, transparent exoskeleton of the first-tailed stage (Figure 11.1).

Adult Circulatory System

Milne-Edwards (1873) and Patten and Redenbaugh (1899) are usually cited for detailed descriptions of the adult circulatory system. The heart is large in comparison with the rest of the body, lying within a wide sinus (the pericardial sac) directly beneath the dorsal axis of the carapace. About half of the heart is within the prosoma, extending from a point between the compound lateral eyes to about one-half the length of the opisthosoma (colorplates 11 and 12). It has the appearance of a jointed tube. Its exterior appears striated due to longitudinal strands of connective tissue. A large median ganglionated nerve and a pair of lateral nerves are conspicuous on the dorsal surface. In cross-section, the heart is somewhat triangular in its middle portion (Figure 11.2).

A cross-section of the heart cuts longitudinally through the vast majority of the muscle fibers, revealing bundles of fibers of varying sizes. Because the fibers cross each other radially, tangentially, and circularly, they form an intricate, loose network. Often many of these fibers divide, subdivide, and then connect again in the formation of sheets of muscle (Jordan, 1917; Fahrenbach, 1999). In the relative simplicity of its striations, *Limulus* muscle seems more like that of vertebrates than that of arthropods. Cardiac and skeletal muscles are very similar, differing

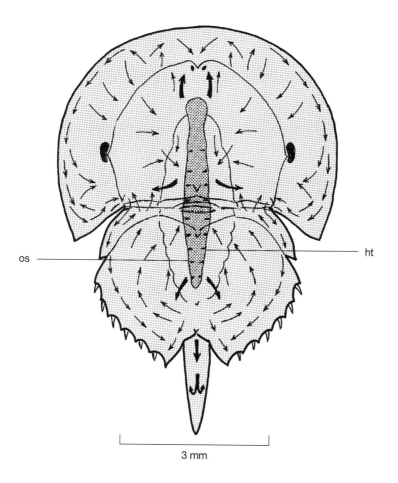

os —

— ht

3 mm

FIGURE 11.1 Blood circulation in the first-tailed (second instar) stage of *Limulus (redrawn after Packard, 1872);* dorsal view, diagrammatic. Heavy lines outline the chitinous carapace; the major suture lines and creases in the carapace are lighter lines. The axial, tubular heart (ht) and its several pairs of ostia (os) are darkly stippled; the rest of the body is lightly stippled. Wide arrows indicate hemolymph flow arterial circulation; narrow arrows indicate venous circulation. The sinuses in the tissues are not shown.

mainly in their microscopic details. Chemically, the organic phosphoric fractions of the skeletal muscle are all of higher levels than in the cardiac muscle (Engel and Chao, 1935). This is related to the physiologically different activities of the two types of muscles. The rapid, strong actions of the skeletal muscle and its heavy energy demand are met by a large reserve of an immediately available source of energy (for example, argininephosphoric acid). A relatively small reserve of this immediate source of energy is sufficient for the requirements of the slow, steady rhythmic contractions of the cardiac muscle.

The largest portion of the heart is just behind its midline, from which it tapers and somewhat flattens in both directions toward its extremities. The hemolymph enters the heart through eight pairs of transverse slits, the ostia on the sides of the heart. Each ostium has a grating of longitudinal connective tissue strands lying across the opening and two semilunar valves that prevent the backflow of the blood that is

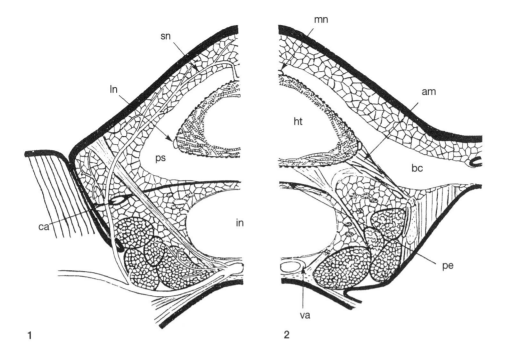

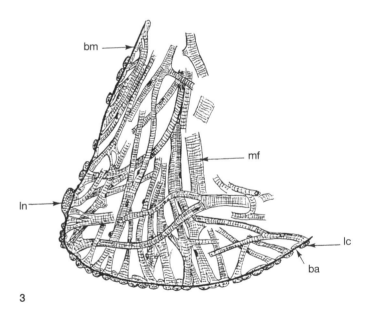

FIGURE 11.2 Diagrammatic representation of vertical sections of the heart. *Redrawn from Patten and Redenbaugh, 1899.* (1) Anterior view through the opisthosoma in the region of the ninth neuromere (nerve ganglion) showing the course of one of the cardiac nerves (scale line, for 1 and 2 = 2 cm). (2) Anterior view of the cross-section through the posterior portion of the ninth neuromere and the anterior part of the tenth neuromere. (3) A cross-section through one corner of the heart of a young *Limulus* showing structural details (scale line = 5 mm). am = alary muscle, ba = basal portion of a heart muscle fiber, bc = branchio-cardiac canal, bm = basement membrane, ca = collateral artery, ht = heart, in = intestine, lc = longitudinal connective tissue, ln = lateral cardiac nerve, mf = heart muscle fiber, mn = median cardiac nerve, pe = pericardial membrane, ps = pericardial sinus, sn = segmental cardiac nerve, and va = ventral artery. Columns of muscles are indicated by long lines (in 1 and 2).

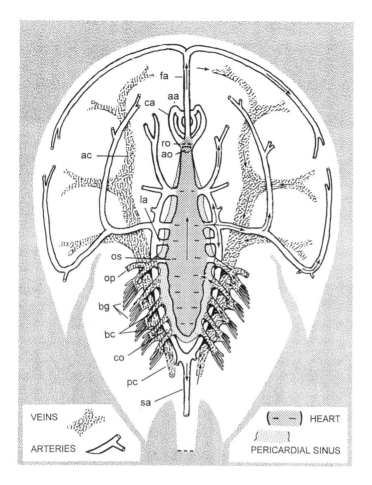

FIGURE 11.3 A diagrammatic representation of the dorsal aspect of the circulatory system of an adult *Limulus,* (not to scale). *Based on Redenbaugh and Patten, 1899; Patten, 1912.* Neither the ventral arteries nor the arteries arising from the vascular ring around the brain and nerve collar are shown. Arrows indicate the direction of blood flow. aa = aortic arch; ac = anterior cardinal vein; ao = aortic valve; ca = cephalic artery; co = collateral artery; fa = frontal artery; la = lateral artery; os = ostium; bg = book gills; sa = superior abdominal artery; pc = posterior cardinal vein; bc = branchio-cardiac canals; op = opercular canal; ro = rudimentary ostium.

pumped anteriorly in the heart. Starting at the posterior end of the heart, the ostia open in sequence, pulling in oxygenated blood that is then moved forward by waves of muscular contraction (a peristaltic movement).

The arterial system is well developed in *Limulus* (Figure 11.3). Eleven arteries carry blood from the heart. Usually three arteries pass blood forward from the anterior end of the heart and four pairs of arteries circulate blood laterally, from the sides of the heart opposite the four pairs of the anterior-most ostia. The arteries discharge hemolymph into sinuses, from which it is moved by pressure from heartbeats and by movement of the body organs and the appendages throughout the tissues before flowing into large, thin-walled veins. After passing through the book gills, the hemolymph passes into the pericardial sac and thence into the heart. Rhythmic movements of the appendages

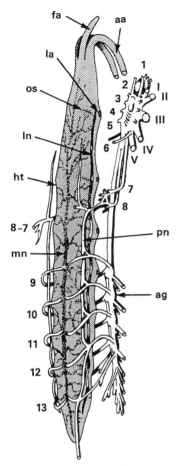

FIGURE 11.4 Diagram of the enervation of the heart. *Modified from Patten, 1912.* The heart was tipped to the right, giving a three-quarter view, to better show the nervous system, especially the median nerve cord as well as the ventral nerves. aa = aortic arch, ag = abdominal ganglion, fa = frontal artery, la = lateral artery, ln = lateral nerve, mn = median nerve cord, os = ostium (one of a pair of valves opening into the heart), pn = pericardial nerve, I–V = nerves to the corresponding ambulatory appendages, 2–8 = cardiac nerves from the brain, 7 & 8 = nerves carrying inhibitory fibers, and 9–13 = cardiac nerves from the thoracic ganglia, carrying accelerator fibers to the heart.

bearing the book gills (the branchia), coordinated by impulses from opisthosomal ganglia, circulate water over the surfaces of the book lamellae (Fourtner et al., 1971). These contractions ventilate the book gill lamella and act as a secondary heart, pumping blood to the pericardial sinus (Watson and Groome, 1989). When the heart muscles relax, elastic ligaments attached to the pericardial wall expand the heart, sucking hemolymph into the heart through the eight pairs of ostia.

Nerve Connections to the Heart

All the nerve tissues lie upon the surface of the heart and can be readily removed without damaging the muscle tissue. A median ganglionated cord on the dorsal surface of the heart connects with a pair of lateral nerves. Some strands form a network from a number of minute nerves that come from the median nerve (Figure 11.4). A series of dorsal segmental cardiac nerves, six through thirteen, convey impulses from the ventral nerve. The heartbeat is regulated by sets of impulses from a pacemaker (at the level of the fifth and sixth ostia, where the greatest number of nerve cell bodies exist), inhibition (impulses from the seventh and eighth nerves), and by augmentation (mainly impulses from the ninth through the eleventh dorsal nerves).

Three sizes of nerve cells are in the nerve clusters (ganglions) on the heart. All the cells have a long axon (process) and from none to many short extensions (dendrites) of the cell body. Giant cells are unipolar (only one long process); somewhat smaller cells with short processes are bipolar, and the smallest cells with many dendrites are multipolar.

Blood Pressure

Redmond et al. (1982) described the sequence of events in the beating of the heart and measured the normal blood pressure within the pericardium, heart, aortas, and prebranchial circulation. The heart beats on an average of thirty-two times a minute, with a maximum of fifty-one beats. Contraction of the heart produces a rapid rise in pressure until the aortic valve and probably the four pairs of lateral arteries open. The rising pressure forces the hemolymph into the aortic arch and frontal artery (see Figures 11.3 and 11.4, aa and fa, respectively). Then, when the heart muscles begin to relax, the pressure falls rapidly in the heart and slowly in the aorta. This systolic/diastolic cycle lasts about 2 seconds and produces average pressures of 26/3—measured in cm H^2O (Redmond et al., 1982) and 19/2 mm Hg (Fahrenbach, 1999). Both heart rate and blood pressure decrease rapidly when *Limulus* is exposed to the air and return rapidly when crabs are reimmersed in seawater.

Heart Physiology

In the early 1900s, the heart of the horseshoe crab was subjected to intensive experimentation with the goal of determining the relative importance of an intrinsic beat in the heart muscle compared to a heartbeat activated by a nerve impulse. It was the anatomical structure, including associated nervous tissue, and the large size of the *Limulus* heart that made it ideal for physiological studies and thus, perhaps, the best known of any invertebrate heart. Many of these investigations were pioneered by Anton Carlson, beginning in 1904, followed by Walter Garrey (1912), Henri Fredericq (1947), Peter Heinbecker (1933), and Iping Chao (1933, 1935). They used an important characteristic of *Limulus:* the nerve and muscle of the heart can be easily separated anatomically. As a result, they could experiment on the normal functioning of the heart, with and without its nerve tissue, as well as on the effects of various chemical and physical stimuli on the heart action. The hardiness and availability of great numbers of *Limulus* seem to have predestined it for these and other laboratory investigations. Most of the discoveries were discussed from the standpoint of comparative heart physiology, especially in reference to the vertebrate heart. Thus, the heart of the horseshoe crab has played a unique role in theoretical discussions of cardiac function in higher animals. The research has also helped to explain the reactions and behavior of horseshoe crabs when physical or chemical aspects of their environment change (see also Chapter 9). The rest of this chapter summarizes examples from three main lines of experimentation.

Demonstration of the Normal Functioning of the Heart

Carlson (1904), using a simple dissection technique, demonstrated that cardiac contractions in the adult *Limulus* are purely nervous in origin. He found that removal of cross-sections of the heart, leaving the nerves intact, did not interfere with coordination of the remaining tissue. Cutting the nerve cords, however, destroyed the coordination. When he removed the nervous elements of the heart, the normal rhythmic activity of the heart stopped immediately and permanently. Yet the heart muscle was still capable of a single contraction for each stimulus, whether pinched or excited by an electrical current. Thus, the rhythmic manner in which the adult heart beats is due to the rhythmical activity of the nerve cells of the elongated heart ganglion. This neurogenic origin of the heartbeat differs from that of mammalian hearts, in which the beat originates in the heart muscle (myogenic hearts). Carlson also found that the degree of the automatism of the different regions of the heart varied with the number of ganglion cells present. Interestingly, these cells were more numerous in the posterior region of the heart. Is that because that is where the first and greater muscular contractions take place that force the blood forward? Perhaps, but pulsations of the branchial appendages (book gills) also aid circulation (Wyse and Page, 1976; Watson and Wyse, 1978; Freadman and Watson, 1989; and Wyse et al., 1980).

Several investigators have studied the pacemaker function of the ganglionated portion of the median nerve cord. Garrey (1930) showed that it is this portion of the nerve cord that initiates the cardiac rhythm. Samojloff (1930) demonstrated the pacemaker characteristic of the heart ganglion by showing that it reacted to extra systoles induced artificially by electrical stimulation. According to Heinbecker (1933, 1936), individual ganglion cells have pacemaker functions. He demonstrated that propagation of a nerve impulse (action potential) in the *Limulus* heart results essentially from the activity of two components: one derived from the axons of the large pacemaker ganglion cells, the other from the axons of smaller motor ganglion cells.

Carlson (1906a) found that the rate of conduction in the intrinsic motor nerve plexus of the heart was slower than that in the peripheral motor nerves—conduction in the lateral nerves of the heart was 40.9 cm per second, 41.0 cm per second in the median nerve, and 32.8 cm per second in an ambulacral (limb) nerve. The optimum temperature of *Limulus* heart muscle was found to be from 10° to 15°C (50° to 59°F), yet the heart ganglion rhythm continues up to 42°C (108°F), at which point it was suppressed until the temperature is lowered. Varying the temperature of the muscles alters the strength but not the beat of the heart, whereas both the rate and intensity of the heartbeat are altered when the nerve ganglion is subjected to temperature variations.

Prosser (1943) was interested in the component parts of the nervous system that resulted in the mass discharge that corresponded to the beat of the heart as well as the activity of single units. In one experiment he isolated the cardiac ganglion of *Limulus* and demonstrated that its rhythmic discharge consisted of a slow wave upon which axon spikes were superimposed. These slow waves of nerve impulses were maximal in the midsegments of the heart and could be greatly increased by isotonic sucrose.

Effect of Chemicals on Heart Action

Various salts affect both the heart ganglia and muscle. When heart preparations are bathed in solutions of higher or lower than normal concentrations of salts, the activity of the ganglia and the muscle tissues are depressed or primarily stimulated, respectively (Carlson, 1906b). In isotonic solutions of sodium chloride, both nervous and muscular components of the *Limulus* heart closely paralleled the regularity of normal rhythm (Carlson, 1907). Various isotonic solutions of sodium initiate a new rhythm and increase this rhythm in a quiescent nerve in the following order: $Na_2SO_4 > NaCl > NaNO_3 > NaBr > NaCNS$. At toxic levels these solutions produced an irregularity and early cessation of rhythm in the reverse order. An increase in potassium or sodium or a decrease in calcium in solutions bathing a heart preparation accelerate the gross heart rate, whereas the opposite in concentrations act in an opposite manner (Prosser, 1943). Drugs affect the heart differently before and after development of the nervous system: the beat of the embryonic heart, before innervation, is unaffected by acetylcholine, but the adult heartbeat is accelerated (Prosser, 1942). There is little effect on the automatic rhythmic activity of the ganglion cells when less-than-toxic concentrations of pH are applied to heart preparations (Chao, 1935).

Watson and Groome (1989) and Groome et al. (1994) explored the regulation of the *Limulus* heart by intrinsic modulators. These substances probably reach the heart through release from the seven pairs of cardioregulatory nerves that arise from the last three pairs of circumesophageal ganglia and from the abdominal ganglion. Several compounds apparently have a role in cardioregulation: acetylcholine (excitatory function); serotonin (inhibitory function); various amines (which increase the contractility of heart muscles); several neuropeptides (which increase the strength of heart contractions without altering the heart rate); and one or more FMRFamide-like peptides (which affect cardioacceleration).

Physical Stimuli Also Affect Heart Action

The deganglionated heart responds to pinching or to electrical stimulus (Carlson, 1904). When compression of nerves of the heart is progressively increased, the contraction of the heart muscle supplied by those nerves is gradually and progressively reduced (Garrey, 1930).

❈　　❈　　❈

While research on the structure and function of the *Limulus* heart held center stage nearly a hundred years ago, today it is the immune system that holds the greater implications for science and medicine. Researchers are continuing to seek answers to questions ranging from how the horseshoe crab protects itself from invasive enemies, particularly pathogenic microbes and multicellular parasites that penetrate its integument, to identifying the medical implications of this defense system. Chapters 12 and 13 outline the results of this highly technical research.

CHAPTER 12

INTERNAL DEFENSE AGAINST PATHOGENIC INVASION: THE IMMUNE SYSTEM

Peter B. Armstrong

The Essence of Immunity

One of the major barriers to success for any individual organism is parasitism and disease. In this context, parasites are species that draw their food resources from the live bodies of a host species, that exhibit some degree of adaptive structural and behavioral modifications to life on or in the host species, and that spend much or all of their life in association with that host species. Parasites may be unicellular or multicellular, prokaryote (see Exhibit 12.1), eukaryote, or virus. It has been estimated that more than half of all species are parasites or commensals (Price, 1980), and parasites affect the longevity and fecundity of practically every free-living organism. A highly efficient defense system that functions to protect the host from parasitic infection is especially necessary for any long-lived free-living species to enable its members to survive the numerous episodes of pathogenic challenge that they will certainly face during the period between conception and the age of reproductive maturity.

Humans are long-lived and the subjects of human medicine, immunology, and epidemiology are well-established disciplines. It is not as well appreciated that many invertebrates are similarly long-lived. In fact, the maximum life span for many invertebrate species is as long as it is for humans in spite of continual challenge from pathogens and without the benefit of medical care (Finch, 1990). The best-characterized immune system of any long-lived invertebrate is that of the horseshoe crab. This chapter outlines our present understanding of the antipathogenic defenses of the horseshoe crab, as an exemplar of the long-lived invertebrate.

The immune system is composed of the array of tissues, cells, and effector molecules that operate to limit the severity of attack by pathogenic parasites that have gained access to the internal milieu. Successful infection by a parasite is facilitated by suites of virulence factors, unique molecular attributes of the pathogen used to enable colonization and infection of the host (Mekalanos, 1992; Lantz, 1997). The immune system typically targets both the parasites themselves and the virulence factors necessary for parasitic invasion and survival in the internal milieu of the host. In higher animals, the majority of the effectors of the immune system are found in the blood or hemolymph, presumably because the blood has ready access to all parts of the body and is best prepared to concentrate defense effectors at a site of pathogenic invasion. The blood-vascular system of *Limulus* consists of a dorsal heart, a system of interconnected blood-filled spaces, and the blood plasma and blood cells (Shuster, 1978).

Most people's idea of the immune system is the adaptive immune system of vertebrates, which involves the actions of an enormous variety of antibody proteins with differing and highly specific recognition capabilities and whose diversity is dependent on the reorganization of the relevant genes in the maturing lymphocyte. The antibody-based adaptive immune system is strictly an invention of vertebrate evolution, and it is completely lacking in invertebrates. Instead, invertebrates depend entirely on an evolutionarily primitive system of immunity, the innate immune system, that is characterized by broad target recognition and that frequently is constitutively expressed. Elements of innate immunity are found also in vertebrates, where it arguably is of equal importance with adaptive immunity in prevention of parasitism and disease. As we will see, some of the elements of the innate immune systems evolved early in the history of animals and have been preserved in the evolution of diverse animal phyla. It is interesting that even long-lived invertebrates are quite competent to fend off pathogenic attack in the complete absence of the antibody-based immune system. Certainly the environments in which invertebrates find themselves are well stocked with microbes and potential eukaryote parasites. Why vertebrates seem to require antibody-based immunity and the immune systems of invertebrates operate perfectly well without this form of adaptive immunity has not received much attention.

Characterization of the immune system of *Limulus* has been facilitated by the relative simplicity of the composition of the blood-vascular system of the animal. It contains but a single type of blood cell, in contrast to the diversity of blood cell types found in most other animals. The single blood cell of *Limulus* is the granular amebocyte (Loeb, 1920; Armstrong, 1985). The fluid phase of the blood, the plasma, also is relatively simple in composition. The inorganic salts of the plasma are identical to the salt content of the ambient seawater, 0.5 M NaCl, 50 mM MgCl$_2$, 10 mM CaCl$_2$, and so on. The protein composition of the plasma is dominated by the respiratory protein, hemocyanin, which is present at 40–60 mg/mL. In *Limulus,* as with many other invertebrates, the oxygen-binding protein of the blood is in solution in the plasma, rather than sequestered in blood cells. There are only two other abundant proteins. The pentraxins are present at 1–3 mg/mL and α_2-macroglobulin is present at 0.2–1 mg/mL (Armstrong et al., 1996a). Characterization of immunity is also conferred by the ease in obtaining large quantities of blood. Blood comprises about one third of the body mass of the horseshoe crab and it is relatively easy to collect 100–200 mL of blood from a large individual (my personal best bleed is 386 mL from a single large female). If treated gently, there is excellent survival of animals following bleeding, and I routinely return them to the ocean following collection of blood.

Elements of the innate immune system identify the presence of foreign cells or potentially toxic products of invading pathogens by reacting to specific categories of microbial products, known as the pathogen-associated molecular patterns (Janeway and Medzhitov, 2002). Components at the surface of the invader, such as lipopolysaccharide (also known as endotoxin) from the outer cell membrane of Gram-negative bacteria, peptidoglycans from Gram-positive bacteria, and (1,3)-β-D-glucans from the cell wall of fungi, are strongly favored for these recognition events. Long-chain sugars are important components of these macromolecules, and sugar-binding proteins are prominent elements of the innate immune system.

As we will see, certain of the immune effectors of *Limulus* are specifically responsive to these particular microbial components. Additionally, *Limulus* has a diverse array of proteins that bind sugars and amino phosphate compounds, presumably to mark and immobilize invading microbial pathogens that display these molecules on the cell surface. Finally, proteases, enzymes that chop proteins into smaller fragments, are important virulence factors of invading unicellular and multicellular pathogens (Armstrong, 2001) and *Limulus* has a variety of protease inhibitors that function to regulate endogenous and exogenous proteolytic enzymes.

Immune System of *Limulus*

As already noted, based on cell morphology, the peripheral blood contains a single cell type, the granular amebocyte (Armstrong, 1985, 1991). In the absence of wounding or challenge with pathogens, this cell is an ovoid, nonadhesive, immotile cell, 15 to 20 μm in its longest dimension. The cytoplasm is packed with secretory granules (Figure 12.1). When bleeding occurs, or when bacteria gain access to the blood, this cell releases the contents of its secretory granules into the external environment (Armstrong and Rickles, 1982) and transforms into an adhesive cell that attaches to other blood cells and to noncellular surfaces (Armstrong, 1980) (Figure 12.2). The activated cell becomes motile and initiates ameboid locomotion if attached to planar surfaces (Arm-

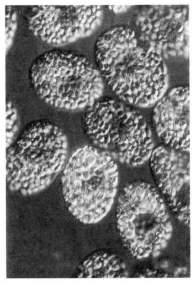

FIGURE 12.1 Granular amebocytes from *Limulus* in the unactivated state. These cells were collected directly into aldehyde fixative to prevent post-extravasation activation so as to show the discoid form of the unactivated cell. Microscopic inspection of several thousand cells prepared in this way from animals freshly collected from the wild shows a population with 100 percent of the cells with this morphology. The cytoplasm is packed with the secretory granules.

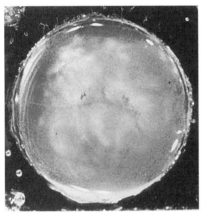

A

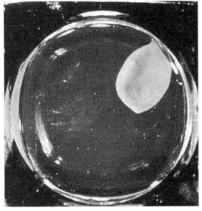

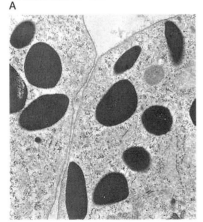

B

FIGURE 12.2 Aggregation of the *Limulus* amebocyte following bleeding. (A) Adhesive mat of "amebocyte tissue" that forms when amebocytes are collected under sterile conditions directly into sterile embryo dishes. Top left dish, the cells settle to the bottom of the dish and aggregate. Top right dish, after several hours, the mat of amebocyte tissue contracts in a manner analogous to the contraction of the activated blood platelets contained in the mammalian blood clot. (B) Electron micrograph of amebocyte tissue showing the area of attachment of two adherent amebocytes.

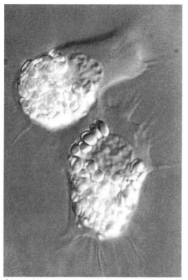

A

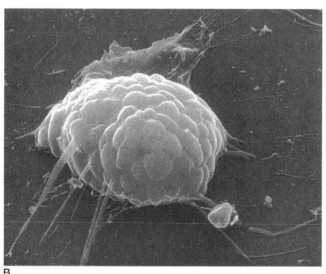

B

FIGURE 12.3 Motility of the *Limulus* amebocyte. The extravasated amebocyte becomes motile when placed in contact with a solid surface, such as a microscope coverglass. The cells extend hyaline pseudopods, P, and filopods, F, in the direction of locomotion and stretch out retraction fibrils, R, which are extended portions of the cell that terminate in residual adhesive contacts with the substratum, at the trailing end of the cell. (A) Light micrograph of living cells migrating on a microscope coverglass. (B) Scanning electron micrograph of a similar cell.

strong, 1977, 1979) (Figure 12.3) and the cells undergo profound morphological transformation (Figure 12.4). The granular amebocyte is an important contributor to immunity, functioning as the repository of the blood clotting system and of a menagerie of antimicrobial effector molecules (Iwanaga and Kawabata, 1998; Iwanaga, 2002).

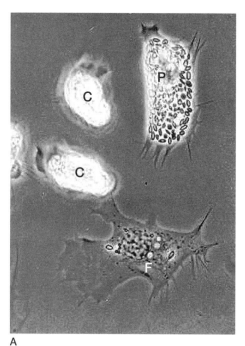

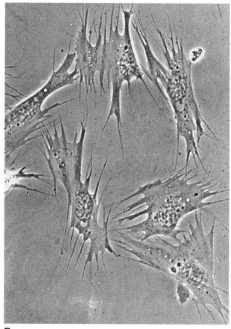

A   B

Immune Attack on the Pathogen

An important strategy to thwart pathogenic invasion is the direct attack of the host's immune system on the invading pathogen. This is well illustrated in *Limulus* by systems that immobilize the pathogen at the sites of initial entry. These systems are tasked with the outright cytolytic destruction of the pathogen and the phagocytic uptake and degradation of invading microbes by the blood cells. These systems are mediated by a combination of agents found in the plasma, in the secretory products of the blood cells, and by the blood cells themselves.

Immobilization of Invading Parasites

By immobilizing and containing the invading parasites to a restricted location close to the site of initial entry, the host prevents their systemic dispersal to all parts of the internal milieu and buys time for other components of the immune system to kill or inactivate the foreign cells. The horseshoe crab accomplishes immobilization by entrapping invading microbes in the fibers of its blood clot. In mammals, the clotting system involves the polymerization of the plasma protein, fibrinogen, into the fibrillar elements of the fibrin clot (Furie and Furie, 1992). In

FIGURE 12.4 *Limulus* amebocytes adopt a variety of different morphologies after they are removed from the animal and allowed to attach to an adhesive surface. (A) Cells in different degrees of degranulation and flattening on the coverglass. The two compact cells, C, are highly motile and have not yet initiated degranulation; the partially flattened cell, P, is less migratory but still contains its complement of cytoplasmic granules; and the highly flattened cell, F, has degranulated completely, save for two remaining granules. (B) Following complete degranulation, cells may adopt bizarre morphologies.

TABLE 12.1 Elements of the coagulin clotting system.

Component	Functional Characterization	References
Coagulogen	Structural protein of the clot	Tai et al., 1977; Takagi et al., 1979
Proclotting enzyme	Proteolytic activation of coagulogen	Muta et al., 1990
Factor B	Proteolytic activation of proclotting enzyme	Muta et al., 1993
Factor C	Proteolytic activation of factor B	Muta et al., 1991
Factor G	Proteolytic activation of proclotting enzyme	Seki et al., 1994
Transglutaminase	Isopeptide cross-linking	Tokunaga et al., 1993a; Tokunaga et al., 1993b

Limulus the soluble precursor protein, coagulogen, undergoes proteolytic modification to coagulin, which then polymerizes to the fibrils of the extracellular clot (Table 12.1) (Holme and Solum, 1973; Tai et al., 1977; Takagi et al., 1979; Srimal et al., 1985). Both the fibrin clot of humans (Rotstein, 1992) and the coagulin clot of *Limulus* immobilize invading bacteria. Coagulin-immobilized bacteria are held so tightly that they lack even Brownian motion (Bang, 1979).

Two of the principal requirements of any clotting system are that it be inactive in the intact animal and that it be responsive to bleeding by a rapid activation that is spatially restricted to the wound site. In the intact animal coagulogen, the soluble precursor of the clot-forming protein, coagulin, is sequestered in the cytoplasmic granules of the blood cells (Murer et al., 1975). Exocytosis of the blood cell granules (Figures 12.5 and 12.6) releases coagulogen into the external milieu. Inadvertent

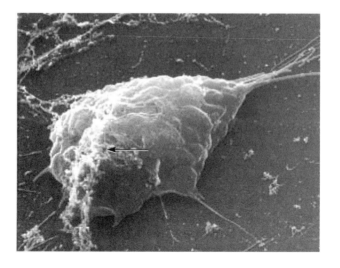

FIGURE 12.5 Degranulation of cultured blood cells is accompanied by polymerization of a coagulin clot, even in a sterile, lipopolysaccharide-free culture environment. The flattened, partially granulated cell in this scanning electron micrograph is associated with a sparse layer of fibrillar material that presumably is coagulin clot. Cultures that have been incubated for a sufficient time for complete degranulation are completely covered by this layer of clotted protein.

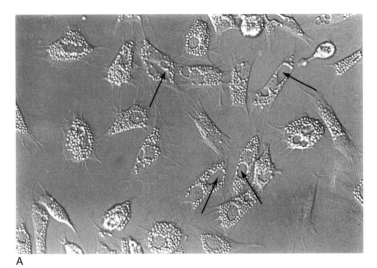

A

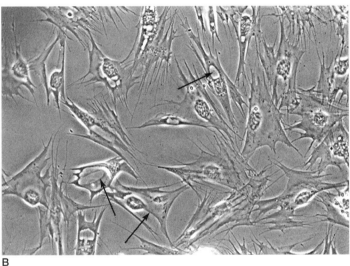

B

FIGURE 12.6 Degranulation of cultured amebocytes can be observed visually in cells cultured on microscope coverglasses. (A) These flattened cells are in the early stages of degranulation, with sites of degranulation of individual cells seen here as islands of granule-free cytoplasm (arrows). (B) Degranulation is complete and sites of massive degranulation remain in some cells as internal lacunae (arrows).

clotting is regulated by two features of the system; clotting requires both the release of coagulogen from the exocytotic granules and the appropriate proteolytic modification of the soluble coagulogen protein, because only the proteolytically modified form is able to polymerize into the fibrils of the clot.

To date, four proteases have been identified that function in activation of coagulogen (see Table 12.1). All of these are released as inactive proenzymes and remain inert in the absence of microbial pattern recognition molecules. Two proteolytic activation cascades have been identified for the activation of proclotting enzyme, which is the ultimate protease in the activation pathway and the protease responsible for proteolytic activation of coagulogen. The two activation cascades are activated by different classes of pathogen-associated molecular pattern molecules (Figure 12.7). Lipopolysaccharide, an important constituent of the outer membrane of Gram-negative bacteria, activates the factor C arm of the pathway, and $(1,3)$-β-D-glucan from the cell wall of fungi activates the factor G arm (Iwanaga et al., 1992). The clotting system responds to these agents as signals, respectively, for infection by Gram-negative bacteria or by fungi. In addition to these microbially activated coagulogen clotting pathways, the clotting of coagulogen can be activated under sterile conditions, in the absence of microbes, when living, degranulated blood cells are present (Armstrong and Rickles, 1982). The enzymatic basis for activation of clotting in the absence of lipopolysaccharide and $(1,3)$-β-D-glucans is unknown.

The endotoxin-activated clotting system has received extensive attention, in part because it is the basis for the commercially important LAL (Limulus Amebocyte Lysate) test for endotoxin. This is a three-enzyme cascade, in which proclotting enzyme is proteolytically activated by factor B, which in turn is activated by a proteolytic cleavage administered by activated factor C (see Table 12.1). Pro-factor C is activated by the presence of endotoxin. This involves the binding of endotoxin to high-affinity sites of the pro-factor C at the amino-terminal end of the protein (Tan, Ng, Yau, Chong, Ho, and Ding, 2000). Recombinant protein containing these domains binds endotoxin with high affinity, which may have application in the removal of endotoxin from biological samples and in the protection of patients from LPS-induced

FIGURE 12.7 Clotting of the blood of *Limulus* involves conversion of a soluble protein, coagulogen, into coagulin, a form that polymerizes to construct the insoluble fibrillar elements of the clot. Conversion is mediated by proteolytic modification of coagulogen by the protease, clotting enzyme. Clotting enzyme is initially present in an inactive form, proclotting enzyme, that itself requires proteolytic modification of activation. Activation of proclotting enzyme is effected by at least two protease cascades that themselves are activated by products of microbes; lipopolysaccharide from Gram-negative bacteria or $(1,3)$-β-D-glucans from fungi.

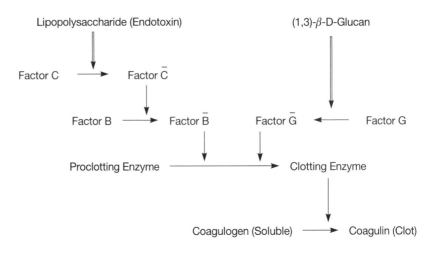

THE IMMUNE SYSTEM

lethality (Tan, Ho, and Ding, 2000). Once endotoxin has bound to pro-factor C, this protein self-activates and initiates activation of the clotting cascade and the ultimate production of the clot. Detection of a clot or of this system of proteases that initiate clotting is an indicator of the presence of endotoxin by the LAL test.

Cytolysis of the Invading Pathogen

Cytolytic systems contribute to immunity by the direct destruction of foreign cells that enter the internal spaces of a host species. The best-characterized cytolytic systems in *Limulus* are a plasma-based system involving the protein, limulin, and a cell-sequestered system using a family of small peptides, the tachyplesins and related peptides. Limulin is a member of the pentraxin protein family, which boasts prominent members of the innate immune systems of vertebrates (Pepys and Baltz, 1983) and arthropods (Nguyen et al., 1986a; Nguyen et al., 1986b; Tennent et al., 1993; Iwaki et al., 1999). *Limulus* has multiple pentraxins, of which limulin (Armstrong et al., 1996b) and *Limulus* forms of C-reactive protein and serum-amyloid protein (Shrive et al., 1999) are the best characterized. Limulin is responsible for a potent cytolytic activity in the plasma that has been assayed by its ability to lyse sheep erythrocytes (Armstrong et al., 1993). In this context, mammalian red blood cells have served in the characterization of a number of cytolytic systems as a surrogate "foreign cell" because they are readily available and their cytolysis is readily quantified by measurement of the appearance of hemoglobin in the bathing saline. Limulin is necessary for hemolysis by *Limulus* plasma because removal of limulin abolishes the hemolytic activity of the plasma. Hemolytic activity can be restored to limulin-depleted plasma by the addition of purified limulin. Limulin is sufficient for hemolysis because purified limulin is hemolytic in a Ca^{+2}-dependent manner at 1–3 nM in the absence of other plasma components. Although the other pentraxins of *Limulus* are not hemolytic (Armstrong et al., 1996b), they show potent membrane-lytic activities for artificial phospholipid membranes. Based on these observations, it is suggested that all of the *Limulus* pentraxins are capable of lysing foreign cells, including the cells of potentially pathogenic microbes.

Cytolytic peptides are a widespread class of antimicrobial defense effectors and are found throughout the animal kingdom (Gabay, 1994). The peptide antibiotics of *Limulus* are composed of the tachyplesins and related molecules. These agents contain seventeen or eighteen residues and are rich in basic and hydrophobic amino acids. In their native state, the peptides are folded to form structures with the basic residues on one face of the folded peptide and the hydrophobic residues on the

TABLE 12.2 Major categories of immune defense proteins and peptides released from amebocytes.

Category of Agent	Examples	Functional Characterization	Reference
Lipopolysaccharide (LPS)-binding proteins	Anti-LPS factor (LALF); tachylectin-4; L6	Bacterial killing; neutralization of LPS	Muta et al., 1987; Saito et al., 1997; Saito et al., 1995
Lectins	L10	Binding to carbohydrates of microbial surface	Okino et al., 1995
Bactericidal proteins	Big defensin, Tachycitin, LALF	Killing of bacteria and fungi	Kawabata et al., 1997; Kawabata et al., 1996
Bactericidal peptides	Tachyplesin; polyphemusin	Cytolysis of bacteria	Nakamura et al., 1988; Miyata et al., 1989
Protease inhibitors	LICI-1; LICI-2; LICI-3, L-cystatin, α_2-macroglobulin	Inactivation and clearance of proteases	Miura et al., 1994; Miura et al., 1995; Agarwala et al., 1996a; Agarwala et al., 1996b; Armstrong et al., 1990; Ikawi et al., 1996

opposite side (Iwanaga et al., 1994). This organization facilitates their interactions with cell membranes. The cytolytic peptides are contained within the secretory granules of the blood cells and are released by exocytosis (Table 12.2).

Membrane-lytic proteins and peptides operate by two different processes for the disruption of the integrity of the plasma membrane. The barrel-stave transmembrane pore mechanism involves the establishment of pores created by the physical insertion of multiple units of the cytolytic peptide or protein across the lipid core of the membrane. Each unit is oriented to present a hydrophilic face to the central channel and a hydrophobic face to the fatty acid chains of the membrane. Limulin apparently operates by this mechanism (Swarnakar et al., 2000). Channel-forming agents provoke cytolysis when water flows into the cell in response to the high internal osmotic pressure generated by the high concentration of macromolecules of the cytoplasm. Macromolecules of a sufficient size in the bathing medium protect cells reversibly from cytolysis. Protection results from the experimental establishment of a concentration of external macromolecules of a size that is larger than the channel pore size, which is equal to the concentration of macromolecular osmolites within the cell, thereby equalizing the osmotic pressure gradient and protecting the cell from osmotic rupture. The minimum size for the external osmolite that is protective yields an estimate of a limulin pore size of 1.7 nm (Swarnakar et al., 2000).

A second mode of action is the carpet model, in which the membrane-lytic agent coats the surface of the plasma membrane and causes the lipid membrane to disrupt into tiny beads, or micelles, resulting in the formation of transient discontinuities and general disruption of membrane integrity. This process, which is found for a variety of antimicrobial peptides, is initiated by a surface binding of the peptide to establish a continuous carpet that is able to alter lipid packing to an extent sufficient to cause the bilayer to break into small lipid micelles (Shai, 1999). The tachyplesins apparently operate by the carpet model. In solution, these peptides are amphipathic with positively charged and hydrophobic faces. During membrane lysis, the peptides lie parallel to the membrane surface with the hydrophobic faces penetrating only slightly into the lipid core (Matsuzaki et al., 1993).

Phagocytosis

One of the most important antimicrobial reactions of the blood cells of higher animals is phagocytosis of small particles and encapsulation of foreign bodies too large to be phagocytosed (Ratcliffe et al., 1982). In the latter reaction, numbers of blood cells attach to the foreign body and enclose it in a many-layered cellular capsule (Salt, 1970). Both phagocytosis and encapsulation require the ability to discriminate between self and non-self. Surprisingly, for so well studied an organism as *Limulus*, little is known of the possible import of these reactions for defense. The amebocyte is definitely capable of phagocytosis (Armstrong and Levin, 1979), but its phagocytic capabilities appear to be relatively feeble. In studies of the phagocytic processing of carbonyl iron particles by *Limulus* amebocytes *in vitro*, I seldom found more than two to five particles in any one cell, and most cells even in our most active preparations contained no particles. It is possible that in this and in the few other studies of phagocytosis in *Limulus*, optimal conditions were not utilized, but present evidence does not suggest that a generalized phagocytic attack by blood cells is very important.

Lectins

Recognition of foreign cells, and a variety of their products, involves binding of immune effectors to the carbohydrates of the foreign cells and molecules. Lectins are proteins that bind, with varying levels of selectivity, the terminal sugars of oligosaccharide chains. These potentially function as recognition receptors for the carbohydrates of foreign cells and their molecular products. The plasma and the secreted products of the horseshoe crab blood cells contain a diverse array of lectins (Kawabata and Iwanaga, 1999). The previously described plasma lectin, limulin, is responsible for the plasma hemolytic system. Limulin is capable of mediating the lysis of foreign cells bearing sialic acid residues at the termini of cell surface oligosaccharide chains. The other members of the pentraxin class of proteins, of which limulin is a member, may have similar activities for those cells to which they show binding capabilities.

Several of the lectins from the secretory granules of the blood cells bind to bacteria and some show antimicrobial activities (see Table 12.2). Tachylectin-1 binds bacterial lipopolysaccharide and shows antibacteriological activity for Gram-negative bacteria (Kawabata et al., 2001). The immune defense functions of the several other lectins of *Limulus* are less well characterized, but they may operate in facilitating the phagocytic recognition and clearance of target cells and additionally may participate in cross-linking microbes to the coagulin clot. In addition to anticellular activities, the lectins may operate in the endocytotic clearance of glycoconjugates secreted by invading pathogens.

Actions of the Immune System against Secreted Products of Pathogens

Pathogens employ suites of secreted virulence factors to assist in the colonization of the host. These may contribute to the disease process itself (Mekalanos, 1992; Lantz, 1997). Additionally, certain of the products of microbial parasites, such as the lipopolysaccharides of Gram-negative bacteria, are both structural elements of the microbe and have potent toxic actions on the host. An important function of the immune system is to inactivate, neutralize, and clear these toxic products with the aim of reducing the success of the infecting parasite and of reducing the severity of the disease symptoms created by that infection.

Lipopolysaccharide-Binding Peptides and Proteins

Lipopolysaccharide (LPS) is the principal agent responsible for disease associated with infection by Gram-negative bacteria. A diverse collection of the bioactive peptides and proteins secreted from the *Limulus* blood cells react with lipopolysaccharide. Several of the blood cell lectins bind to sugar residues of LPS (see Table 12.2). As noted previously, binding of LPS by factor C of the clotting cascade initiates the blood-clotting cascade by initiating its autoproteolytic activation (see Figure 12.7). The core lipid A portion of endotoxin is a target for binding of the blood cell protein, anti-LPS factor (LALF). This 101-residue basic protein (Muta et al., 1987) binds and inactivates LPS (Warren et al., 1992) and shows bacteriolytic activity for Gram-negative bacteria. LALF and shorter segments of the intact protein are under investigation for the neutralization of LPS in clinical settings, in which the management and reversal of septic shock syndrome would represent a major therapeutic advance (Kloczewiak et al., 1994; Ried et al., 1996).

Protease Inhibitors

Proteases free in the blood and tissue spaces have the capacity for considerable mischief. Endogenous proteases left over from blood clotting or the activities of inflammatory blood cells contribute to a number of diseases in humans. Proteases are also important virulence factors for microbial and multicellular parasites (Armstrong, 2001). In response, animals have evolved a variety of protease inhibitors capable of inhibiting this class of enzymes (see Table 12.2). The protease inhibitors are of two basic functional classes, the active-site inhibitors, which bind to and inactivate the active site of the target protease, and the α_2-macroglobulin class of inhibitors, which physically entrap the target proteases in an internal pocket in the α_2-macroglobulin protein. Typically the active site of the trapped enzyme is unaffected and can hydrolyze synthetic amide and ester substrates small enough to diffuse into the α_2-macroglobulin cage, but the enzyme loses its ability to react with protein substrates that are too large to enter the cage. The *Limulus* immune system features examples of both classes of inhibitors.

The α_2-Macroglobulin Family of Proteins

The α_2-macroglobulin family of proteins includes complement components C3, C4, and C5 (Tack, 1983; Sottrup-Jensen, 1987, 1989), which are important elements of the vertebrate innate immune system (Law and Reid, 1995), and α_2-macroglobulin homologues found in the plasma of arthropods (Quigley and Armstrong, 1983; Armstrong et al., 1985; Levashina et al., 2001), mollusks (Armstrong and Quigley, 1992; Bender et al., 1992; Thogersen et al., 1992), and all classes of vertebrates (Starkey and Barrett, 1982).

The best-characterized function of the α_2-macroglobulins of vertebrates (Van Leuven, 1984) and arthropods (Melchior et al., 1995) is the binding and clearance of proteases from the internal milieu. Most prokaryote and eukaryote parasites that have been adequately studied use proteases as essential elements of their armamentarium of virulence factors for invasion and survival. Proteases function in invasion of the parasite across the integuments and in evasion of host immune effectors of the host; and, in some parasites, they operate directly as toxins. In this context, it has been suggested that one important function for protease inhibitors is the inactivation and clearance of the proteases of invading parasites (Armstrong, 2001), as a strategy to reduce pathogen survival and virulence. α_2-macroglobulin is uniquely adapted to perform this function because, unlike the other classes of protease inhibitors, α_2-macroglobulin is able to bind and inactivate proteases of all classes and enzymatic mechanisms, and from all sources (Quigley and Armstrong, 1983; Quigley and Armstrong, 1985). The promiscuous character of α_2-macroglobulin makes it an ideal scavenger for the diverse array of proteases produced by the variety of potential parasites that an animal might encounter in its life.

The basic function of α_2-macroglobulin in protease clearance is to convey bound proteases to a receptor-mediated endocytotic clearance system. In this manner, the proteases (Figure 12.8A) and protease-re-acted α_2-macroglobulin (Figure 12.8B) are cleared from the blood of the horseshoe crab within half an hour. The blood cells appear to mediate the clearance, because labeled protease-reacted α_2-macroglobulin could be isolated from detergent extracts of the blood cells at the very stages that protease-reacted α_2-macroglobulin was being cleared from the plasma. In mammals, the cell surface receptor involved in binding of protease-reacted α_2-macroglobulin to macrophages and other endocytic cells has been identified as low-density, lipoprotein, receptor-related protein-α_2-macroglobulin receptor (LRP/α_2M-R), a member of the low-density lipoprotein receptor family (Strickland et al., 1990; Kristensen et al., 1990). In this pathway, α_2-macroglobulin functions as a kind of opsonin that promotes the binding and endocytosis of a molecularly diverse array of endopeptidases to cell surface receptors. The receptor recognizes as its cognate ligand a structure unique to the protease-reacted form of α_2-macroglobulin rather than the protease itself. Following binding, α_2-macroglobulin and its cargo of bound protease are degraded in secondary lysosomes (Melchior et al., 1995).

The internal defenses against microbial pathogens can be divided into the adaptive responses and the innate immune responses. The adaptive immune responses involve the activities of T- and B-lymphocytes, which identify foreign molecules with specialized receptor proteins on the cell surface. During lymphocyte maturation, the genes that encode these receptors undergo multiple rounds of chopping and splicing of the DNA such that each individual maturing lymphocyte ultimately expresses its own unique form of the receptor protein with the ability to bind a limited and unique target macromolecule. The entire population of lymphocytes, each with its own unique target recognition capability, allows the animal to respond to a large array of diverse molecules,

THE IMMUNE SYSTEM

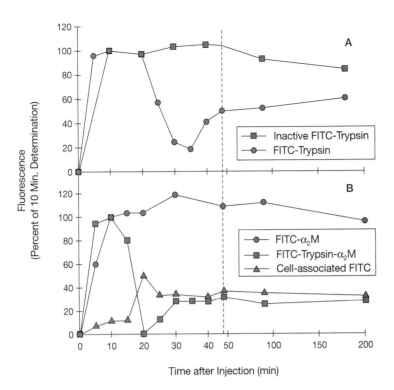

FIGURE 12.8 Role of α_2-macroglobulin in the clearance of proteases from the blood of *Limulus*. In this experiment, trypsin or *Limulus* α_2-macroglobulin was labeled with the fluorescent dye FITC (flourescein isothiocyanate), which becomes covalently linked to lysine residues in the proteins. The labeled protein was then injected into the heart and blood was collected at multiple later times from a distant site (the blood spaces of the legs). After removal of the blood cells, the amount of fluorescence present in the plasma (A, B) or with the blood cells themselves (B) was determined with a fluorometer. This allowed quantification of the amount of fluorescent protein in the blood at various times after its introduction. The abscissa is the amount of fluorescence, expressed as percentage, and normalized to the amount measured in the plasma at 10 min. after injection, a time of complete mixing of injected protein and plasma. Trypsin (A) and trypsin-conjugated α_2-macroglobulin (B) are cleared with similar kinetics: an early phase of increasing concentration as the labeled protein is pumped out of the heart and into the peripheral circulation is followed by a period of rapid clearance from the blood, followed by a perod of reappearance of fluorescence in the blood. Clearance is specific for enzymatically active trypsin (inactivated trypsin is not cleared [A]) and for protease-reacted α_2-macroglobulin (unreacted α_2-macroglobulin is not cleared [B]). Fluoresceinated protein accumulates in the blood cells ("Cell-associated FITC," B) when it is cleared from the plasma ("FITC-Trypsin-α_2M," B), suggesting that the blood cells are taking up the trypsin-conjugated α_2-macroglobulin. The fluorescence that reappears in the blood at later times (t > 40 min, A) is present on low molecular mass peptides because this material readily passes through microporous membranes that retain all proteins larger than 10,000 molecular mass. These observations are interpreted to indicate that α_2-macroglobulin is the major pathway for clearance of proteases from the blood, that the blood cells participate in clearance, and that the fate of the protease-α_2-macroglobulin complex is proteolytic degradation, with the proteolytic fragments being released back into the blood.

many of which are important constituents of invading parasites. Adaptive immune systems have the characteristics of highly specific target recognition, amplification of the response upon challenge, and immunologic memory.

In contrast, the innate immune systems show more broadly defined target recognition capabilities and lack immunologic memory, and the effectors may be constitutively expressed. Examples of innate immune systems and mediators include the cationic antimicrobial peptides and proteins of the secretory blood cells (for example, the defensins of insects and mammals; Dimarcq et al., 1990; Gabay and Almeida, 1993; Gabay, 1994), the several inducible antibacterial peptides and proteins of insects (Dunn et al., 1994; Hoffmann and Hoffmann, 1990), and the tachyplesins of *Limulus* (Nakamura et al., 1988; Miyata et al., 1989), lectins (Sastry and Ezekowitz, 1993; Kawabata and Iwanaga, 1999), the complement system of vertebrates (Law and Reid, 1995), and the varied blood clotting systems employed by different animals (Bohn, 1986; Iwanaga et al., 1992).

Although some of these systems are restricted to particular taxa, others are found in phylogenetically diverse groups of animals. These latter must have appeared early in animal evolution and can be presumed to be important for survival when they can be shown to have been retained during the evolution of a diverse array of species with very different body plans, habitats, and basic physiologies. The α_2-macroglobulins and the pentraxins are in fact present, often at high abundance, in the plasma of animals as diverse as humans and horseshoe crabs. α_2-macroglobulin functions in protease clearance both in mammals and in *Limulus*. The best characterized of the *Limulus* pentraxins, limulin, functions as the principal cytolytic protein of the plasma. Undoubtedly other functions will be discovered for both of these proteins in the identification and clearance of potentially harmful cells and molecules that have entered the internal milieu.

Typically, the innate immune system has been under-appreciated and less thoroughly studied than the adaptive immune system. The understanding of immunity in invertebrates is likely to find practical application in improved methods for veterinary care of invertebrates important for aquaculture and may prove useful for the development of methods for biological control of invertebrates that are agricultural pests or vectors for human diseases. At least some of the many pathogens that are responsible for vector-borne diseases are harmful for their invertebrate intermediate hosts as well as for humans. An improved understanding of the immune defenses that the invertebrate hosts mount against these human pathogens may offer strategies for manipulation of the intermediate host to render it immune to infection by the parasite for which it is the transmission agent for infection of humans. And *Limulus* will continue to play an important role in this enterprise, possessing, as it does, the best-characterized immune system for any long-lived invertebrate.

EXHIBIT 12.1

amphipathic Regions of a protein that show adjacent patches that are highly soluble in water, due to the presence of electrically charged amino acids, and other patches that are water-insoluble and lipid-soluble, due to the presence of amino acids with hydrophobic groups.

coagulogen A soluble form of the protein that is the structural protein of the blood clot in *Limulus*. Coagulogen is contained within secretory granules of the blood cells and is released when these cells release the granule contents. Coagulogen is then modified by a protease, clotting enzyme, and the modified form of the protein is now able to polymerize into the fibrils of the blood clot.

cytolysis The killing of a cell due to damage inflicted on the cell membrane. The cytolytic destruction of invading pathogens is an important process in immunity because it eliminates the parasite before it has the opportunity to cause disease.

cytolytic See Cytolysis.

defensins A class of peptides that have antimicrobial activities and that are elements of the innate immune systems of vertebrates and many invertebrates.

eukaryote Member of the large class of organisms whose cells have a well-structured nucleus. This class includes all animals, plants, and fungi.

exocytosis Process by which cells secrete the contents of secretory vesicles. Many cell types contain membrane-bound secretory vesicles in the cytoplasm that, upon the receipt of the proper signal to the cell, fuse the vesicle membrane with the cell's plasma membrane and discharge the vesicle contents into the external environment of the cell.

fluoresceinate The attachment of fluorescent dye molecules to larger macromolecules, usually proteins. Because fluorescent dyes can be seen even at very low concentrations, fluoresceination enables the investigator to see the labeled proteins even in situations in which the protein is present at low concentration. Fluorescence microscopy is one of the preferred methods for localizing dye-labeled proteins by microscopic inspection of biological samples.

glycoconjugates Macromolecules, such as proteins or lipids, that have covalently bound sugars. Many extracellular proteins are heavily glycosylated, with multiple long-chain sugars attached to the protein backbone.

hemolytic The cytolytic destruction of red blood cells. A widely used strategy to identify effector molecules of the immune system with cytolytic activity is to screen for molecules with a hemolytic action. The mammalian red blood cell is used here as a model foreign cell whose lysis is easily measured by the release of its content of hemoglobin into the bathing medium. The ultimate goal is to find molecules that can lyse invading pathogens, and experience has shown that the ability to lyse red blood cells is often a direct reflection of the ability of the molecule to lyse foreign cells in general, including the cells of invading pathogens.

hydrophobic Insoluble in water. Molecules that are hydrophobic are insoluble in an aqueous environment but are soluble in the lipid core of cell membranes.

isopeptide bonding Covalent linking of two protein chains. Isopeptide bonding is used to impart mechanical strength to the fibrillar blood clot of mammals and is used to link proteins to the fibrils of the blood clot of *Limulus*.

lipopolysaccharide The toxic product of Gram-negative bacteria and the agent of disease caused by Gram-negative bacteria. Lipopolysaccharide (LPS) is a ubiquitous component of the outer leaflet of the outer membrane of Gram-negative bacteria. The membrane anchor of LPS is lipid A, a central phosphodisaccharide unit that is attached to multiple fatty acid chains. The central phosphodisaccharide unit is also attached to a long oligosaccharide chain containing the novel sugar, 3-deoxy-D-manno-octulosonic acid (KDO).

lymphocyte Class of cells of the vertebrate immune system that use antibodies or antibody-like molecules in their several roles in immunity.

lymphopoietic The generation of lymphocytes by proliferation of stem cells.

α_2-**macroglobulin** A protein of the innate immune system that binds all classes of proteases and that serves as an agent for the removal of proteases introduced by invading pathogens.

micelles Aggregates of lipid molecules. Because lipids have long hydrophobic fatty acid chains, they tend to aggregate when placed in an aqueous environment so as to allow the fatty acid chains to associate with one another and to escape the aqueous environment.

osmolite Molecule in solution that contributes to the osmotic pressure of that solution.

pentraxins Class of plasma proteins that play important roles in immunity in vertebrates and in *Limulus.* In *Limulus,* the best-characterized pentraxin is limulin, which is responsible for the hemolytic actions of *Limulus* plasma.

peptidoglycans Complexes of the disaccharide, N-acetyl glucosamine-N-acetyl muramic acid, and amino acids (usually L-alanine, D-alanine, diamino pimelic acid, and D-glutamic acid) that are cross-linked to constitute an important structural element of the cell wall of eubacteria. Peptidoglycans are recognized by elements of the immune system as indicators of the presence of bacteria.

phagocytosis Process by which particles are engulfed by a cell. During phagocytosis, processes of the cell surround the particle and eventually the membrane contacting the particle reorganizes to establish a vacuole positioned within the cell and containing the particle. In some higher animals, phagocytosis is an important process for the killing and destruction of microorganisms. Phagocytosis is also a mechanism for capture of food particles for many protozoans and for the digestive cells of some multicellular animals.

phospholipid The major class of lipids of cell membranes. Phospholipids most frequently contain two long fatty acid chains, which are positioned in the interior of the membrane, linked through glycerol phosphate to one of a variety of polar head-group molecules, which are positioned at the surface of the membrane.

polymerization Assembly of subunits to form a larger polymer. Linkage of subunits may be covalent, as is the linkage of monosaccharides in a larger oligosaccharide chain or of amino acids in a protein, or may be noncovalent, as is the linkage of coagulin protein subunits to form the fibrillar structures of the *Limulus* blood clot.

polyphemusin A class of antimicrobial peptides found in arthropods.

prokaryote Member of the class of organisms whose cells lack a well-defined nucleus. The prokaryotes are composed of the eubacteria and the archeobacteria.

proteolytic Ability to sever the peptide bonds that link amino acids in a protein. Proteolytic enzymes, proteases, play important roles in the invasion of pathogens into potential hosts and endogenous proteases of the host play essential roles in activating various host immune defense molecules. An example of this latter function is the involvement of the protease, clotting enzyme, in the proteolytic modification of the soluble protein, coagulogen, so it can self-polymerize to form the coagulin fibrils of the *Limulus* blood clot.

tachycitin An antimicrobial protein from *Limulus.*

tachylectin A lectin, a class of sugar-binding proteins, from *Limulus.* The tachylectins are proposed to play roles in the recognition and elimination of microbes that have invaded *Limulus*.

tachyplesins A class of antimicrobial peptides from *Limulus*.

transglutaminase An enzyme that catalyzes the formation of isopeptide bonds linking the amino acid glutamine of one protein with the amino acid lysine of a second protein. Transglutaminase-catalyzed cross-linking is active in the blood clots of vertebrates and *Limulus*.

CHAPTER 13

CLOTTING CELLS AND *LIMULUS* AMEBOCYTE LYSATE: AN AMAZING ANALYTICAL TOOL

Jack Levin, H. Donald Hochstein,
and Thomas J. Novitsky

Levin and Bang (1964a) discovered a substance from the blood cells of *Limulus* that clots the plasma when exposed to gram-negative bacteria. Almost immediately after Levin and Bang (1968) subsequently prepared *Limulus* amebocyte lysate (LAL) from washed blood cells, the significance of this substance became apparent, sparking avid research on the mechanisms involved in its activity. The intensity of this research resulted in more information on the blood cells and medical uses of horseshoe crabs than on any other aspect of their biology. To review this vast research field, we divided the subject into three parts. These parts, in sequence, describe the many steps in research required to bring a new biological material to market—from the discovery (Levin), through the development of standards and regulations of a biologic (Hochstein), to the commercialization (Novitsky) of LAL made from the blood cells of *Limulus polyphemus.*

Discovery of a Clotting Mechanism in the Blood Cells

Perhaps the earliest description of the coagulation of *Limulus* blood was that of W. H. Howell (Figure 13.1A) of Johns Hopkins University in 1885. In a report in the *Johns Hopkins University Circulars,* he concluded: "The striking resemblance in the method of union of the corpuscles to form the clot, in mammalian blood and the blood of *Limulus,* for in-

FIGURE 13.1 Pioneers in the study of horseshoe crab amebocytes. (A) William H. Howell in 1903. *Courtesy of the Archives of the Johns Hopkins University School of Medicine and Hospital.* (B) Leo Loeb. From *Biological Memoirs 35:205–251, 1961. New York: Columbia University Press, with permission of the National Academy of Sciences.*

stance, and the close similarity in chemical properties of the fibrin produced, suggest that the formation of fibrin in the two cases may result from an essentially similar series of changes." This was a remarkable insight because, as our knowledge of *Limulus* coagulation has advanced, we have learned just how closely it resembles mammalian clotting.

In the same decade, the Marine Biological Laboratory (MBL) was established in 1888 at Woods Hole, Massachusetts. This laboratory was near breeding grounds of *Limulus* and so the species was common in the area. Among the early studies, those by Leo Loeb (Figure 13.1B) on the blood cells and coagulation were classics. The American horseshoe crab was soon recognized as an animal well suited to morphological and physiological research, which has long been associated with the Marine Biological Laboratory. The role of *Limulus* in immunological studies began when H. Noguchi (1903), a young professor of pathology at the University of Pennsylvania, injected horse red blood cells into *Limulus* and found that the substance produced was a powerful agglutinin.

Discovery of the Activation of Blood Coagulation in *Limulus* by Gram-Negative Bacteria

Frederik Bang began his remarkable study of clotting and immunology in *Limulus* at the MBL. His investigation of how horseshoe crabs and other marine invertebrates responded to injection of bacteria arose out of his medical interest in immunity and from his speculation that species of ancient origin might reveal primitive immunological functions.

During the summers of 1950 and 1951, taking a cue from Noguchi's report on the reaction of *Limulus* to the injection of equine red cells, Bang injected various bacteria into the circulatory system of horseshoe crabs. No agglutinins formed, but a ubiquitous marine pathogen, *Vibrio*, a gram-negative bacterium (GNB), produced an infection in the crabs. While this pathogen grew rapidly in juvenile horseshoe crabs, it did not invariably kill them. However, large doses injected into adult crabs were lethal. Bang's (1953) landmark study revealed the following:

- The infection produced intravascular clotting.
- GNB and a heat-stable extract of these bacteria were equally effective in producing coagulation.
- Gram-positive bacteria failed to promote a response (Bang, 1956).

Subsequent studies by Levin and Bang of *Limulus* indicated that a substance from the blood cells was extremely sensitive to endotoxin. This discovery led to medical and commercial applications of the substance, *Limulus* amebocyte lysate (Levin, 1985; Levin, 1987).

Collaboration between Jack Levin and Frederik Bang

Bang put this investigation aside until 1963, when he discussed his observations on *Limulus* blood with C. Lockard Conley, director of the Hematology Division at The Johns Hopkins University School of Medicine and Hospital, where Bang was also a faculty member. They decided that further investigation by a hematologist would result in an effective collaboration (Figure 13.2). Jack Levin, a Fellow in hematology, at that time was studying the Shwartzman reaction, a mammalian clotting phenomenon associated with GNB. The first experiments of the collaboration, beginning in the summer of 1963 at the MBL, were designed to learn more about *Limulus* coagulation and to determine similarities between amebocytes and platelets.

Levin and Bang (1964a, 1964b) quickly learned that cell-free plasma from *Limulus* would not clot, but studying coagulation proved difficult because samples of blood often clotted spontaneously. Because of his earlier work with bacterial endotoxin, a component of the cell wall of all GNB, Levin suspected bacterial contamination of the blood samples. When blood samples were drawn into sterile and pyrogen-free (in other words, endotoxin-free) containers, the blood remained liquid. This finding enabled Levin and Bang to demonstrate that the coagulation system came from the amebocytes, the cells found in *Limulus* blood; and that after release of the coagulation factors from amebocytes,

plasma could be gelled by bacterial endotoxin. The realization that an extremely sensitive material was produced by the blood cells of *Limulus* led to the medical and commercial applications of amebocyte lysate.

Preparation of LAL

Before they could reliably make LAL, Levin and Bang first had to find a way to stabilize amebocytes after they were exposed to air. Levin learned that N-ethyl maleimide (NEM) would stop cellular adhesion and aggregation. Armed with this new tool, he was able to prepare NEM-stabilized amebocytes that could be washed free of plasma components. The intracellular fluid was then harvested by osmotic lysis of the amebocytes and subsequent separation from the cellular debris; this extract is now known as LAL. Modern-day preparation of LAL reagent still follows the basic steps outlined by Levin.

Levin and Bang (1968) recognized the marked sensitivity of the system and its applicability for assaying bacterial endotoxin. Levin demonstrated how the reaction could be adapted to a convenient *in vitro* test for endotoxin. When LAL reagent is mixed with an equal volume of a solution containing endotoxin, an opaque gel forms. There is an inverse relationship in which the amount of time required for the reaction increases as the endotoxin concentration decreases. The time required for formation of a firm gel in a 10×75 mm test tube was used to estimate the concentration of endotoxin. Alternatively, an estimate of endotoxin could be determined by making serial twofold dilutions of endotoxin and incubating for a fixed period of time. In this application, the highest dilution that formed a firm gel could be used to calculate endotoxin concentration. Today, with the availability of incubating spectropho-

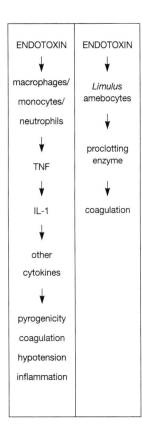

FIGURE 13.3 Mammalian (left) and *Limulus* (right) response to endotoxin *(courtesy of Dr. J. F. Cooper).* IL-1 (interleukin); TNF (tissue necrosis factor).

TABLE 13.1 Functions of *Limulus* amebocytes.

Hemostasis

Blood coagulation

Contain entire coagulation mechanism

Respond to bacterial endotoxin

Phagocytosis

Antibacterial activity

tometers, assays employing the kinetics of gel clot formation (turbidity), similar to that described by Levin and Bang (1968), have increased the quantitative precision of the LAL test.

What Is the Similarity between *Limulus* and Mammalian Responses to Endotoxin?

The response of mammals to endotoxin is more complex than in the horseshoe crab (Levin, 1988). Nevertheless, there are similarities in the effects of bacterial endotoxins in mammals, including man, and in invertebrates, of which the horseshoe crab is an example.

Biological Effects of Endotoxin in Mammals

Endotoxemia produces a complex pattern of systemic toxicity in mammals that ranges from fever induction to life-threatening effects such as hypotension and shock. When endotoxin enters the cardiovascular system in man, it is transported by binding proteins to macrophages, monocytes, and neutrophils—which, in turn, release a number of pro- and anti-inflammatory mediators, such as tissue necrosis factor (TNF) and interleukin-1 (IL-1) (TNF is the primary cytokine stimulated). These cytokines act synergistically to induce a complex biological response that includes pyrogenicity, shock, activation of blood coagulation, and inflammation. For example, under certain conditions, TNF that is released from activated macrophages induces, via other factors, fatal disseminated intravascular coagulation. Also, TNF and IL-1 induce the production of prostaglandin E_2, which then acts on the hypothalamus to cause fever (Pearson, 1990). Figure 13.3 is a simplistic scheme that compares mammalian and *Limulus* responses to endotoxin.

The Response of Amebocytes to Endotoxin and Glucan

In contrast to the varied leukocyte responses in mammals, the amebocytes of the horseshoe crab respond to endotoxin only with aggregation and subsequent release of coagulation and antibacterial factors. Chapter 12 describes the biochemical mechanisms of coagulation in horseshoe crabs. *Limulus* hemolymph contains only one type of cell, called an amebocyte or granular hemocyte. The principal functions of amebocytes are to provide hemostasis and cellular defense against invading bacteria (Table 13.1). Amebocytes contain two types of secretory granules, and the contents of these granules are released in response to invading bacteria. All of the coagulation enzymes are serine protease zymogens that constitute a coagulation cascade activated by endotoxin (Iwanaga et al., 1998). Ultimately, the activated clotting en-

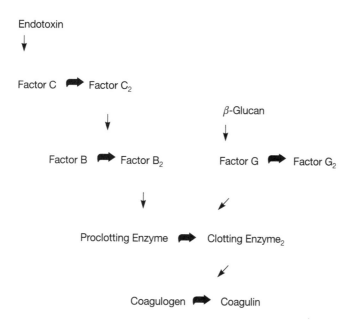

FIGURE 13.4 Biochemistry of the LAL cascade showing endotoxin and glucan pathways.

zyme converts coagulogen (the clottable protein) to an insoluble coagulin gel.

This cascade is the basis of the LAL gel-clot test, more commonly known in the pharmaceutical industry as the Bacterial Endotoxins Test (BET). As seen in Figure 13.4, the LAL reaction contains an "alternative pathway" that is activated by $(1,3)\beta$-D glucan. Since β-glucans are a major component of the cell walls of fungi, it is not surprising that a modified LAL reagent has led to the development of a new fungal diagnostic. Morita et al. (1981) was the first to show that LAL responded to water-soluble glucans. Later, Miyazaki et al. (1992) and Obayashi et al. (1995) refined the LAL reagent to react only with glucans and applied the new reagent (G Test) to the diagnosis of Candidiasis and invasive deep mycoses. Today the G Test is in clinical use as a fungal diagnostic in Japan (Seikagaku Corporation, Tokyo) and is in clinical trials (as Glucatell) in the United States (Associates of Cape Cod, Inc.).

What Is the Principal Biomedical Application of the *Limulus* Endotoxin Test?

After Levin and Bang demonstrated the practical nature of the LAL test, they achieved some success in testing for endotoxin in urine, cerebral spinal fluid, genitourinary exudates, and endotoxemia (Levin, 1987; Prior, 1990). When Levin applied the new test to detection of human endotoxemia (Levin et al., 1970a, Levin et al., 1972), he found powerful plasma inhibitors to the reaction between endotoxin and LAL (Levin, 1970b). Alterations in test methods were required to inactivate inhibitors by protein-denaturing steps and to reduce false positives or negatives (Pearson, 1990). For numerous reasons, the LAL test never gained official recognition as a diagnostic for human disease except in Japan. Levin and others, however, continued a lively research interest in the application of LAL to the study of endotoxemia and sepsis and the test still serves as the gold research standard for the detection of endotoxin in human and animal tissues (Novitsky, 1994, 1999).

Today, while LAL is still used in biomedical research, its major use is to monitor endotoxin contamination in parenteral drugs, medical devices, and biologics. The earliest use of intravenous therapy for humans was complicated by unpredictable occurrence of pyrogenic (fever-causing) reactions due primarily to endotoxin contamination. Subsequently it was found that ubiquitous endotoxins are the most common pyrogens. This is a result of contamination of water and other substances used for intravenous therapy by gram-negative bacteria. While solutions can be sterilized with steam under pressure (autoclaving), endotoxins are not destroyed. Thus, concern for endotoxin contamination in injectable (parenteral) drugs led the industry to seek an assay. In mammals, fever is perhaps the easiest to measure of all endotoxin/pyrogen responses. After Siebert (1923) demonstrated that measuring the febrile response of a rabbit could monitor contaminated intravenous solutions, an assay was developed using rabbits whose rectal temperatures were monitored following injection of a fixed volume of test solution into the marginal ear veins. In the early 1940s, this "Pyrogen Test" was adopted by the U.S. Pharmacopoeia (USP) (U.S. Pharmacopoeia XIII, 1945) and remained the official test for pyrogens until its replacement with the LAL assay in the late 1970s and early 1980s (Federal Register Notice, November 4, 1977; U.S. Pharmacopoeia XXI, 1985).

Collaboration between Jack Levin and James F. Cooper

The collaboration of Levin and Cooper led to the LAL application that completely revolutionized the way that we test parenteral products for endotoxin (bacterial pyrogen). The studies of Levin had shown that LAL was exquisitely sensitive to endotoxin. The next step was to determine whether the new test had sufficient sensitivity, specificity, and other attributes to replace the existing rabbit pyrogen test. The advent of short-lived radiopharmaceuticals for diagnostic imaging created a demand for pyrogen testing that could not be met by the rabbit assay.

Dr. Henry N. Wagner, chairman of the Division of Nuclear Medicine, The Johns Hopkins Medical Institutions, suggested a collaboration to explore the *Limulus* test as an alternative to the rabbit test. A graduate student at the time, Cooper proposed to investigate the relative sensitivities of the two tests and the applicability of the *in vitro* test to radioactive drugs. The study commenced in 1969, with Levin and Wagner as advisors.

Experiments were performed to determine the threshold pyrogenic dose of endotoxin for rabbits using lipopolysaccharide (LPS) derived from *E. coli* and *Klebsiella*. There was an excellent correlation between rabbit febrile response (fever index) and LAL reactivity (gel time), which indicated that the *in vitro* test was indeed an indicator of a biological response (Cooper et al., 1971). To compare sensitivity, rabbits were injected with geometrically decreasing doses of LPS, on a microgram-per-kilogram basis, until the threshold pyrogenic dose was found. LAL prepared at Woods Hole by Levin was used to determine the gel time of endotoxin solutions used in the comparison study. The LAL would gel (positive reaction) at a concentration of endotoxin tenfold lower than that required to produce a pyrogenic response in rabbits, as determined using what was the official pyrogen test. This high level of sensitivity led to further comparison studies of rabbit and *Limulus* tests in biomedical research and the pharmaceutical industry.

Limulus Test as an Alternative to the Rabbit Test for Pharmaceuticals

The rabbit test was fully entrenched in the regulatory process as a screening test for pyrogenic levels of endotoxin in injectable drugs. For the new test to replace the old, the drug industry and regulators would need to validate that the LAL test was more sensitive and specific and that there were no significant nonendotoxin pyrogens in pharmaceutical products that would be missed by LAL assays. Large studies were needed to document the efficacy of the LAL test. For standards, endotoxin was purified by removal of protein to yield LPS. To define a biological activity of endotoxin, 5 endotoxin units (EU) were assigned to 1 ng of an *E. coli*–derived LPS that was initially used as an endotoxin standard. A number of large studies found that approximately 1 ng/kg, or about 5 EU/kg, was the threshold sensitivity of rabbits to LPS (Dabbah et al., 1980). For a 10 mL dose of an infusion fluid for a rabbit, the sensitivity would be approximately 0.5 EU/mL. In contrast, most LAL preparations at this time detected as little as 0.1 EU/mL, or at least a fivefold advantage in sensitivity.

The question of nonendotoxin pyrogens was a big concern for the drug industry and the Food and Drug Administration (FDA). It was critical to show that LAL testing did not fail to detect any pyrogens of importance to pharmaceutical quality control (QC) testing. Baxter, a major producer of intravenous fluids, reported the results of 143,196 LAL tests and 28,410 rabbit tests performed on intravenous fluids and medical devices (Mascoli and Weary, 1979). Their data confirmed that rabbits often gave equivocal results; that endotoxin was the only detectable pyrogen in parenteral products; and that LAL was the more sensitive test. These data confirmed the advantages of the new test, alleviated the concern for nonendotoxin pyrogens, and influenced the FDA to approve LAL as an alternative to the rabbit test for testing parenteral products.

What Was the Benefit of LAL Testing to the Health Care System?

The impact of LAL testing on the improvement of quality and safety of sterile products in the United States was dramatic. For example, an outbreak of pyrogenic (febrile) reactions attributable to endotoxin occurred in 1974. The drug in question was normal serum albumin (25 percent), which is used to replace human albumin in cases of trauma and severe disease states. This incident prompted an investigation by the Centers for Disease Control (CDC) (Hochstein et al., 1979). They used the new LAL test to show that 45 percent of albumin supplies on the market at that time contained levels of endotoxin capable of producing fever in patients receiving this protein. Albumin producers were subsequently required to use the LAL test extensively in process testing, with the result that pyrogenic reactions were eliminated by 1978 (Steere et al., 1978). The CDC reported two additional outbreaks of culture- and LAL-positive drugs and devices during this period (Highsmith et al., 1982). Influenza vaccines have also been improved by the introduction of similar in-process testing by LAL methods to eliminate endotoxin content.

The route of injection is critical to the response to drugs contaminated with endotoxin. In 1972, an outbreak of meningeal reactions occurred that were attributed to drugs that were injected into the cerebrospinal fluid for nuclear imaging (Cooper and Harbert, 1975). The drugs had passed sterility and rabbit pyrogen testing. In contrast to the albumin-induced reactions previously described, resulting from hundreds of milliliters being infused for therapeutic effect, only 1 ml of the endotoxin-contaminated drugs injected into spinal fluid was capable of inducing life-threatening aseptic meningitis. Following this episode, the

FDA set rigid endotoxin limits (by LAL) for intrathecal drugs, and LAL replaced the insensitive rabbit pyrogen test.

In summary, the LAL reagent became a unique example of a successful *in vitro* replacement of an animal toxicity test because it correlated with an important mammalian response to toxicity. The goal of an *in vitro* alternative test to an animal assay is to isolate specific mechanisms and responses of cell systems in the whole animal. If the *in vitro* test addresses a critical feature of *in vivo* toxicity, there is a tendency to become reliant on the new test, which then gains universal acceptance (Flint, 1994). Such is the case for endotoxin measurement by LAL, which is now the test of choice for screening of injectable drugs for endotoxin. The pioneering work of Levin and Bang has revolutionized the way we test for endotoxin and made an enormous contribution to patient safety.

Symposia Highlight Scientific Interest in Amebocyte Lysate

Excitement emanating from research on the many properties of horseshoe crab blood led to six international symposia and publication of their proceedings during the period of 1979 to 1991. Elias Cohen and other scientists (1979) organized the first symposium at the Marine Biological Laboratory. Five others followed in quick succession (Watson et al., 1982; ten Cate et al., 1985; Watson et al., 1987; Levin et al., 1988; Sturk et al., 1991).

Elias Cohen and Carl Shuster had been fellow graduate students learning the techniques of systematic serology in the laboratory of Alan Boyden at Rutgers University. There they shared the same equipment, with Cohen concerned with snake systematics and Shuster (1962) with the serological relationships of three species of horseshoe crabs, as determined by antibodies produced in rabbits injected with antigens (hemocyanin) from the crabs. Ultimately, Cohen studied lectins in *Limulus*. The initial section in Cohen's book introduced several aspects of the biology of *Limulus*. That introduction proved provident, because none of the subsequent symposia considered the animal itself. Actually, it was the 1980 "*Limulus* Expedition" at the Marine Biomedical Center and the Marine Laboratory, Duke University (Bonaventura et al., 1982) that compiled basic information on the physiology and biology of horseshoe crabs. Later, Prior (1990) also included a chapter on the natural history and biology of *Limulus* in a book on clinical applications of LAL.

The first session of the 1978 symposium was notable in other ways. It was there that Shuster first met Koichi Sekiguchi and one of his students, Koichiro Nakamura, and at which a graduate student, Mark Botton, reinforced his interest in *Limulus*—an interest that continues today. In discussions with Elias Cohen and Sidney Galler, Shuster decided that a manuscript he wrote (1949) on the circulatory system and blood of *Limulus,* which had been deposited at the Marine Biological Laboratory library and had been consulted by several of the lysate pioneers, should be distributed separately because it was deemed too long for a chapter in the book that resulted from the first *Limulus* symposium. Sid Galler, as Deputy Assistant Secretary for Environmental Affairs, Department of Commerce, offered to have it printed, but it was accepted for publication by the Department of Energy, Federal Energy Regulatory Commission (FERC). Senator William Proxmire selected the publication (Shuster, 1979) for a "Golden Fleece" investigation, and that investigation prompted the Chairman of FERC to order the publication's destruction.

Role of the FDA

The mission of the Food and Drug Administration includes the assurance of safety and efficacy for drug products. The Division of Biologic Standards (DBS), later known as the Bureau of Biologics (BoB) and now a part of the Center for Biologics Evaluation and Research (CBER), was formerly located on the campus of the National Institutes of Health (NIH). The DBS had specific responsibility for the quality of biological drugs derived from blood products. An integral part of quality assessment at DBS was the rabbit pyrogen test (U.S. Pharmacopoeia XX, 1980). Pyrogen testing required considerable resources in the form of animals, facilities, and trained personnel. Briefly, the rabbit pyrogen test consists of the intravenous injection into three rabbits of a sample of each final container lot of an injectable product. The test dose for each rabbit must be at least 1 ml, with a maximum of 10 ml/kg of body weight. This dose is determined relative to the human dose. Following injection of the product to be tested, the rectal temperature is determined hourly in each of three rabbits for 3 hours and any temperature rise recorded. After taking the last hourly reading, the sum of the temperature rises in the three rabbits is determined. If the value is less than 1.4°C, the product passes; if the rise is 1.4°C or more, the test is continued in five additional rabbits. Then, if the sum of the total rise in the eight rabbits is less than 3.7°C, the product passes; if not, the product fails and cannot be released.

Although pyrogen testing employing rabbits is now semiautomated and utilizes rectal temperature probes linked to computers, technicians who cradled the rabbits in their laps while they inserted a rectal thermometer every hour during the test performed the original tests. These technicians became quite adept at handling the rabbits, so much so that new testers often adversely affected the outcome of the test by making the rabbits anxious. This cradling of the rabbits, known as the "lap test," was a hallmark of the FDA prior to lysate acceptance.

Testing of Biological Products for Endotoxin Using Both LAL and Rabbits at the FDA

Edward Seligmann of the Division of Biologic Standards (DBS) accepted the offer of James Cooper for collaboration as a way to evaluate the merits of LAL as a new test for bacterial pyrogen. At that time, LAL was believed to be capable of detecting nanogram quantities of endotoxin, the ubiquitous pyrogen found in the cell wall of gram-negative bacteria. Laboratory personnel were performing the rabbit pyrogen test on at least four licensed biological products per day. These products included 5 and 25 percent normal serum albumin (Human NSA), purified protein factor (PPF), immune serum globulin (ISG), and antihemophilic factor (AHF) (Hochstein et al., 1979).

The initial part of the evaluation of LAL at DBS was to incorporate the new test into the daily test routine and gain information through parallel testing. From LAL prepared by Cooper at Chincoteague, Va., aliquots of buffered LAL were lyophilized in single-test ampules providing a stable and convenient preparation of lysate (Figure 13.5). An initial study of 155 radiopharmaceutical and biological products established the feasibility of LAL as an alternative test for endotoxin (Cooper et al., 1972). However, evidence of interference with the LAL test by proteins indicated the need for further studies.

A

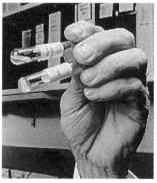

B

FIGURE 13.5 (A) Donald Hochstein at the Bureau of Biologics, NIH, Bethesda, Md., comparing LAL tests. (B) The clear fluid in the top vial indicates no reaction, while the cloudy gel in the bottom vial denotes a positive test.

Bleeding Horseshoe Crabs at the NASA Base in Chincoteague, Virginia

The initial parallel study determined that the *in vitro* test was of genuine value in detecting endotoxins. Therefore, it was clear that the BoB needed to be self-sufficient in its supply of LAL to continue the comparison studies. Because the BoB laboratory was located about 150 miles from the Atlantic Ocean, horseshoe crabs were difficult to obtain, so it was decided to have a crew spend one week each year, for the next several years, at the National Aeronautical and Space Agency (NASA) base in Chincoteague, Va., bleeding adult horseshoe crabs. After bleeding about 800 to 1,000 crabs each week, the group would return to the laboratory with several liters of lysate and 200 to 300 live crabs.

Holding Horseshoe Crabs in Tanks for Bleeding Experiments

These crabs were placed in a three-tier fiberglass lobster tank unit. The tanks, from bottom to top, were six, five, and four feet wide. This arrangement provided total visibility of the interior of the bottom two tanks and steps for reaching the top tank. Each tank contained 7 inches of synthetic seawater (Instant Ocean). A pump connected to the bottom tank pumped the salt water to the top tank. From there, it flowed through an overflow into the middle tank and then back to the bottom tank. Each tank held between 70 and 100 crabs, depending on their size.

When the tanks were full and there were extra crabs, they were put in plastic boxes and placed in a refrigerator at 4° to 8°C. If the crabs were kept moist, they would survive in the refrigerator for at least 6 months without food. The crabs did not move in this cold environment as long as the light was off; but when the light in the refrigerator was turned on, the crabs would respond to this stimulus by moving their telsons. The crab tanks were housed in the rabbit holding room. Because rabbits had to be sacrificed after pyrogen testing, their livers were removed, rinsed in tap water, sliced, and fed to the crabs, which thrived on this diet. Each crab in the tanks was numbered so a bleeding record could be maintained. The crabs were bled on a semiannual basis. It was noted that, with more frequent bleeding, the crabs produced hemolymph depleted in amebocytes, indicating that it took at least six months for these captive horseshoe crabs to regenerate their normal amebocyte count.

Regulation of the Lysate Test

The LAL test began to be used in the medical community in 1972. The FDA received complaints that lysate was being used to diagnose septicemia, meningitis, and urinary tract infections. Therefore, the FDA decided to regulate the sale and use of this product. Because LAL was a blood product, and because the expertise for LAL was in the Division of Control Activities, it was decided to assign LAL to the BoB.

In the *Federal Register Notice* of January 12, 1973 (Federal Register Notice, 1973), the FDA announced that LAL derived from the circulating blood cells (amebocytes) of the horseshoe crab, *Limulus polyphemus,* was a biological product. As such, it was subject to licensing requirements as provided in Section 351 of the Public Health Service Act (42 U.S.C. 262). This notice also stated that drugs that are subject to a pyrogen test required the rabbit test and that the LAL test could not be substituted as a final pyrogen test.

A Standard Endotoxin Needed for Determining LAL Sensitivity

It soon became evident that a reference standard endotoxin (RSE) was needed to determine LAL sensitivity. In 1974, FDA contracted with the University of Montana to prepare 30 grams of purified endotoxin from the common intestinal bacterium, *E. coli* O113:H10:K negative (Rudbach et al., 1976). This bulk endotoxin was labeled EC (for *E. coli*) and, due to its extremely stable nature, was simply stored in a desiccating jar at room temperature. Not surprisingly, this original batch of endotoxin is still in use today and should last for many more years.

In 1976, a portion of this bulk EC endotoxin was used to prepare the first lot of RSE, EC-1. Before freeze-drying this small lot, normal serum albumin (human) at 0.1% concentration was added as a stabilizer. A second lot was prepared like EC-1 and designated as EC-2; it consisted of 1,500 vials that contained 1.0 μg of EC endotoxin. Lot EC-2 was found to be extremely stable. Sufficient potency data was also collected to support assigning a unit of activity value to the preparation. Most control authorities agree that it is better to use units of activity rather than a concentration based on dry weight in expressing the strength of standard preparations. Based on a collaborative study involving FDA and several licensed LAL manufacturers, a value of 5.0 endotoxin units per nanogram (EU/ng) was assigned to Lot EC-2 (Rastogi et al., 1979). The only criticism of Lot EC-2 was that it contained normal serum albumin (human) that might bind endotoxin. To eliminate this possible problem, Lot EC-3 was prepared as a pilot lot with no fillers or stabilizers. Following several months of use, this lot appeared to be satisfactory. Lot EC-4 was prepared in the same way. After EC-4 was freeze-dried with no filler or stabilizer, the vials appeared empty and it was quite difficult to dissolve all of the endotoxin into the water used for reconstitution. This lot was depleted in 1981.

Each new lot of endotoxin created problems in that the in-house standards had to be revalidated to reflect the new endotoxin. To minimize the need for frequent recalibration, a large lot of endotoxin (Lot EC-5) was prepared that was intended to last 10 to 15 years. Lot EC-5 was prepared from the EC bulk powder. The FDA/BoB contracted with a licensed lysate manufacturer to freeze-dry 30,000 vials from a single bulk. The protocol used for Lot EC-5 was jointly prepared by the USP and the FDA, using lactose and polyethylene glycol as stabilizers (Hochstein et al., 1983).

This preparation was stored at $-20°C$ at the USP facility in Rockville, Md.; it was distributed by the agency as Endotoxin Standard Lot F. Several hundred vials of this material were stored at the CBER facility in Kensington, Md., and distributed upon request to licensed manufacturers as the U.S. Reference Endotoxin Lot EC-5. A rabbit pyrogen study indicated that about 10 EU/kg of EC-5 was needed to elicit a $1°C$ rise in a rabbit. A similar study was done in man to determine the threshold pyrogenic dose to EC-5 (Hochstein, 1994). At the Clinical Research Center in New Orleans, La., human male volunteers were divided randomly into groups of twelve. Each group was given one intravenous injection of Lot EC-5 at a dose of 0, 2, 4, 8, or 16 endotoxin units (EU) per kilogram of body weight. Oral temperatures were taken and recorded every 15 minutes for 8 hours. The pyrogenic response to U.S. Standard Endotoxin in humans during the test period was determined. The results indicated that there was a direct correlation between EU/kg administered and the temperature rise. The threshold pyrogenic dose ($1.0°F$ rise in 50 percent of the volunteers) in this study was approximately 4.1 EU/kg.

This was the first report (Hochstein et al., 1994) to describe the human dose response to an intravenous administration of a pyrogen-free water control and four dose levels of U.S. Standard Endotoxin, Lot EC-5. With this information, the FDA guideline on the use of the lysate test stated that the maximum human dose of endotoxin should be 5 EU/kg. This dose might cause a slight fever in about half of the patients, but it should not cause shock or death. This was the best endotoxin lot. However, by 1995, it was rapidly being depleted and, accordingly, a replacement was needed quickly. USP stated that if CBER furnished the bulk EC powder and some expertise for the preparation of EC-6, they would have it prepared at the National Institutes of Biological Standards and Controls (NIBSC). Therefore, in late 1996, about 60,000 vials were prepared, with each containing 10,000 EU. This lot was labeled EC-6 (Poole et al., 1997). Following an extensive collaborative study involving more than 20 worldwide testing facilities, this

CLOTTING CELLS AND *LIMULUS* AMEBOCYTE LYSATE

lot was accepted as the World Health Organization (WHO) standard endotoxin and has been distributed by USP as Lot G endotoxin. This lot should last about 10 years.

What Is Required for a License to Produce and Distribute *Limulus* Amebocyte Lysate?

The FDA is responsible for licensing all *Limulus* amebocyte lysate manufacturers as well as any lysate made from the other species of horseshoe crabs, such as *Tachypleus*. To be licensed, a manufacturer must complete Form 3439 (Application to Market a New Drug, Biologic or Antibiotic Drug for Human Use) and Form FDA 2567 (Transmittal of Labels and Circular). The manufacturer then prepares three lots of lysate. Each lot must be tested for potency using the U.S. Standard Endotoxin EC-6, which is identical to USP Endotoxin Lot G. Samples of the three lots must then be submitted to the CBER for FDA testing. The samples are distributed to specific laboratories for moisture, sensitivity, and sterility testing.

Once the paper review and sample testing are complete, an on-site inspection of the manufacturing facility is scheduled. The prelicensing inspection is scheduled so the appropriate personnel will be present for answering questions during the inspection. One of the inspectors is usually a committee member with expertise in the testing of the product. The inspection involves not only checking the physical plant but also reviewing daily records, including standard operating procedures for each production step. The validation records of equipment such as freeze dryers, airflow hoods, and autoclaves are also reviewed. If any exceptions to the federal regulations are noted during the inspection, the inspector discusses them with the management before leaving the facility. The manufacturer then has the responsibility to reply in writing to each observation.

The licensing committee reviews all data and submits a recommendation to the CBER Director regarding licensure. Recently the FDA has implemented more stringent requirements for the LAL manufacturers, elevating the production of a lab reagent to that of an injectable drug. While stricter regulations seem commendable, the additional burden on manufacturers may result in more expensive lysate. It should be noted that Levin prepared lysate in a marine biology research laboratory, under conditions of high humidity, with open windows and no temperature control; and Associates of Cape Cod started in the basement of a residential home. LAL from both these sources was adequately sensitive and stable.

Licensed LAL Manufacturers

In 1976, two manufacturers applied for a license to produce and market LAL. They completed all necessary requirements for a license to manufacture biological products and were licensed in 1977. Since 1977 additional companies have been licensed to produce lysate, including one foreign company, Seikagaku Corporation of Tokyo, Japan. Due to an increasing regulatory burden, however, companies have come and gone or have combined. Today, only five companies, out of a total of nine licensed, remain. In order of licensure, these are:

- Associates of Cape Cod in Massachusetts, now owned by Seikagaku Corporation
- BioWhittaker in Maryland, now Cambrex Bio Science Walkersville, Inc., owned by Cambrex of New Jersey
- Baxter Healthcare in South Carolina
- Haemachem Division, Wako Chemicals USA, Inc. (the LAL product line of Haemachem in Missouri was recently sold to Wako Chemicals USA, in Richmond, Va., and is now the Haemachem Division, Wako Chemicals USA, which has a new LAL production facility on Cape Charles, Va.
- Endosafe in South Carolina, now owned by Charles River Laboratories of Massachusetts

Seikagaku Corporation relinquished its license to manufacture *Tachypleus* lysate when it acquired Associates of Cape Cod.

Why Were FDA Guidelines Needed for Amebocyte Lysate?

Following the *Federal Register* notices of January 12, 1973 (the amebocyte lysate of the horseshoe crab is a biological product), November 4, 1977 (describing conditions for use of LAL as an end-product test), and January 18, 1980 (announcing the availability of a draft guideline for end-product testing), a booklet entitled "Guideline on Validation of the *Limulus* amebocyte lysate Test As an End-Product Endotoxin Test for Human and Animal Parenteral Drugs, Biological Products, and Medical Devices" was published in December 1987. It did not take long for biomedical investigators and the users of lysate to realize that endotoxin prepared from different genera of bacteria and even different lots prepared from the same genus gave different results with the same lot of lysate on a per weight basis. Endotoxin lots varied in potency as much as 100 to 1,000 times. This meant that the sensitivity of each lot of LAL had to be determined with a standard endotoxin and a standard lysate lot so another laboratory could repeat the results. The FDA determined that, to control lots of LAL, both a standard endotoxin and a standard lysate were needed. This prompted the FDA to purchase a large lot of lysate from a licensed manufacturer and label it as U.S. Reference Lysate Lot 1. The large lot of endotoxin was Reference Endotoxin Lot 6.

A separate set of guidelines were issued in 1987 under section 10.90(b) (21 CFR 10.90[b]) of the FDA's administrative regulations that provide for use of lysate by pharmaceutical manufacturers. In this way, manufacturers of human drugs (including biologicals), animal drugs, and medical devices were informed of procedures that the FDA considered necessary to validate the use of LAL as an end-product endotoxin test. Although these guidelines are not legal instruments, a manufacturer that adheres to the guideline would be considered in compliance with relevant provisions of the applicable FDA Current Good Manufacturing Practice regulations (CGMP) for drugs and devices and other applicable requirements. As provided in 21 CFR 10.90(b), persons who use methods and techniques not provided in the guideline should be able to adequately ensure, through validation, that the method or technique they use is adequate to detect the endotoxin limit for the product.

This guideline also describes acceptable conditions for use of the *Limulus* amebocyte lysate test, including procedures for using this methodology as an end-product endotoxin test for human-injectable drugs (including biological products), animal drugs, and medical devices. The procedures may be used in lieu of the rabbit pyrogen test. In the guideline, the terms "lysate" and "LAL reagent" refer only to *Limulus* amebocyte lysate licensed by the Center for Biologic Evaluation and Research, FDA, and "official test" means that a test is referenced in a USP drug monograph, a new drug application, new animal drug application, or a biological license.

How Is Commercial Production of LAL Monitored?

Once a manufacturer is licensed, a sample of each LAL lot must be submitted to CBER for release and must be accompanied by two forms. One is a protocol that lists all the manufacturer's test results from sterility to chemical and physical data. The second form is the raw data potency sheet that shows the test results on the content of four final containers in parallel with four replicates of the FDA reference lysate. Both lysates are tested with the reference standard endotoxin. Because the sensitivity of the reference lysate and the potency of the reference standard are known, the testing laboratory can determine whether the samples submitted by a manufacturer meet the standards. Both the reference lysate and standard endotoxin are supplied to LAL manufacturers for their in-house release testing. FDA laboratories test the sensitivity of each lot of lysate before it is released. If the test value is within a twofold dilution of the manufacturer's sensitivity test, the lot is released. If the testing is outside of the twofold range, four more samples are tested; if it is still outside the range, the manufacturer is notified of the problem and requested to retest more samples.

If the manufacturer is unable to match the FDA results, the lot is rejected and the manufacturer is notified that the lot cannot be released. The testing of samples and the review of the product protocol usually take 2 to 3 weeks. The number of lots released each year has increased and so has the number of *Limulus* bled. For example, 127,000 crabs were bled in 1991, 163,000 in 1995, 247,000 in 1996, 257,000 in 1997, 280,000 in 1998, and 240,000 in 1999. A study showed that when crabs are returned to their natural habitat within 48 hours following bleeding, the mortality rate is quite low (Rudloe, 1983). That is why the FDA guidelines require that the crabs be returned to water as soon as possible following bleeding.

Inspection of Manufacturing Facilities

Each manufacturing facility (both the horseshoe crab bleeding area and the processing/manufacturing area) is inspected at least every 2 years. If any exceptions are noted during the inspection, an annual inspection will be performed until these exceptions are eliminated. During these on-site visits, the inspectors:

- Review many standard operating procedures (SOPs) and current good manufacturing practices (CGMPs).
- Interview persons directly involved with daily activities.
- Review books containing data of laboratory testing of the product.
- Review all temperature charts for deviation from set ranges.
- Check to be sure that all products ready for release are segregated from unreleased products.
- Review records indicating that the crabs were returned to their natural habitat within 48 hours after bleeding.
- Check to see whether all processing steps have been dated and initialed as completed.
- Check to see whether any new senior personnel have been employed, equipment obtained, or renovations taken place since the last inspection and whether these were reported to the FDA as major changes.

Commercialization of *Limulus* Amebocyte Lysate

Apart from the discovery at the Marine Biological Laboratory in Woods Hole of the clotting mechanism of the horseshoe crab extract and the LAL reagent with subsequent research into its clinical application by Jack Levin, other local scientists became interested in applying LAL to their research. The story of the commercialization of lysate takes us back to the early 1970s, only a few years after Levin and Bang's discovery. Another scientist, Stanley W. Watson (Figure 13.6), a microbiologist at the Woods Hole Oceanographic Institution (a separate entity from the Marine Biological Laboratory where LAL was discovered), decided to try LAL as a means of monitoring the purity of cell membranes he was preparing from marine nitrifying bacteria. Watson reasoned that the LAL test would distinguish between membrane preparations containing endotoxin (cell envelope or wall) from those lacking endotoxin (plasma membrane). Because LAL from Levin's lab was in short supply but horseshoe crabs were not, Watson set up a small production operation in his laboratory. Unfortunately he found his reagent (as well as that of Levin) rather insensitive for his purposes, and so he set out to improve the reagent. This project ultimately proved successful, and researchers as well as pharmaceutical companies sought Watson's reagent.

FIGURE 13.6 Stanley W. Watson, a pioneer in the commercialization of LAL.

The numerous papers employing LAL to detect endotoxin that appeared in the early 1970s and the interest of the FDA's Bureau of Biologics in the reagent for use as a replacement for the pyrogen (rabbit) test (Cooper et al., 1970, 1971; Levin and Bang, 1968) piqued the interest of pharmaceutical manufacturers. Ever mindful of new research developments as well as their regulators, the FDA, the large industry leaders began their own lysate studies. Many of these companies contacted Levin for reagent but were usually turned away because Levin only had enough for his own research. Horseshoe crabs, however, were easily shipped from Woods Hole. Soon the supply department at the Marine Biological Laboratory found itself shipping crabs all over the United States, but especially to the large midwestern pharmaceutical laboratories. However, unlike Watson, these laboratories found that producing LAL themselves was a less-than-rewarding experience.

Begging Levin or the supply department at the MBL to produce reagent, even for a fee, was met with rejection. But there was another source—Watson. As it turned out, the head of the MBL supply department, John Valois, was married to Watson's chief technician, Frederica (Freddy) Valois. Faced with relentless requests for reagent, John began to refer inquiries to Watson. Soon Watson found himself supplying free reagent and consulting for a number of pharmaceutical manufacturers as well as a few companies interested in commercializing LAL.

Founding of Associates of Cape Cod

Faced with a shortage of reagent for his own use, Watson approached the Woods Hole Oceanographic Institution and the Marine Biological Laboratory with the prospect of setting up a small laboratory to supply LAL commercially. Today this sort of arrangement would undoubtedly be encouraged; but in the early 1970s, it was practically unheard of. Watson, nevertheless, was convinced that LAL could be commercially viable; and, after failing to convince the Woods Hole research institutions and several established companies to produce LAL, he set up his own company. He was also able to build on his earlier work with LAL and earned, with colleague James Sullivan, U.S. Patent No. 4,107,077 "*Limulus* Lysate of Improved Sensitivity and Preparing the Same," thus ensuring some protection against potential competition.

Watson had actually formed a company, Associates of Cape Cod (ACC), in 1972 with his wife for the purpose of selling real estate. This corporation served as a convenient platform for the LAL business and was expanded as such in 1974. The early business, affectionately referred to as the "crab lab," was operated out of Watson's home in Woods Hole. Crab bleeding was done in the garage and processing of the reagent and research conducted in a basement laboratory. In 1977, Watson moved to a rented facility in the town of Falmouth and shortly thereafter was granted the first FDA license to produce and market LAL. As predicted, Watson quickly had competitors; but with his patent, which was granted in August of 1978, the company survived and grew. Thomas Novitsky became a member of the staff of ACC in 1978. Watson died in 1995, but not before seeing his company become a model for the successful commercialization of a basic research product. Soon after, ACC merged with Seikagaku Corporation, its Japanese counterpart in the LAL business. Seikagaku is credited with commercializing this technology using the Asian horseshoe crab, *Tachypleus tridentatus*.

Acceptance of LAL in the Marketplace

A *Federal Register Notice* announcing the licensing of a commercial manufacturer and the requirements for substitution was supported by the FDA in 1977 (FR 1977). Although LAL was now allowed, with conditions, for use as an alternative to the pyrogen test, industry acceptance was slow. Most of the large injectable drug manufacturers such as Abbott and Travenol had already been experimenting with LAL for years. In fact, Travenol went ahead and became a licensed manufacturer of LAL for internal use only.

Many companies, however, had years of experience with the rabbit pyrogen test. While this test was expensive compared to LAL, it was a tried-and-true procedure. In practice, the pyrogen test was pass-fail with minimal control and no standard for comparison. The LAL test, on the other hand, required standards and positive and negative controls, and it was seen as a more sensitive test and more prone to sample interference than the rabbit pyrogen test. Companies feared that more of their products would be rejected by the LAL test and, even worse, that the FDA would compare the concentration of endotoxin in similar products from different manufacturers. This subpyrogenic level of endotoxin, while passing the traditional rabbit pyrogen test, could easily become a focal point for additional FDA regulation. In addition, the FDA was requiring comparative data, pyrogen test versus LAL, prior to allowing substitution.

In retrospect, the industry was slow to adopt what on the surface appeared to be an ideal example of technological progress. However, companies seldom miss an opportunity to save money and therefore soon started using the LAL test in an unregulated manner to monitor their production water and raw chemical materials. Because the LAL test was quick and easy to perform and was, most importantly, much more sensitive than the pyrogen test, a manufacturer could be assured that raw materials that tested either negative or very low with LAL would, when formulated into a final product, always pass the pyrogen test. This finding encouraged the use of LAL as an in-process control and created a significant market long before LAL replaced the pyrogen test as the official release test. Even medium to small manufacturers saw the wisdom in this approach and the massive amounts of data, generated and discussed in scientific meetings and papers, convinced the FDA to remove the comparison restrictions on substitution of the LAL test in 1987.

How Is LAL Currently Manufactured?

Horseshoe crabs are collected from several shallow water coastal areas along the eastern seaboard of the United States. Although horseshoe crabs occur in many coastal locations from Maine into the eastern part of the Gulf of Mexico, major LAL collecting areas are located near the production facilities of the LAL manufacturers and are in areas of high

FIGURE 13.7 Fisherman collecting horse-shoe crabs from Pleasant Bay, Cape Cod, Mass.

densities of crab populations. Thus, Massachusetts/Rhode Island, New Jersey/Delaware/Maryland/Virginia, and the Carolinas are prime collecting areas. Crabs are collected by hand or are taken in nets (Figure 13.7) and delivered "fresh" to a processing facility. In most cases, the processing facility is separate from the production facility, due to a combination of incompatibility with the procedures used to produce the final product and the remoteness of the collecting areas from the pharmaceutical industry.

The original licensed manufacturer, Associates of Cape Cod, still conducts all operations at one site in Massachusetts. Collected animals are delivered fresh and moist to the processing facility, where they are cleansed by rinsing and brushing. The crabs are then flexed along the hinge that separates the forepart from the mid-part and immobilized in a V-shaped trough, or "bleeding rack." To collect the "blood," or hemolymph, a large-gauge needle is inserted into the heart or pericardial sinus that lies beneath the almond-shaped arthrodial membrane that separates the body halves. Hemolymph then flows by gravity through the needle into a collecting bottle or bag (Figure 13.8). Bleeding is completed in only a few minutes. The animal is not ex-

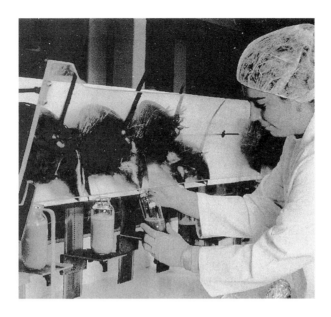

FIGURE 13.8 Technician collecting hemolymph from a series of horseshoe crabs.

sanguinated due to the design of the circulatory system—only the large blood vessels, including the pericardial sinus and connected adjacent arteries, are emptied. Most of the hemolymph (70 percent) remains in the spongy tissues of the prosoma (cephalothorax) and diffuses slowly into the larger vessels. The best part of the bleeding process is that it does not appear to inflict any long-term harm to the animals. Shortly after bleeding, the crabs are released back into their environment. Studies have shown only a small difference in mortality between collected animals that were bled or unbled and released. A study in Florida waters reported a difference of approximately 11 percent, which is well below the mortality in other fisheries considered to have sustainable yields (Rudloe, 1983). Horseshoe crab mortality associated with bleeding for lysate production has recently resurfaced as an issue due to concern over continued viability of horseshoe crab populations, in part due to the mortalities occurring in some of the commercial fisheries. However, recent studies either confirm Rudloe's findings or show even less mortality during processing in the laboratories. It should be noted that the lysate industry has always been a champion of horseshoe crab protection.

Isolation of Amebocytes, Extraction, Formulation, and Packaging

Commercial production of lysate basically follows the techniques described by Levin, with the exception of scale and quality control. For example, numerous crabs are bled simultaneously, producing liters of raw product per day. At Associates of Cape Cod, during the height of production, as many as 1,200 crabs are bled per day. Amebocytes are separated from the blue-colored plasma in banks of centrifuges in a controlled environment. Some manufacturers deviate from Levin's procedure by first freezing the cells, followed by repeated freezing and thawing to lyse the cells. Other manufacturers follow the original procedure by suspending and lysing the cells in hypotonic solution or by mechanically rupturing cells with a homogenizing device.

Once raw lysate is produced, all procedures are carried out in Class 100 or 1000 clean rooms rather than on a research lab bench (Figure 13.9). In contrast to Levin's procedure, all currently marketed LAL is lyophilized in either multi- or single-test vials. LAL manufactured in the United States is provided in standard, stoppered serum vials or cartridges, while that in China is packaged in single-use sealed glass ampoules. Commercialized LAL also differs among manufacturers in LAL formulation, in other words, components added to raw LAL to affect stability, solubility, sensitivity, and specificity. Common ingredients in-

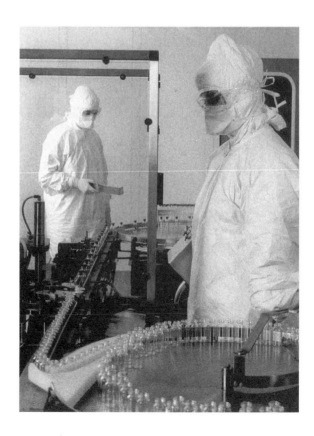

FIGURE 13.9 Bottling finished LAL (*Limulus* amebocyte lysate).

clude mono- and divalent cations, buffers, sugars, proteins, and detergents. Although all commercial LAL preparations are roughly equivalent when tested with the reference endotoxin standard in water, formulation differences become apparent during testing for naturally occurring endotoxins in pharmaceutical products.

The way the LAL test is performed today differs only slightly from that described by Levin, i.e., equal volumes of LAL and the sample to be tested (0.1 ml each) were mixed in a small (12 or 10 × 75 mm) glass test tube, which was incubated at 37°C in a water bath for some time. Current regulations have fixed volumes, time, and temperature for the reaction as well as the number of replicates and controls to be included. In addition to the original gel-clot method, two additional methods are currently in use, the turbidimetric and the chromogenic methods. However, the gel-clot assay is still the most commonly used methodology, accounting for more than half the number of LAL tests performed worldwide.

Current Gel-Clot Assay

This is essentially a serological type test, in other words, a series of dilutions are tested with the LAL reagent until an endpoint (highest dilution to form a solid gel-clot) is achieved. The test is conducted by adding an equal volume of reconstituted LAL to the sample to be tested (0.1 ml each) in a small, endotoxin-free 10 x 75 mm glass tube. The mixture is then incubated at 37°C for 60 minutes in a water bath or dry-block heater. The test is read by slowly rotating the tube through 180 degrees. It is a positive test if the gel-clot remains in the bottom of the tube. All other tests are considered negative, although an experienced technician will recognize +/− tests as borderline. The test becomes quantitative through use of known concentrations of standardized endotoxins. The most sensitive gel-clot LAL reagent can detect about 0.03 EU/ml (3 picograms).

Turbidimetric Method

The turbidimetric method is a machine-read version of the gel-clot assay. Because the formation of turbidity or cloudiness accompanies the formation of a gel-clot, the rate of increase in turbidity can be used to measure the concentration of endotoxin, as originally described by Levin (1968). In this assay a nephelometer, spectrophotometer, or optical reader is used to measure either the total amount of turbidity present after a set incubation period, or the rate of increase of turbidity over time. The first method is referred to as the endpoint and the latter as the kinetic turbidimetric assay. For the kinetic assay, an incubating reader must be used as well as a computer to continually record and eventually analyze the data. Two manufacturers, Associates of Cape Cod, Inc. (Falmouth, Mass.) and Wako Pure Chemical Company (Osaka, Japan), have developed instruments to specifically perform the kinetic turbidimetric LAL assay. Microplate readers (spectrophotometers) are also used.

Chromogenic Method

In the late 1970s, the Seikagaku Corporation of Tokyo, Japan, commercialized a LAL test that used a synthetic peptide coupled with a chromophore (chromogenic substrate) to provide a test with an easily visualized (colored) endpoint. Currently this assay has evolved to include both endpoint and kinetic versions, using the same machines as the turbidimetric assay (as long as the wavelength used is low enough to "see" the chromophore, in other words, a wavelength of about 405 nm). An additional endpoint version of the chromogenic assay uses a chemically converted chromophore, which can be read at a higher wavelength (545 nm). This assay is referred to as the diazo-coupled chromogenic test. The basic test is composed of a LAL reagent and a chromogenic substrate. More recent versions contain both components in a single vial.

To perform the test, a small amount of the reconstituted reagent is mixed with the sample in a tube or microplate and incubated. The incubation times for the chromogenic (as well as the turbidimetric) assays are generally shorter than the gel-clot assay for similar sensitivities. Most commercially available chromogenic assays require 30 to 45 minutes. If the substrate and LAL are combined initially, the assay is stopped after the incubation period and read in the spectrophotometer. If they are separate, the substrate is added after the first incubation period and allowed to incubate a few minutes longer. Adding a small amount of acid stops (fixes) the reaction. At this point the assay can be read as a yellow-colored endpoint (paranitroanilide) or converted to a magenta-colored azo-dye endpoint. The diazo conversion is simple chemistry and takes less than 5 minutes. The diazo-coupled version is more sensitive and is less likely to be interfered with by yellow/straw-colored samples than the original method. The chromogenic assay can also be read kinetically by using an incubating spectrophotometer coupled with a computer. However, the diazo variation can only be used in the endpoint version.

Kinetics and Automation

Although most pharmaceutical quality control and assurance laboratories still use the gel-clot methodology, more and more companies, especially the larger ones, are opting for the kinetic turbidimetric and chromogenic assays. Although great care must be taken in handling the liquids and labware used with these tests (endotoxin is ubiquitous in the environment and contamination of this test is quite easy), automation is attractive and feasible.

What Is the Current Status of the LAL Test?

In 1984, the FDA published concise guidelines for the use of LAL both for in-process testing and final release of product. The USP also embraced the LAL test and in 1985 published the Bacterial Endotoxin Test (BET) and began systematically to incorporate the LAL test in various drug monographs.

Europe and Japan initially lagged behind the United States in acceptance (although Japan has allowed the LAL test to be used as a diagnostic for endotoxemia, while the United States has not), but now acceptance is universal as a replacement for the rabbit pyrogen test in pharmaceuticals.

Although the LAL test is internationally recognized and accepted, no one as yet agrees on a single methodology for each assay version (gel-clot, chromogenic, turbidimetric) or on a single standard. In part, this is due to the commercial supply of LAL, with each manufacturer touting its method as best and, to a greater degree, each country or regulatory agency believing that its standard is the best. In 1997, the FDA, USP, and European Pharmacopoeia (EP) agreed to accept FDA's current standard. Japan, however, has not yet agreed. From a practical standpoint, multiple standards create confusion in the marketplace. It is hoped that science and international politics will soon find some common ground on this issue.

Why Hasn't the LAL Test Been Approved for the Diagnosis of Endotoxemia or Other Bacterial Diseases?

When the FDA first approved the LAL reagent, it was with the caveat "not for use in the diagnosis of endotoxemia in humans." This was because the early scientific literature was not in agreement. Even those studies that concluded that endotoxin in human blood or plasma could be detected with the LAL reagent found that the correlation with other clinical indicators, such as bacterial infection, fever, mortality, and morbidity, was inconsistent and often less than 90 percent. Although proponents of a LAL/endotoxin diagnostic argued that improved sensitivity, assay design, and endotoxin recovery (from blood or plasma) would prove the utility of this test, this remains a controversial and unsettled issue.

A recent study (SEPTEST trial) concluded that there was no clinical utility of the test under stringently controlled conditions, even though the assay—in this case, a diazo-coupled chromogenic test—consistently recovered endotoxin from samples that contained endotoxin (Ketchum et al., 1997). In contrast, Hurley (1994, 1995a, 1995b) and Hurley and Levin (1999) found a general concordance among endotoxemia, bacteremia, gram-negative sepsis, and mortality when numerous published studies were compared by meta-analysis. On the other hand, very good evidence demonstrates the clinical utility of LAL for the detection of gram-negative bacteruria and meningitis (Prior, 1990). Unfortunately, the LAL test is relatively expensive for routine screening (such as urine analysis) and is very easily contaminated (yielding false positives). Therefore, it seems unlikely that the LAL test in its current versions will be approved (or demanded) as a clinical diagnostic for these indications.

What Is the Future of the LAL Test?

The LAL market has never been large—perhaps US$50–75M worldwide. This is probably the reason why only a few companies remain interested in this market. Three of the original commercial companies (Associates of Cape Cod; Microbiological Associates, now Cambrex Bio Science Walkersville, Inc., and Haemachem, now Haemachem Division, Wako Chemicals USA) remain, and Endosafe (now Charles River Laboratories, Inc.) entered the market in 1989. Some dozen others have come and gone. Although a Chinese market exists, its size is unknown as the Chinese have their own industry based on lysate from the Asian *Tachypleus* species. The number of lysate manufacturers in China is at least five. Because the pharmaceutical market in the United States and Europe is considered mature with respect to controlling endotoxin contamination, growth of the LAL market is low, probably less than 10 percent per year. Although developing nations continue to improve their pharmaceutical industries, LAL is relatively expensive compared to the traditional rabbit test in most countries outside North America, Europe, and Japan.

Supply of Horseshoe Crabs
Although the market for LAL is small, the world's pharmaceutical manufacturers have come to depend on LAL, just as the LAL manufacturer depends on the horseshoe crab. Currently, the U.S. LAL industry has an adequate supply of crabs, but the Japanese industry has encountered se-

rious shortages. Chinese supplies seem adequate; but the same development, accompanied by the destruction of beaches and pollution, which doomed the Japanese crabs, may soon seriously reduce their supply. In the United States, the use of horseshoe crabs for eel and more recently conch (whelk) bait is currently a threat to the survival of the LAL industry as well as, possibly, to that of the animal. For almost two centuries, the horseshoe crab has been used at many places along the Atlantic coast for fertilizer, bait (mainly eels), and livestock food, generally at a nonthreatening level. *Limulus* survived large harvests from Delaware Bay for over a century, when thousands of tons were collected annually and ground up for fertilizer. The horseshoe crab fertilizer industry ended in the 1960s, but the population has never recovered to the level of the immense numbers seen in the nineteenth century. Now that conchs have been added to eels in the demand for bait, bait fisheries have expanded and horseshoe crab populations up and down the Atlantic coast are declining. Many populations are noticeably smaller than in former years and their occurrence has become more spotty or scattered. In recent years, prodded mainly by the American Audubon Society and the American Bird Conservancy, state governments and the Atlantic States Marine Fisheries Commission have developed a coast-wide horseshoe crab management plan (see Chapter 15).

Recombinant, Chemical, and Other Biochemical Alternatives

It is surprising that a simple chemical method or at least an alternative biological one has not yet replaced the LAL test. Probably the complex, enzymatic nature of the mechanism of the LAL test has been its savior—it is one of the most sensitive bioassays known. This is because the multicascade nature of the LAL reagent acts as a biological amplification system. With special care, the endotoxin associated with one *E. coli*–like bacterium per ml of water can be detected with LAL. The closest chemical method, using highly sophisticated machines (gas chromatography coupled with nuclear magnetic resonance spectrometry), is thousands of times less sensitive. Yet a genetically engineered version of the LAL test should be possible. To date, the substrate has been synthesized chemically. The first enzyme of the cascade has been produced in yeast from a horseshoe crab gene (Ding et al., 1997). Although the cost to produce commercial quantities of even this first enzyme (pertinently, three enzymes are present in the natural product) is extremely high, Cambrex Bio Science Walkersville, Inc. (formerly BioWhittaker) has recently introduced BioWhittaker™ PyroGene™ Endotoxin Recombinant Factor C Assay, an Endotoxin assay based on a recombinant Factor C (rFC) (See Figure 13.4).

Another line of research using another protein from the amebocytes of the horseshoe crab has shown some success. In this research a small peptide molecule, endotoxin-neutralizing protein (ENP), which specifically binds endotoxin with a high affinity, was isolated, characterized, and produced recombinantly (rENP). This recombinant peptide is capable of being inexpensively produced in large quantities and its use as an endotoxin assay has been patented (Wainwright et al., 1990). Unfortunately, the sensitivity of either the rFC or rENP-based assay is not as great as the LAL (0.01 EU/ml for the recombinant assays vs 0.001 EU/ml for LAL). These or possibly some other assay may replace LAL in the future.

KING CRAB FERTILIZER: A ONCE-THRIVING DELAWARE BAY INDUSTRY

CHAPTER 14

Carl N. Shuster, Jr.

The peak years of exploitation of the Delaware Bay population of horseshoe crabs are about 150 years apart and represent two entirely different fisheries. The first peak was during the mid-1800s when a few million horseshoe crabs were harvested annually and converted into fertilizer. At the same time there was a small fishery that caught horseshoe crabs for eel bait. The commercial fertilizer industry ended by 1970. In its place a substantial new industry has emerged, from the late 1970s to the present time, using increasing numbers of horseshoe crabs as bait for large snails (conchs to the fisherman, whelks to the scientist). A summary of coast-wide commercial landings of horseshoe crabs, 1965 to 1997, although incomplete and probably understated, indicates that the first landings of over one million pounds per year occurred in 1976 and increased almost steadily to six million by 1997 (Atlantic States Marine Fisheries Commission, 1998).

Historically, farmers and watermen from Florida to Massachusetts used small numbers of horseshoe crabs for fertilizer, eel bait, and occasionally as food for poultry and hogs (Goode, 1887). But only in Delaware Bay was their capture for fertilizer a well-defined industry, due to the great abundance of the crabs on the beaches (Smith, 1891; Shuster 2001). That industry provides a record of what occurred in a fishery that peaked with annual harvests of millions of these crabs and yielded only 100,000 per year at its close in the 1960s. It is also a basis for consideration of what may happen to such a population under a similar prolonged and heavy harvest, as in the current bait fishery.

King Crab Fertilizer Industry of Delaware Bay

Limulus was called a king crab by colonials because, at that time, horseshoe crabs were thought to be crustaceans and they were the largest crabs known. This common name was used in all the early writings about the fertilizer industry based on *Limulus* and has been perpetuated by the watermen.

It has always been easy to collect horseshoe crabs during their spawning migrations to beaches of Delaware Bay in the spring and early summer. When the practice of collecting horseshoe crabs for use as fertilizer began is not known, but it may date back to colonial times. In the early 1800s, in his accounts of the shorebirds on Delaware Bay, Alexander Wilson noted that "the dead bodies of the [crabs] themselves are hauled up in wagons for manure, and when placed at the hills of corn, in planting time, are said to enrich the soil, and add greatly to the increase of the crop" (Brewer, 1840). To varying degrees, small-scale hand harvesting has persisted ever since; even into the late 1950s, the elderly Conover brothers, Louis and Clinton, were still collecting crabs for fertilizer. Russell Garrison, another farmer, also collected a few hundred horseshoe crabs each year and fed them to poultry and hogs or dried and ground them up for fertilizer. Once *Limulus* eggs were so numerous they could be shoveled from the shore by the wagonload and used as feed for poultry (Cook, 1857). Although horseshoe crabs were harvested from both sides of Delaware Bay, the larger and longer lasting fishery was that on the New Jersey side (Figure 14.1).

King Crab Industry in New Jersey

Historically, the largest numbers of the horseshoe crabs caught in New Jersey were collected in fishing pounds, which were first used around 1870. By 1880, nine of them were in operation (Earll, 1887) and there were still nineteen pounds in the 1930s, when the fishery was declining. While there may have been more pounds during the years of abundant harvests, the locations along the shoreline probably were similar, depending on accessibility to spawning areas, including land ownership. The only known record of the exact location of nineteen king crab pounds resulted from a mapping exercise of the Cape May coastline of Delaware Bay, 1935–1936, between Fishing Creek and Goshen Creek (see Figure 14.1), roughly corresponding to the area where the tidal flats along the peninsula are the widest (Works Progress Administration, 1938). These pounds ranged in length from 150 to 2,500 ft.

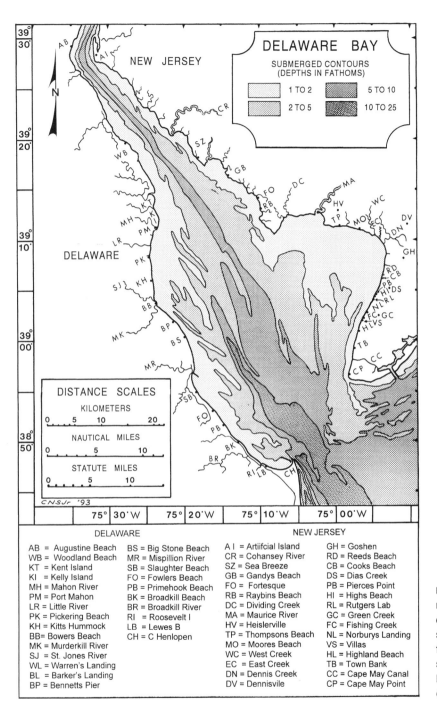

FIGURE 14.1 Delaware Bay and the continental shelf at the mouth of the bay. Many of the sites related to horseshoe crab spawning areas and past and present fisheries are identified. The extensive shallow water areas on both sides of the bay provide a large nursery for the young crabs.

DELAWARE BAY
SUBMERGED CONTOURS
(DEPTHS IN FATHOMS)

1 TO 2	5 TO 10
2 TO 5	10 TO 25

DISTANCE SCALES

KILOMETERS
0   5   10   20

NAUTICAL MILES
0   5   10

STATUTE MILES
0   5   10

C N S Jr '93

DELAWARE

AB = Augustine Beach
WB = Woodland Beach
KT = Kent Island
KI = Kelly Island
MH = Mahon River
PM = Port Mahon
LR = Little River
PK = Pickering Beach
KH = Kitts Hummock
BB= Bowers Beach
MK = Murderkill River
SJ = St. Jones River
WL = Warren's Landing
BL = Barker's Landing
BP = Bennetts Pier

BS = Big Stone Beach
MR = Mispillion River
SB = Slaughter Beach
FO = Fowlers Beach
PB = Primehook Beach
BK = Broadkill Beach
BR = Broadkill River
RI = Roosevelt I
LB = Lewes B
CH = C Henlopen

NEW JERSEY

A I = Artiifcial Island
CR = Cohansey River
SZ = Sea Breeze
GB = Gandys Beach
FO = Fortesque
RB = Raybins Beach
DC = Dividing Creek
MA = Maurice River
HV = Heislerville
TP = Thompsons Beach
MO = Moores Beach
WC = West Creek
EC = East Creek
DN = Dennis Creek
DV = Dennisvile

GH = Goshen
RD = Reeds Beach
CB = Cooks Beach
DS = Dias Creek
PB = Pierces Point
HI = Highs Beach
RL = Rutgers Lab
GC = Green Creek
FC = Fishing Creek
NL = Norburys Landing
VS = Villas
HL = Highland Beach
TB = Town Bank
CC = Cape May Canal
CP = Cape May Point

Design and Construction of a Pound

The expansive, sandy intertidal flats along the Cape May shoreline were ideal for the construction of pounds. Smith (1891) described the fishery in detail and illustrated the main features of pounds and weirs that, although similar in design, differed in the materials used to construct them. The pounds extended out several hundred yards from the shore onto the flats. Poles made of small trees or saplings were driven deeply into the sand about 10 ft apart. At low water, they rose some 7 or 8 ft above the flats. A leader, also called a hedge, led from the beach to the trap. Poultry wire with about a 2 in mesh (sometimes a 4 or 6 in mesh) was used as netting. Depending on the length of the pound, several large traps (bowls or pockets) were arranged along its axis, with a bayward terminal trap and two or three more, equally spaced, toward the shore. Each pocket had at least two parts: an outer one for fish and an inner one (nearer the shore) for the horseshoe crabs. Wings, extending landward in a V-shaped manner from the opening into the bowl, flanked the inner pockets. An inclined plank enabled the horseshoe crabs to get into but not escape from the inner bowl. Several stout wires separated the inner and outer bowls and kept the crabs from entering the outer bowl and crushing the fish. A heart-shaped bay was characteristic of the pounds of Selover and the Conover brothers in the 1960s. Access doors were on the south side of the pound because that was where the crabs would be harvested during an ebbing tide.

Examination of the sketch by Russell Conover (Figure 14.2) suggests a poignant question—how many horseshoe crabs were caught in the outermost trap compared with the inshore trap? During the 1940s when Conover was most engaged in the fisheries, he recollects that about equal numbers were caught in both pounds at the times of maximum spawning activity and that, during those years, each pound usually caught between 40,000 and 70,000 crabs annually. This observation raises more questions. For example, what does this somewhat equal distribution suggest? Do the crabs move more or less evenly into the shallow waters from the beaches or do they fan out as they leave the beaches? If the latter, the pounds may have trapped crabs that spawned on neighboring beaches.

Winter ice damage usually made construction of pound nets an annual necessity. Building one was hard work and was often conducted under stressful weather conditions. Cutting poles, if they were not already available, usually began in late winter. The messy chore of dipping wire in tar to retard corrosion was usually assigned to the youngest members of the crew. The younger men also loaded and manned the scows that brought the poles to the men who were setting them. In the

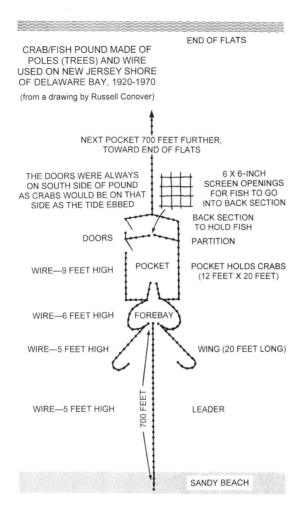

FIGURE 14.2 A bird's-eye view of the essential parts of a New Jersey pound, based on a pencil sketch by Russell Conover (not to scale). His family and others used this kind of trap along the Cape May shore from 1920 to 1970.

early years, setting the poles was arduous. Teams of about seven men set the poles by hand, working them into the flats by yawing them back and forth. A team could set the poles for a pound net in about a week. When small pumps became available in the 1940s, the poles were jetted into the bottom by a team of about five men. Setting the poles by jetting took about half as long as when they were set by hand. In the fall, salvageable poles were sometimes removed.

Weirs (Stake Nets) Weirs were usually set on muddy bottoms in a cove (Smith, 1891). They consisted of poles or stakes driven into the mud to form a leader, wings, and a bowl. Poles from 3 to 6 ft long were employed in the wings and leader. Those poles constituting the bowl were

8 to 10 ft long, and their diameters were 2 to 4 in at their larger ends and 1 to 1.5 in at their smaller extremities. These poles were spaced from 1.5 to 3 in apart to permit the water to wash through. Each bowl was somewhat semicircular or heart-shaped, the extremities of the brushwork joining the wings about midway through their length. A wedge-shaped platform of boards was in the space between the wings as they approached the leader from the door (entrance) into the pound.

The platform was inclined at a gentle slope, and was about 5 ft long, projecting about a foot over a support. It was important that the pitch of the platform was not too great and that the floor was not too smooth; otherwise the crabs could not, or would not, walk up it. The floor of the bowl was made of cheap boards to prevent the crabs from scratching into the mud and loosening the poles. Such traps were most commonly used by farmers, who cut poles from their nearby woodland. One person, using a boat, could fish two to four weirs per tide. When the weirs were set on sandy bottoms, they were fished by horse and wagon; a person could fish more sand-bottom weirs than those on muddy bottoms. The catch per weir was equivalent to that of a pound.

Fishing the Pounds Both boats and wagons were used in tending the nets, although the wagons were considered more convenient and were more extensively employed. At low tide, the pound was fully exposed, and workers could retrieve the captured crabs with pitchforks or with a crab-spear consisting of a single piece of sharp-pointed metal mounted on a long handle. The catches varied with the year, month, and tide. Usually, the nets were fished twice a day during low tide. Seventy-eight-year-old Samuel Compton of Dias Creek was one of the last of the old-timers to operate a pound and tend it by wagon (*The Sunday Bulletin,* Philadelphia, 1953). He covered his horse, Harry, with burlap to protect him from the biting insects. During the last few years that Compton operated a pound off the north side of Pierces Point, Russell A. Conover and his father and brother helped build the pound. In September and October, they cleaned the wire and made the pocket smaller and put smaller mesh screen on the pocket to catch eels during November and early December.

Sometimes, skiffs or barges were used instead of wagons. After collection from the pounds or weirs, horseshoe crabs were hauled to the beach, where holding pens were erected high on the bank beyond the high tide level. These pens accommodated 25,000 to 300,000 horseshoe crabs at a time (Figure 14.3). There, the crabs began to dry out and usually died within 2 to 3 days. Soon the crabs were riddled by maggots of carrion flies and a pungent, nauseating stench emanated from them.

A

B

FIGURE 14.3 Two views of a large holding pen on the Delaware Bay shore of New Jersey. (A) Side view. (B) Looking down on the pen. This pen, estimated as possibly 1.8 m high x 3 m wide x 24 m long, could have held as many as 20,000 adult horse-shoe crabs. The crabs were held in many such pens, of all sizes, until dry, prior to being ground into a meal that was added to a fertilizer mix. *From Fowler (1908), plates 64 and 65, courtesy of the New Jersey Museum.*

Milne and Milne (1947) gave an excellent description of the trapping of horseshoe crabs in the 1940s and illustrated a special kind of wooden car that ran on rails from the shore down to the water, where the crabs were unloaded from a barge into the car for transport to the pens on shore. These cars had a V-shaped cross-section and could be tipped to either side on their chassis to unload.

Cancerine Factories in New Jersey

At first the harvesting of horseshoe crabs for fertilizer was primarily by individuals for use on their property. The crabs were either dried or composted before use. Over the years, as the acreage of farmed lands increased, the need for fertilizer also increased. Commercial fertilizer factories were built to produce a dried, ground-up material that was marketed as cancerine. This was a viable industry due to the abundance of *Limulus*. Records indicate there were at least five cancerine fertilizer factories in New Jersey, but it is not likely that more than two or three of these plants were operating at the same time.

The earliest cancerine factories in New Jersey were the Ingram and Beesley plant near Goshen (ca. 1855–1880) and the Vincent Miller plant in Dennis Township (ca. 1865–1880). The Kirby and Smith plant on West Creek (ca. 1870–1890s) was perhaps the largest cancerine factory of all (Beers, 1872). This plant alone processed 1 to 2 million crabs annually (Smith, 1891). In 1890, 275,000 king crabs from New Jersey were transported to a plant at Billingsport; another 225,000 crabs from New Jersey were transported to Baltimore, MD (Smith, 1891). As late as 1907, most of the crabs penned in New Jersey were also shipped to Billingsport (later called Paulsboro) for processing at the end of the harvest season (Fowler, 1908).

The Camp family operated the last horseshoe crab fertilizer factory on the bay shore from the 1920s to 1930s (Reed, 1986). It was located at King Crab Landing just south of Pierces Point (Figure 14.4). This property was adjacent to the present location of Limuli Laboratories, a horseshoe crab bleeding facility owned and operated by Benjie Lynn Swan. The Camp fertilizer plant was a large wooden building on timber piling. The Espoma Company of Millville, New Jersey, was probably the last commercial producer of cancerine in New Jersey, processing fewer than one hundred thousand crabs a year from around 1930 to 1966 (see Figure 14.5). The explanation that accompanied the photographs in this figure stated:

The horseshoe crabs we use are taken from the shores of Delaware Bay during late May, with the exact time being determined by the moon, tides, and wind. A strong northeast wind creating rough water

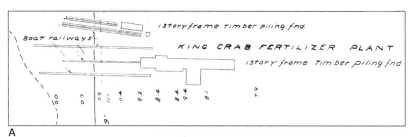

A

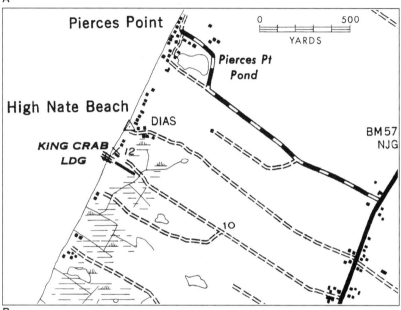

B

C

FIGURE 14.4 The Camp fertilizer plant, located at King Crab Landing, just south of Pierces Point, probably was the last king crab fertilizer plant on the Cape May shore of Delaware Bay. (A) It was a large wooden building on timber piling (from sheet 20 of a series of 1936 WPA maps; courtesy of Aerial Photo and Map Library, New Jersey Tidelands Management Bureau). (B) This king crab plant was also shown on a National Ocean Survey chart (1977) and the U.S. Geological Survey map (1972). (C) The only known view of Camp's plant is from the beach, on a glass slide by T. C. Nelson *(courtesy of the Haskin Shellfish Research Laboratory, Rutgers University, Port Norris, NJ).*

FIGURE 14.5 Several steps in the processing of the crabs. (A) Crabs were delivered to the fertilizer plant. (B) Inside the plant, crabs were transferred to a crusher. (C) The crusher consisted of a large metal hopper with slanted sides down to a rectangular trough in which worm gears crushed the dried crabs. (D) The pulverized material passed down a tunnel into a large spigot that controlled bagging. *Reproduced from a set of 1958 photographs, printed with the permission of the Espoma Company, Millville, N.J.*

TABLE 14.1 Production of horseshoe crab meal during the last years of processing *Limulus,* 1959–1966, at the Espoma Company, Millville, NJ

Year	Number of Crabs	Cost per 1,000	Supplier	Pounds	Tons	Crabs per Ton
1959	10,000	$18.00	Crowley	*	9.64	2,594
	15,000	$18.00	Leroy Selover	19,275		
1960	64,500	$18.00	Crowley **	52,475	26.24	2,458
	58,700	$18.00	Selover **	45,600	22.80	2,573
1961	6,300	$20.00	Hayes	*		
	46,600	$18.00	Selover	43,915	21.96	2,409
	76,126	$18.00	Crowley	*	34.01	2,562
	11,000	$20.00	E. Lee	68,025		
1962	51,500	$18.00	Selover	*	36.63	
	35,000	$18.00	Wilson Eldredge	3,265		2,352
1963	35,000	$18.00	Selover	*		
	15,000	$18.00	Eldredge	43,125	21.56	2,319
1964	98,000	$18.00	Selover/Eldredge	79,575	39.79	2,463
	6,050	$18.00	Sam Peterson	4,200	2.10	2,881
1965	51,000	$18.00		38,400	19.20	2,656
1966	35,000	$18.00		23,775	11.89	2,944

* Combined totals, listed opposite the second supplier.
** Crabs were delivered by Kenneth Shellenberger.
Note: Under cost per 1,000, the $20.00 was for delivered crabs.

will delay the time of our catch. The crabs are usually caught after they have spawned and started to return to the bay.

The principal supplier was Leroy "Tokey" Selover, who operated a pound just south of the Green Creek sluice during the 1960s. Since then, a cancerine meal produced in Virginia from the offal from blue-crab shucking houses has supplanted king crab meal in Espoma fertilizers, and the company continues producing natural organic-based specialty lawn and garden plant food for home use. For production numbers during the years 1959 to 1966 and a chemical analysis of the Espoma Company horseshoe crab meal, see Tables 14.1 and 14.2, respectively.

At first, cancerine production used horseshoe crabs that were dried in holding pens. Whereas composting used most of a crab, the air drying of crabs was less efficient because maggots and other carrion species, including beetles, ate much of the organic matter. In later years, newly harvested horseshoe crabs were steamed prior to being ground up. This produced a higher-quality meal, which was also used to feed

TABLE 14.2 Analysis No. 271550 (September 22, 1956) by Gascoyne and Company, Inc. (Baltimore, Maryland), of a sample of the Espoma Company's (Millville, NJ), horseshoe crab meal.

	Percent
Moisture	6.32
Available phosphoric acid	1.21
Insoluble phosphoric acid	0.14
Total; phosphoric acid	1.35
Nitrogen	8.60
Potash	0.54
Acid-base equivalent	(265.97-A lbs. calcium carbonate per ton)
Calcium (Ca)	0.7034
Magnesium (Mg)	0.2340
Iron (Fe)	0.1581
Copper (Cu)	0.0078
Manganese (Mn)	0.0015
Boron (B)	0.0246
Sulphur (S)	0.4440
Molybdenum (Mo)	0.0092
Zinc (Zn)	0.0044

livestock, particularly hogs, and poultry. As late as 1948, these fresh (usually designated as green) horseshoe crabs were steamed and processed for chicken mash and pig meal (Dickinson, 1958). Dickinson bought thousands of the crabs from the local farmers and used them as feed for broilers and young pigs. Due to the strong flavor imparted by the crabs, the chickens' feed was changed at least 2 weeks prior to the use of the chickens' eggs or meat. Following World War II, when the old farms were being sold for housing developments in Cape May County, the Board of Health no longer permitted the storage and drying of horseshoe crabs in pens because these pens were deemed to be a menace to public health (Dickinson, 1958).

King Crab Industry in Delaware

Quantities of king crabs were taken at only a few localities on the Delaware shore and the fertilizer factories were usually nearby. Unlike New

Jersey, there are no records indicating that either pounds or weirs were ever used. Rather, the fishery was confined to the shore and the immediate shallow water and was conducted from boats or wagons, according to whether the men were fishermen or farmers (Smith, 1891). The fishery was chiefly in the vicinity of Bower's Beach and the crabs were transported to Warren's and Barker's Landings on the St. Jones River (Figure 14.6).

In 1888, six lighters (skiff-like boats), from 12 to 16 ft long, were deployed from Barker's Landing near a fertilizer factory. When the wind was offshore, these boats plied up and down along the shore, receiving the crabs collected from the beach and shallow water. Once these boats were full, they were off-loaded into two large scows, 30 and 42 ft long, that were anchored offshore. The scows transported the crabs to the storage site near the factories. At Bower's Beach there were only four professional horseshoe crab fishermen in 1888, but they were joined by thirty-five others, including farmers, farmhands, and wood-choppers (Smith, 1891). These fishermen employed 25-ft scows. Semi-professional fishermen and others in the area made half of their collections in boats and half in wagons. In addition to the quantities sold for fertilizer production, about 25,000 crabs were fed to hogs annually. Vessels sent out from factory operators transported a portion of the Delaware catch to New Jersey.

The hand fishery during the king crab spawning season continued into the 1930s. Men and boys walked through the water along Bowers Beach at night, towing small scows and, upon feeling or seeing a crab, tossed it into the scow (Federal Writers' Project, 1938). A thousand crabs brought $5 to the collector. A night's catch was put into wire-netting pens in the water until taken by scows to Warren's Landing. There, on a long wharf made of cedar poles, they were stacked by the thousands (see Figure 14.6) to dry in the sun. In October, they were pulverized or broken into small pieces in crushers. Sometimes small carts on rails were used to move the crabs from the scows to storage in pens. When used for feed, the crabs were smashed while alive and thrown to hogs or chickens.

Cancerine Factories in Delaware

In some years, then and now, horseshoe crabs were more abundant in Delaware waters than along the New Jersey shore (Fowler, 1908). A cancerine factory was established in Delaware to minimize shipping necessitated by the varying abundance of horseshoe crabs on each side of the bay.

Warren's Landing on the St. Jones River may have been used for almost one hundred years, from the 1840s until the 1930s, as a principal

A

B

C

D

FIGURE 14.6 Warren's Landing on the St. Jones River. (A) and (B) Views to the east. (C) and (D) Views to the west. (A) On June 28, 1928, the family of William E. DeWitt was on the dock. (B) DeWitt with two large female crabs (left to right, approximately 39.5 cm wide = 15.5 in and 36 cm wide = 14+ in). (C) The men are standing on a cedar pole matting covering the marsh bank. The crabs nearest the large scow have just been unloaded; later they will be neatly stacked in rows along the planking to shed water. (D) A closer view of the king crabs; the white house in the background is the same one seen in (C). *Photographs from the Delaware Board of Agriculture Collection, courtesy of the Delaware State Archives.*

storage site for king crabs. By 1844, Jehu Reed shipped cordwood from his pine forests and many tons of fine peaches to Philadelphia and New Jersey from his port at Warren's Landing (Montgomery, 1974). Reed used king crab fertilizer in his peach orchards and established a fertilizer plant at the landing. For many years the crabs were stacked high at the landing during the summer and crushed for fertilizer in the fall. What happened to Reed's 1840s plant at Warren's Landing is not known; but there was a king crab plant operating there in the late 1930s (Federal Writers' Project, 1938), at which time it took about 1,500 crabs to make a ton of fertilizer. During the late 1920s and early 1930s, Mrs. Pauline Parker recalled that, as a child, she picked up horseshoe crabs at two for one cent, earning fifty cents to a dollar in an afternoon (Montgomery, 1974).

Smith (1891) reported that two factories designed to prepare king crabs for fertilizer were in operation in Delaware in 1887 and 1888. A small factory at Banckenburg Creek (Byles, 1859) used a furnace to dry about 200,000 crabs per year. A factory near Barker's Landing had a steam mill in which the crabs were ground while green and then mixed with sodium sulfate. There, as many as 100,000 crabs were used in some seasons; but in 1887 and 1888, only 50,000 and 30,000, respectively, were used. Nearly the entire output was sold in Delaware and the rest in the Delmarva Peninsula for $25 to $30 per ton.

Impact of the King Crab Fertilizer Industry

In the early years, the number of crabs must have seemed inexhaustibly large. For example, Cook (1857) reported that a resident of Town Bank (lower Cape May County) collected 750,000 crabs on a half-mile beach in 1855, and 1.2 million on a 1 mi stretch in 1856. Cook (1857), in noting the vast numbers of horseshoe crabs in Delaware Bay, concluded:

> If the number is not materially diminished, the manufacture [of cancerine] could be extended so as to produce many thousand tons every year, but there is so little knowledge of the habits of the king crab, that no judgment can be formed as to the effect that will be produced on a coming year's supply by the destruction of great numbers of those which come to the shores to lay their eggs.

Even as early as the 1880s, catch-per-effort was decreasing in the Delaware Bay king crab fishery (Smith, 1891). For example, New Jersey fishermen caught 1.6 million horseshoe crabs in 1890 compared to 1.3 million in 1887, a 19 percent increase. However, over that same time interval, the fishing effort (quantity of gear used) increased by 163 percent, causing Smith to deduce the following:

> It cannot be denied, as shown by reliable returns, that in some localities there has not only been a maintenance of the supply but even an increase; but the general trend is and has been toward a decrease . . . the catch has only been maintained by an increase in the amount of apparatus.

Because pounds or other trapping devices were not employed in Delaware, the decrease there was not quite as pronounced as on the New Jersey side of the bay, but the catch was probably "influenced by the great drain on the species in New Jersey" (Smith, 1891). Smith also pointed out that Rathbun's (1880) warning on the decrease in the abundance of horseshoe crabs was timely, and considered that the diminution of the king crab population was no doubt due to the practice of capturing the crabs during the spawning season, before all of the eggs were laid.

Can a Population Recover?

While the total population of horseshoe crabs was never determined, U.S. Fisheries Statistics from Delaware Bay show that average annual harvests of about 1.5 million crabs were sustained for most of the years from the 1870s through the 1920s. The big question is, What percentage of the population was harvested each year? In the year 1890, pounds and weirs took two-thirds of the 1.7 million horseshoe crabs caught in New Jersey (Smith, 1891). Eighty-eight percent of those pound/weir catches were taken in four areas: Dennisville (120,000), Dias Creek (250,000), Green Creek (401,000), and Fishing Creek (180,000), while another 600,000 crabs were collected by hand. Many fewer horseshoe crabs were either seen or taken from Delaware Bay during the 1950s and 1960s (Figure 14.7).

With a population as numerous as that in the Delaware Bay area, a relatively rapid recovery may occur within two decades when there is no substantial fishery. This is suggested by isolated observations over the years on one stretch of beach, from Highs to south of the Rutgers Oyster Research Laboratory to Green Creek on Cape May, NJ (see HB, RL, and GC, Figure 14.1). During the 1951 spawning season, the daily maximum was 232 spawning adults on a 0.47 km stretch of Highs Beach (= 494 per km; Shuster, 1997). At the Green Creek end, during the last 8 years (1959 through 1966) of the king crab fertilizer industry, the Espoma Company converted an average of 77,000 crabs to fertilizer

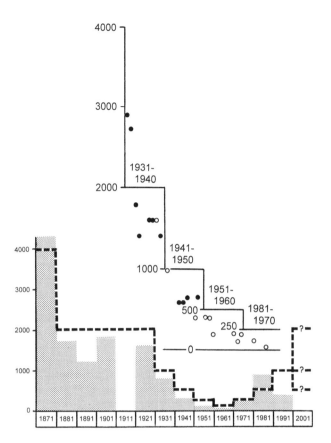

FIGURE 14.7 A summary of the horseshoe crab fisheries catch statistics for Delaware Bay during the years of the king crab fertilizer industry, collected by various federal fisheries agencies for the decades 1871 through 1961 (based on Table 1, Shuster and Botton, 1985). The average number of crabs harvested, per decade, is given in thousands and the heavy dashed line represents a decrease or an increase of 50 percent at each step. The averages for the decades of 1971 to 1991 are from spawning survey data (Swan et al., 1999). The upper graph provides the available annual data points, 1930 to 2000: the solid circles represent the total Delaware Bay harvest and the open circles when only New Jersey harvests were reported (the scale is twice that of the bottom graph).

per year (Dean R. Sanders, 1996, personal communication). By 1977, during peak spawning on 2 km of beach south of the Rutgers Laboratory, there were an estimated 63,000 males and 12,600 females. That represented a great increase in the spawning population during a time in which commercial fishing pressure for fertilizer and bait was minimal. Even after the bait fishery was well under way, Botton et al. (1988) routinely counted over 100 crabs per 15 m on this beach during the 1986 breeding season (= 6,667 per km).

The king crab fertilizer industry has passed into history. Its decline was probably due to a combination of several factors: competition with chemical fertilizers, complaints about offensive odors, and the relatively low abundance of *Limulus*. A century later, with the resurgence of a fishery for horseshoe crabs in Delaware Bay and elsewhere, other concerns are being voiced, including adverse impacts on migratory shorebirds and species preservation. Now the stakes are much higher and more geographically widespread (Berkson and Shuster, 1999), as described in Chapter 15.

HORSESHOE CRAB CONSERVATION:
A COAST-WIDE MANAGEMENT PLAN

Carl N. Shuster, Jr., Mark L. Botton,
and Robert E. Loveland

After centuries of use, *Limulus* has recently become the center of attention as a fisheries resource requiring management. This recognition arose over a contention—whether commercial fishing activities would diminish the quantity of horseshoe crabs eggs on the beaches of Delaware Bay and elsewhere in the springtime. Thus, our first and last chapters share a common bond—the migratory shorebird. Also, there continues to be a concern about decreases in horseshoe crab populations elsewhere in North America and in the Indo-Pacific.

Current Uses of Horseshoe Crabs

The many roles of *Limulus* in nature and in human activities mark it as an important marine resource (Berkson and Shuster, 1999; Walls et al., 2002). In addition to the LAL industry (see Chapter 13), horseshoe crabs are also extensively used for bait in the eel and whelk (conch) fisheries. Ecotourism associated with the horseshoe crab and shorebird migrations is significant in the Cape May, N.J., area and elsewhere along Delaware Bay. Manion et al. (2000) analyzed the monetary importance of activities dependent on horseshoe crabs. In one approach they measured net changes in economic well-being or the economic value generated by the provision of goods or services. Their second consideration was the economic benefits created throughout a region by an industry. From these assessments, they estimated that over $260 million annually was generated by the use of horseshoe crabs in three major industries. The relative economic importance of these, out of a possible total score of 20, was biomedical (17), bait (2), and ecotourism (1). While you might question the precision of these rankings, there is little doubt that the biomedical industry is the most important economic use of horseshoe crabs.

Biomedical Uses

Horseshoe crabs are commercially important in the production of *Limulus* amebocyte lysate (LAL). Manion et al. (2000) estimated the value of LAL to regional economics to be over $222 million annually. Although a few hundred thousand horseshoe crabs are bled annually to produce lysate, the LAL industry is based on blood donation procedures. Bled crabs are either returned to the sea with attendant mortalities as low as 4 percent or used for bait under an Atlantic States Marine Fisheries Commission (ASMFC) interstate quota system (more on this later). Research is under way on a lysate alternative that will decrease the dependence of the biomedical industry on horseshoe crabs. Ding and Ho (2001) reported a procedure that may eventually eliminate, or at least reduce, the reliance on whole horseshoe crab blood as a raw material for LAL. They produced genetically engineered Factor C, an endotoxin-sensitive protein in the cascade of reactions that leads to the formation of the clot (see Chapter 12). Thus, another assay for endotoxin is possible (based on recombinant Factor C) that might be at least as sensitive as the present LAL procedure (Pepe and Chen, 2002).

Horseshoe Crab Bait Fishery

Harvesting of horseshoe crabs as bait for American eels *(Anguilla rostrata)* began as a cottage industry in the nineteenth century (Goode, 1887). Beginning in the 1980s, estimated harvests of some 350,000 adult crabs annually from the Middle Atlantic states corresponded with an expanding export market for eels and the lack of economically competitive baits (Botton and Ropes, 1987a). In contrast to the indiscriminate taking of males and females by the fertilizer industry, the eel bait fishery concentrates on the egg-bearing females. Before state and federal catch restrictions, horseshoe crabs taken by trawl were usually brought ashore, sorted, and then delivered to freezer storage facilities. But, in the past decade, when the states began to regulate daily catches, mostly females were kept due to their higher value to the fisher.

The fishery for whelk (*Busycon carica* and *Busycotypus caniculatum*) has also expanded rapidly in recent years. Concurrently, the commercial harvest of horseshoe crabs increased on the Atlantic coast. From Florida to Maine, the average annual landings of 44 million pounds (1971–

1975) increased to 1,589 million pounds annually during 1991–1995, and 3,520 million pounds annually by 1997 (National Marine Fisheries Service, 1998). By 1998–1999, the estimated value of the whelk (71 percent) and eel pot (29 percent) fisheries was about $28 million annually (Manion et al., 2000). Anecdotal information from fishermen in the Delaware Bay area indicates that the eel fishery has remained steady and represents no more than 10 percent of the bait used for whelk. Another distinction between the fisheries is that egg-laden crabs are the choice bait for eels, whereas both male and female horseshoe crabs are used to bait whelk pots.

One indication of the quantity of horseshoe crabs taken as bait for eels and whelks comes from an observation in May 1992 by Botton and Loveland. A major hand-harvesting operation at Moores Beach, N.J., filled at least 150 large burlap sacks with horseshoe crabs that were taken directly off the beach as they came ashore to spawn. Given a conservative estimate of 10 crabs per sack, this one group removed 1,500 crabs per day for about a 10-day period. Many people who visited Moores Beach, Thompsons Beach, Reeds Beach, and other New Jersey beaches in the early 1990s to observe shorebirds witnessed similar hand harvesting. A crisis was created when their outcries about conservation issues and the impacts on the migratory shorebirds were joined by others (Berkson and Shuster, 1999). In 1996, officials from New Jersey and Delaware reported an annual catch of over 800,000 horseshoe crabs from the Delaware Bay area (Peter Himchak and Stewart Michels, personal communications). This is a level of harvest approaching that recorded during the king crab fertilizer industry a hundred years ago (see Chapter 14).

Estimating Delaware Bay Bait Use Munson (1998) used information collected by Delaware and New Jersey fisheries agencies and the Delaware Bay Waterman's Association from the licensed fishermen to estimate the number of horseshoe crabs used as bait in the Delaware Bay eel and whelk fisheries. Using only the numbers that Munson believed to be realistic, the annual harvests in 1999 and 2000 could have been close to 2 million crabs ± 500,000 per year. Demand for horseshoe crabs also depends on the abundance of blue crabs. If blue crabs are abundant, fishermen harvest them instead of whelk. When the water warms up and the blue crab catch declines, the fishing for whelk picks up. Depending on the supply, time of year, and use, a single horseshoe crab is worth from $0.40 to $1.25, so this fishery can be quite lucrative.

Reducing Dependence on Crabs as Bait Efforts are under way to develop artificial baits and reduce the dependence on crabs as bait in other fisheries. Nancy M. Targett and colleagues at the University of Delaware, for example, have isolated and characterized the chemical attractant in crab eggs and are experimenting with various carriers that will hold and exude the substance over a suitable period of time in the marine environment (Ferrari and Targett, 2003). In addition, studies conducted at Associates of Cape Cod have established that the hemolymph, a by-product of lysate production, is an attractant (patent pending; Novitsky, 2002 personal communication). Independently, Frank "Thumper" Eicherly IV, a commercial fisherman from Bowers Beach, Del., has found that bunkers (menhaden) injected with *Limulus* hemolymph are attractive whelk bait.

Bait bags have been used in other fisheries for many years. Their use in the whelk fishery first gained attention at an ASMFC workshop in October 1999, when Frank Eicherly described how he reduced the need for horseshoe crab bait by up to 75 percent. He constructed the bags of plastic netting and suspended them in a whelk pot by a bungee cord. This prevented undesirable species from devouring the bait. Since then, Eicherly has modified the bait container and made whelk pots with a flap at the bottom for easier placement of the bait (personal communication, 2002).

Soon after the ASMFC workshop on alternative baits and trap design, other participants actively promoted the use of bait bags. Ecological Research and Development Group (ERDG) initiated a study to test the effectiveness of bait bags in reducing the demand of horseshoe crab bait in the whelk fishery. Working with the Virginia Institute of Marine Science and Virginia whelk fishermen, ERDG demonstrated that bait needs could be halved without a measurable loss in catch. Reduction of dependence on large quantities of horseshoe crabs was particularly needed in the Virginia conch fishery after the state quota was diminished by the ASMFC; use of the bait bags is now mandated by the state. ERDG has continued its goal to introduce bait bags throughout the whelk industry, from Virginia to Massachusetts. In 2001 and 2002, with support from the National Oceanic and Atmospheric Administration, National Marine Fisheries Service, ERDG distributed over 14,000 bait bags free of charge to whelk fishermen.

Horseshoe Crabs and Ecotourism

Ecotourism activities provide many hours of pleasure in the sighting of the crabs and the migratory shorebirds that feed on their eggs. Each spring, thousands of ecotourists travel to Delaware Bay to observe the horseshoe crabs and migratory shorebirds. Kerlinger and Wiedner (1991) estimated that the ecotourism associated with birding brought $5.5 million into the Cape May economy in 1988, with much of that linked to the horseshoe crab–shorebird phenomenon. The number of visitors has definitely increased since that time; various environmental groups, including the New Jersey Audubon Society and the Nature Conservancy, now lead regular tours of the area. The annual economic activity in the Cape May region associated with horseshoe crab–dependent tourism now ranges from $7 to $10 million. If we assume a similar but lesser pattern in Delaware and the rest of the Delmarva Peninsula, the total horseshoe crab–generated ecotourism could be between $10 and $15 million annually (Manion et al., 2000).

Horseshoe Crabs in Education and Research

Limuli became public relations stars soon after the British Broadcasting Corporation filmed them spawning on a Cape May shore in 1977 for a segment in the *Life on Earth* television series (Attenborough, 1979, 1986). This population of *Limulus* soon became the most filmed on the Eastern seaboard, attracting worldwide attention. Since that initial cinematic event, dozens of crews have filmed the spawning event and the foraging of the shorebirds. Amid strong expressions of concern about the fate of horseshoe crabs, numerous books, articles in newspapers and magazines, and web sites have chronicled the story of horseshoe crabs and their biological and ecological significance. This heightened public awareness has also resulted in increased funding of research on the population biology of the horseshoe crab, its relationship to the expanding commercial fishery, and a coast-wide management plan.

In comparison to the bait fisheries, the quantities of crabs displayed at aquaria and used in teaching, in research, and in preparation of biomedical products (because the crabs were returned to the sea after bleeding) hardly register a blip in the total harvest records. No one has estimated the local economic value of research on horseshoe crabs, but it is supportive of and basic to other uses of horseshoe crabs. Often many years pass before basic scientific research results in tangible economic benefits. Significantly, research on horseshoe crabs over the years has increased the understanding of human anatomy and biological processes, including vision, immunity, and the nerve-muscle complex of the heart. *Limulus* amebocyte lysate and other discoveries described in this book have also had a significant role in human health.

The horseshoe crab–migratory shorebird connection provides an educational experience in the vicinity of New York City (Riepe, 2001). In an educational project called *Green Eggs and Sand,* middle-school science teachers from Delaware, New Jersey, Pennsylvania, and Maryland have participated in workshops conducted each spring since 2000.

Discussions with experts and field trips provide information used in developing teaching modules on the shorebirds and the ecology of horseshoe crabs (Oates, 1999a, b; Kreamer, 2000). Also in Delaware, ERDG (www.horseshoecrab.org) started a conservation program based on the slogan "Just Flip 'Em." Besides rescuing overturned crabs, flipping them reduces the number of dead, odiferous crabs that accumulate on beaches. Other ERDG projects include poster and writing contests for school children and constructing a horseshoe crab museum and research facility in Milton, Del.

Horseshoe crabs have never enjoyed any popularity as seafood in North America, except perhaps among Native Americans, whereas horseshoe crabs are an exotic dish in Asia and, with improvements in transportation, are increasingly eaten despite the danger of a tetrodotoxin (Kungsuwan et al., 1987). *Carcinoscorpius rotundicauda* is much esteemed as food in portions of Southeast Asia (Figure 15.1). In China, more *Tachypleus* spp. are now probably used for food than for lysate (TAL) production (Morton, 1999).

FIGURE 15.1 A horseshoe crab perched on a basket of clams at a market in Bangkok, Thailand. The source of the photograph is unknown, but it was among those obtained at a photographic supply store around 1954 by C. N. Shuster, Jr. Food poisoning occasionally occurs from eating these crabs *(Kungsuwan et al., 1987)*.

Monitoring Horseshoe Crab Activities

While there are a number of approaches that can be used and questions that can be asked in studying horseshoe crab populations, we will briefly consider only some that concern the migratory shorebirds, spawning sites, and the intensity of spawning, migrations, and population sizes.

What Should the Primary Research Focus Be?

If the fate of migratory shorebirds in Delaware Bay is the central concern, should we be placing our maximum effort on a spawning survey or on enumeration of eggs? Or would studies on the efficiency of the birds' digestive processes or on alternative foods also be valuable? Because the migratory shorebirds are abundant only in areas where horseshoe crabs abound, during the early part of the crab spawning season, should this be a factor in how long and to what extent either egg counts or spawning counts are made?

How did this dependence on the eggs arise? We surmise, from our observations of small, outlying flocks of migratory shorebirds, that the dependence on horseshoe crab eggs in the Delaware Bay area is because the eggs attract many more birds than can be supported by other food items. In other areas, where no horseshoe crabs exist, small flocks feeding on intertidal mud flats in tidal creeks on the Delmarva Peninsula appear to be healthy and feeding voraciously. Thus, in selecting the huge food resource of horseshoe crab eggs of Delaware Bay, the migratory birds are ecological opportunists (see Chapter 1). If this food resource disappeared, would migratory birds even out their numbers along all present stopover sites along the Atlantic coast where other food is available, albeit in lesser amounts? How long would it take for this to occur? How detrimental would it be to the birds in the long run?

Spawning Activity

A number of factors that affect spawning have been considered in the annual Delaware Bay survey, including the impact of weather, especially wave amplitude, on the numbers of spawners (Shuster, 1955, 1982); same tide stage observations of spawning (based on the slack water sampling method initiated by the Delaware Water Pollution Commission, State of Delaware, 1959); the window of time when spawning peaks during a period of high tides (Barlow et al., 1986, and local observations); the need for Bay-wide synoptic observations (Shuster and Botton, 1985; Finn et al., 1991); changes in spawning intensity throughout the season (Barlow et al., 1986; Swan et al., 1991–2001); and the tendency for spawning to peak during twilight and night hours (Barlow et al., 1986, and local observations). A counting frame was designed to count hordes of spawners (Figure 15.2).

Smith et al. (2002b) considered these and other factors. Their report on the initial results of a statistical framework for future spawning surveys along the sandy shores of Delaware Bay provides statistical substantiation for what was already known from earlier surveys—the

FIGURE 15.2 The Finn/Shuster counting cage. In 1990, spawners occupied bands 3 m wide in many places along a beach strand some 1,000 m long. Because most of the animals were covered by water, a 1 by 2 m counting cage was used in 1991 to count those spawners. Carl N. Shuster III used tubular steel to frame the cage and covered the sides with a 2 in chicken wire (later replaced by plastic mesh). The cage was light enough to be handled by one person. After 1991, it was no longer needed due to the marked decrease in spawners.

HORSESHOE CRAB CONSERVATION

greater the number of beaches surveyed, the greater the statistical certainty of the numbers of spawners and the greater the amount of time and number of people, the more reliable the survey. Questions remain, however, about priorities. Should more be done? In designing a spawning survey, does it make any difference if an individual spawner is counted several times each year? Are the ages and conditions of the crabs critical to successful spawning (Brockmann, 2002)? If so, how should these factors be monitored? Will the impacts of rapid changes in environmental conditions on spawning, as represented by the matrix of temperature, salinity, and wave height, be accounted for in the statistical model? We agree that *Limulus* is a species that should be managed; but in the Delaware Bay area it is equally important to monitor the feeding success of the migratory shorebirds during their stopover, as is being done by Delaware and New Jersey agencies.

Migrations and Population Sizes

Adult horseshoe crabs have been tagged to determine their distribution and to estimate the size of a population (Shuster, 1950; Baptist et al., 1957), mortality rates of bled crabs (Rudloe, 1983), and the longevity of adult crabs (Shuster, 1950; Ropes, 1961; Swan, personal communication). The largest tagging program was initiated in 1988 in the Delaware Bay area by James Finn, a lysate producer (Finn-Tech Industries), and continued by Benjie Lynn Swan after Finn's death in 1991 (Shuster, 1997a, b). After spawning, the crabs may move up and down along the same shoreline, on either the Delaware or New Jersey side of the Bay (Table 15.1). More distant ranging takes several weeks and includes travel across the Bay and between the Bay and the continental shelf. One of the conditions in the ASMFC management plan is the requirement that lysate producers tag crabs.

Stock Assessment Surveys

In response to questions about the efficacy of finfish collecting gear to collect horseshoe crabs, an ASMFC task force considered appropriate gear and what data to collect. This effort resulted in the decision to mimic commercial trawling. During the fall of 2001, exploratory trawling was conducted for ASMFC, in the vicinity of the horseshoe crab refuge off Delaware Bay by the Horseshoe Crab Research Center, Virginia Tech (Dave Hata and Jim Berkson, personal communications).

Data on other bottom-dwelling marine arthropods may have been skewed because they were studied as if they were fish (blue crabs—Cronin, 1998; lobsters—Corson, 2002). We wonder, therefore, whether only trawling or dredging should be used in the study of *Limulus* on

TABLE 15.1 A temporal and spatial summary of the recovery of tagged horseshoe crabs in Delaware Bay *(courtesy Benjie Lynn Swan, Limuli Laboratories, 2002)*. Time of recovery is shown in days (within 1 week), months (from 1 week to the next spawning season), and years (the second spawning season and afterward). Crabs tagged in the Bay and recaptured within the Bay were either on the same side of the Bay or had crossed from one side of the Bay to the other. Crabs also moved in and out of the Bay to or from the continental shelf.

Time	Same Side of Bay	Cross-Bay	Out of Bay onto Shelf	From Shelf into Bay	Totals
Years	158	69	28	58	313
Months	35	2	26	8	71
Days	414	0	4	0	418
Totals	607	71	58	66	802

the shelf? Or should alternative sampling/observation methods also be used, such as a benthic sled equipped with video gear (Munson and Oates, personal communication, 1999). Would scuba gear be useful? Robert Diaz (the Virginia Institute of Marine Science, personal communication), while employing a sled mounted with TV equipment, often encountered horseshoe crabs during a benthic study off Fenwick Island, Del. Jon Hulburt (personal communication), an accomplished amateur diver, has sighted and filmed *Limulus* during dives on the continental shelf and at the entrance to Delaware Bay, and Kurtzke (2001) has demonstrated the usefulness of scuba gear and videography in observing horseshoe crabs in depths up to 45 feet in Jamaica Bay, Long Island, New York.

The Impetus for Management

By the 1990s the need for the management of horseshoe crabs in the Delaware Bay area was precipitated by the perception that the migratory shorebirds were being impacted by the increasing commercial fisheries for the crabs and by the loss of spawning habitat. The pertinent states responded by restricting the harvests. When fishermen went elsewhere to obtain the crabs, the impact on other horseshoe crab populations was soon felt up and down the Atlantic coast. Thus, the status of the horseshoe crab population in the Delaware Bay area has generated much controversy, with conflicts escalating among various user groups, particularly between the bait fishermen and conservation organizations. The central arguments involve the reliability of information on the number of crabs in the population and the level of stress that a reduction in the number of horseshoe crab eggs places on the migratory shorebirds. Neither question has been fully answered but several kinds of relevant information are available. Yet, if it were not for the concern about the fate of the migratory shorebirds that flock to the Delaware Bay area each spring, there probably would not be a coast-wise horseshoe crab management plan today (Eagles, 2001).

Although regulations to manage the horseshoe crab fishery in Delaware Bay and elsewhere have been in existence only a short time, the biological information on which they are based has been accumulating for several decades (Loveland, 2001). Annual spawning and trawl fisheries surveys and egg counts have indicated a downward trend in horseshoe crab abundance. Starting with the annual Delaware Bay shoreline count of spawners, peak numbers of crabs in 1990 and 1991 were around 1,300,000 each year. This was followed by a sharp decline, and by 1996 there were at least 435,000 spawners. Lower numbers, as in 1995, may have been due, at least in part, to unfavorable weather conditions.

The overall Delaware Bay spawning population has markedly decreased since 1991. Through 1995, Michels (1997) reported that the number of adult horseshoe crabs counted in trawl surveys on the Delaware side of Delaware Bay paralleled the trends in the spawning survey. Richard Weber (1997 and 1998, personal communication) and Botton and Loveland (personal communication) have reported changes in the quantities of crab eggs in Delaware Bay beaches. The latter documented a decrease in the numbers of horseshoe crab eggs in comparing 1990 egg counts (as a baseline) with 1996 and 1997 counts that occurred at two sampling depths: 0–5 cm representing the egg resource available to foraging shorebirds (Figure 15.3A; Botton et al., 1994), and 15–20 cm, a depth at which a large fraction of the egg-laying effort occurs (Figure 15.3B).

Despite these data on the declining abundance of horseshoe crabs in Delaware Bay during the 1990s, watermen believed that the crabs were just as numerous as they had ever been. During these years, states reduced the number of permits issued for the harvesting of crab. It seems, however, that the remaining fishermen compensated by increasing their fishing effort because total landings continued to increase. This is reminiscent of the increase in fishing effort that occurred during the decline associated with the king crab fertilizer industry. Population trends are further obscured by the longevity of horseshoe crabs, which tends to buffer sudden changes in population size. The lack of definitive answers strongly indicates that more extensive studies of the horseshoe

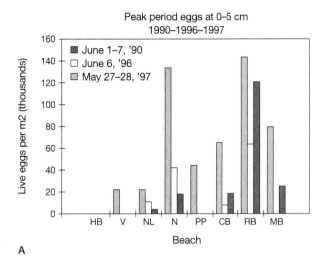

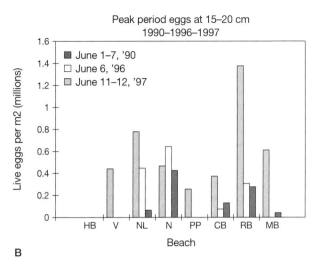

FIGURE 15.3 The abundance of horseshoe crab eggs on the New Jersey shore of Delaware Bay in late May–early June 1990 (solid bars), 1996 (stippled bars), and 1997 (open bars). Samples were taken every 3 m along a transect from the high tide line to the foot of the beach and summed along the entire transect. Refer to Botton et al. (1994) for sampling methodology. HB = Higbees Beach (not sampled), 1996 or 1997, V = Villas, NL = Norbury's Landing, N = North Beach, PP = Pierces Point (not sampled 1996 or 1997), CB = Cooks Beach, RB = Reeds Beach, MB = Moores Beach. (A) The abundance of eggs in surface (0–5 cm) sediments. (B) The abundance of eggs at 15–20 cm sediment depths.

crab population are needed. Early in the controversy, restrictions were placed on the taking of horseshoe crabs by New Jersey, Delaware, and Maryland—those states most concerned about the Delaware Bay stock of horseshoe crabs. State regulations are varied, numerous, and continually being modified (Appendix 13 in Farrell and Martin, 1997; Eagles, 2001).

Issues in the Conservation of Horseshoe Crabs

Conservation strategy is based on the fact that *Limulus* has a 20-year life cycle, that it can be highly migratory, and that multiple age-classes of adults participate in spawning. After spawning on beaches in Delaware Bay, many of the animals disperse over the inshore continental shelf from New Jersey south to Delaware, Maryland, and Virginia (Botton and Haskin, 1984; Botton and Ropes 1987a, 1987b; Swan and Giordano, 2001). It is quite likely that most of the horseshoe crabs collected by trawlers in the middle Atlantic region are part of the Delaware Bay spawning population. For these reasons, the coordinated regional approach advocated by Loveland et al. (1997) was adopted in the ASMFC (1998) management plan. Reports on the status of small populations include those in Pleasant Bay, Mass. (Carmichael et al., 2003) and along the coast of Maine (Schaller, 2002).

Fishery Impacts on a Small Population
The report by Widener and Barlow (1999) documents the impact of a fishery on a local population of horseshoe crab, in the absence of any

other demonstrable factors. In 15 years, from 1984 to 1999 at Mashnee Dike on Buzzards Bay, Mass., the horseshoe crab population declined more than 80 percent and its spawning activity decreased 94 percent. Their data substantiating the decline was an unexpected outcome of the annual quantification of spawning dynamics within three 10 m–square quadrats set at intervals along a 50 m transect at the water's edge. The spawning crabs were then counted in each quadrat every 30 minutes for a period of 2 to 3 hours as the tide flooded and ebbed. One of the changes in the spawning activity, before and after commercial harvesting of the crabs, was a marked contraction of the spawning season, from 59 days in 1984 to 11 days in 1999. Evidence that the population was local was based on the recovery of crabs, a year or more after tagging, and scuba observations of tagged crabs in amplexus during the winter. Thus, harvesting large numbers of horseshoe crabs from a small population can have a significant impact on its size that is much more obvious than in a large population such as that in Delaware Bay.

Morton (1999) reported the decline of populations of *Tachypleus tridentatus* and *Carcinoscorpius rotundicauda* in the vicinity of Hong Kong, due to overfishing and the destruction and pollution of breeding areas, and argued that a ban on fishing is necessary. The situation is especially severe in Japan, where *T. tridentatus* is officially an endangered species; dikes and polders have encroached upon and eliminated prime spawning sites (Botton, 2001). This situation has engendered special attention. The Japanese Society for the Preservation of Horseshoe Crabs is very active in promoting the conservation of the horseshoe crab. In Kasaoka City, the Horseshoe Crab Museum supports public education and runs a breeding program that is helping to restore natural populations along the Seto Inland Sea.

Crabs on the Continental Shelf

The trawl fishery on the continental shelf, with vessels based in Delaware, New Jersey, Maryland, and Virginia ports, has landed increasingly large numbers of horseshoe crabs during the past two decades. In the years 1994 and 1995 alone, the trawlers accounted for about 65 percent of all horseshoe crabs landings in New Jersey (Himchak, 2001). More recently, state and federal regulations, as described later in this chapter, greatly curtailed that fishery. Harvesting of crabs from the continental shelf in the vicinity of Delaware Bay is notable because these crabs are clearly part of the Bay population (Botton and Haskin, 1984; Botton and Ropes, 1987b). This migration in and out of Delaware Bay has been indicated by recovery of 802 tagged horseshoe crabs (out of some 30,000 tagged between 1986 and 2001) (Benjie Lynn Swan and Tom

O'Connell, personal communications). Horseshoe crabs tagged and released on the beaches and on trawlers have been recovered in Delaware Bay and up and down the coast from Staten Island, N.Y., to Cape Charles, Va. (see Table 15.1). These tagging results suggest that adult horseshoe crabs in the Delaware Bay area have localized movements within the Bay during the spawning season (recovery within days) and longer migrations within the Bay or between the Bay and shelf (recoveries months and years later). Specific routes and the duration of stops at particular sites are still unknown, particularly after the crabs have left the Bay. This may soon be clarified because researchers are beginning to use acoustic, radio, and other monitoring techniques.

John Ropes, based on his earlier experience in reporting on the Plum Island Sound population (Baptist et al., 1957), helped to develop the procedures used during the collection of data on the horseshoe crabs encountered in National Marine Fisheries Service (NMFS) finfish trawl and clam dredge surveys on the Middle Atlantic shelf (Ropes et al., 1982; Botton and Ropes, 1987a, 1987b). The bulk of the Middle Atlantic population resides in relatively shallow waters (less than 30 m) between New Jersey and Maryland (Botton and Ropes, 1987b). Moreover, large numbers of immature horseshoe crabs are being captured on the shelf by the smaller trawlers (Shuster, unpublished observations, 1988–2002). This observation raises several questions, such as how soon do the juveniles arrive on the shelf from the bays, at what size (age), and is there a sex difference in crabs caught at different locations? Because the larger, subadult juveniles are among those being harvested, what is the magnitude of the impact on the local population? Further, when discerning age groups in the harvests (young, mid-ages, and old adults), the young are sometimes in the majority. Fisheries managers need to consider the impacts of removing the youngest reproductively active adults from the population.

Harvesting Two Portions of the Same Population

Hand harvesting of horseshoe crabs from their breeding beaches has been the most controversial aspect of the Delaware Bay fishery because of its visibility to the public and the potential disturbances to foraging shorebirds. But the trawl fishery on the continental shelf may actually be impacting the Delaware Bay spawning population more heavily.

During the autumn of 1998, observations of the trawl fishery on the continental shelf revealed an undetected critical impact on the horseshoe crab. Shuster observed that a sizable number of large subadult females and virgin adult females were being collected. These juveniles become sexually mature adults when they molt in late summer and early autumn (Figure 15.4). Virgin females are clearly identifiable as

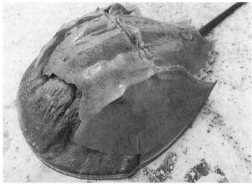

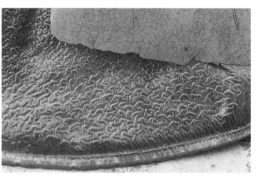

FIGURE 15.4 Three views of a large juvenile female premolt carapace (about 23 cm wide = 9 in wide) captured by a trawl on the continental shelf off Maryland. It was in the process of molting and the thin, brittle, outer, chitinous old shell was fractured during the trawl. These views show the soft, pliable, and leathery character of a large juvenile crab at the time of molting. The pleats in the new carapace are best seen in the bottom illustration. *Photograph by C. N. Shuster, Jr., courtesy of Captain J. Eustler.*

they have ultra-clean, lustrous shells and no mating scars (Figure 15.5). Thus, depleting the large subadult and recently matured females can have a serious impact on one or more year classes, decreasing their numbers before they have a chance to spawn even once. Harvesting of these females may remove up to 8 years of future reproductive output (Botton and Ropes, 1988).

Environmental Considerations

Some of the reported decreases in horseshoe crab populations may be attributable to factors other than commercial fishing. Botton et al. (1988) and Botton (2001) reported that horseshoe crabs are sensitive to

FIGURE 15.5 Unloading a small catch of horseshoe crabs from state-controlled waters off Ocean City, Md., in October 1999. These catches are restricted as well as monitored and data are gathered by the Maryland Department of Natural Resources Fisheries Services. The six crabs on the loading escalator are first-year females with clean shells and no mating scars (see Figure 3.1B). *1998 photographs by C. N. Shuster, Jr., courtesy of Captain J. Eustler.*

habitat quality with maximum numbers of spawners occurring on undeveloped sandy beaches. Coastal zone development, particularly the groins and bulkheads used to protect waterfront houses and roads, are destructive because they block access to spawning sites. Monitoring efforts must be cognizant of the influence of beach quality (for example, slope, width, sediment diameter) on horseshoe crab spawning (Smith et al., 2002a). Habitat loss has already taken place along the New Jersey shore of Delaware Bay, where Botton et al. (1988) estimated that equivalent amounts of disturbed and optimal horseshoe crab spawning habitats existed. Pressure to protect shoreline structures is likely to increase as global warming exacerbates coastal erosion.

Accelerated coastal erosion could deplete horseshoe crab spawning habitat. In the short term, erosion of sand could expose peat layers, which would reduce the suitability of a beach for spawning (Botton et al., 1988). Under natural conditions, sandy beaches, dunes, and coastal wetlands retreat landward as sea level rises (for example, Hull and Titus, 1986; Oerlemans, 1989; Titus, 1990; Titus et al., 1991). However, armored coastlines will behave very differently. Construction of bulkheads and other fixed structures makes an area unsuitable for horseshoe crab spawning. We might learn from the experience in Japan, where *Tachypleus tridentatus* once bred in abundance (Botton, 2001). Once-productive spawning beaches were replaced with vast stretches of concrete embankments and polders that were constructed to reclaim land for industry or agriculture. Tatara Beach, considered one of the prime sanctuary areas in all of Japan, is little more than an 18 by 28 m length of sand where two concrete walls meet at a right angle. In July 1994, we counted a maximum of four mated pairs per day on this beach (Botton et al., 1996).

It has been postulated that beach replenishment of Delaware Bay shores may create or enhance habitats beneficial to horseshoe crabs and shorebirds (U.S. Army Corps of Engineers, 1997). The results of a multiyear project on an isolated Delaware beach in a wildlife refuge strongly indicate that enhancement does occur within a few years following beach replenishment (David B. Carter, personal communication). Carter has expressed guarded optimism that beach replenishment could be a viable approach to improving horseshoe crab nesting habitats, pending the evaluation of several factors, including the abundance of horseshoe crab nests (eggs) and the impact of groins on prevailing currents.

Pollution has been implicated as a major factor in the decline of *Tachypleus tridentatus* in Japan (Itow et al., 1991), but investigations with *Limulus* in the United States do not suggest the same phenomenon. Botton et al. (1998a, 1998b) found that horseshoe crab embryos and larvae were highly tolerant of heavy metals (copper, zinc, and tributyltin) in comparison to similar stages in marine crustaceans. The high tolerance of horseshoe crab embryos and larvae to these metals is consistent with earlier reports showing resistance to oil (Laughlin and Neff, 1977; Strobel and Brenowitz, 1981), PCBs (Neff and Giam, 1977), and the pesticide diflubenzuron (Weis and Ma, 1987). At sublethal levels, heavy metals (particularly mercury and tributyltin) induce embryonic malformations (Itow et al., 1998b) and suppress limb regeneration (Itow et al., 1998a). Interestingly, malformed *Limulus* embryos are rare (< 0.6 percent) in Delaware Bay and Sandy Hook Bay, N.J. (Itow et al.,

1998b), but up to 20 percent of *Tachypleus tridentatus* embryos are abnormal in some areas of coastal Japan (Itow et al., 1991). Whether this indicates a heavier pollutant load in Japan, a greater susceptibility of the Japanese horseshoe crab to pollution, or both is not yet known. Loss of spawning habitat appears to be a more important environmental impact than water pollution, at least for *Limulus* in the mid–Atlantic area.

Developing a Coast-Wide Fisheries Management Plan

Except for the removal of horseshoe crabs from areas of soft clam production in Massachusetts, little was done at the state level prior to the 1990s to manage the crabs. State management of horseshoe crabs began in the mid–Atlantic region: in Delaware and New Jersey in the early 1990s and in Maryland in 1996. After first considering horseshoe crabs as part of the American eel fishery management plan, because the crabs are used as eel bait, the Atlantic States Marine Fisheries Commission (ASMFC) initiated steps in 1997 to develop a coast-wide horseshoe crab management plan.

A task force was organized in 1998 with members mainly from state and federal fisheries agencies, the commercial fisheries industry, and representatives from conservation organizations. Briefly, three committees (stock assessment, technical, and advisory) collaborated in the preparation of reports (for example, ASMFC, 2001a) and recommendations that were usually reviewed by a peer panel (for example, ASMFC, 1999) prior to submission to the ASMFC for action. There, the Horseshoe Crab Management Board, composed of several ASMFC commissioners, reviewed the reports before they were submitted to the entire body of ASMFC for final action.

After reviewing existing studies on tagging, spawning intensity, egg abundance, and mortality from bleeding, the ASMFC task force recommended increasing the intensity of those studies and investigating ways to ascertain the numbers of crabs on the continental shelf and the coast-wide standardization of data collection in fourteen areas (ASMFC, 1998). These, in order of priority, were (1) surveys of female spawners to provide an index of spawning population size; (2) mandatory catch and effort reporting for horseshoe crab, conch, and eel fisheries; (3) surveys of shorebird abundance and surveys of the number of horseshoe crab eggs eaten by the shorebirds; (4) a trawl survey specifically designed to sample horseshoe crab populations to estimate numbers of recruits, age structure, sex ratios, and so on; (5) egg count surveys (conducted in concert with spawning surveys) to provide needed stage-specific mortality information and data on the availability of eggs to shorebird populations; (6) tagging studies (mark-capture) to estimate mortality, the incidence of repeated spawning, and dispersal parameters; (7) genetic studies to identify any regional population structure; (8) estimates of parameters required to develop biological reference points (such as natural mortality rates, growth rates, and fecundity); (9) surveys of the habitat preferences of horseshoe crabs during all life stages including the amount of preferred habitat available to horseshoe crabs (to evaluate the effects of habitat destruction on horseshoe crab populations); (10) sampling of catch to provide information including sex ratio, size distribution, and age distribution; (11) estimates of the proportion of subtidal spawning; (12) mortality estimates during the entire biomedical collection process, from capture to port-return; (13) estimates of fishing discard numbers and associated mortality rates; and (14) the development of artificial bait to replace horseshoe crabs.

Research has been initiated on most of these recommendations, as evidenced by a number of current and recently completed studies that were displayed in a special horseshoe crab poster session (American Fisheries Society, 2002): aerial surveys (to monitor Delaware Bay spawning; to monitor a die-off in Florida); aging (lipofuscin concentration in juveniles' brains; appearance of the carapace of adults); chemical attractant (for use in artificial bait); habitat studies (microtidal system; using GIS [geographic information systems]; beach erosion); hemolymph (total volume; impact of bleeding; parameters); molecular biology (DNA markers; new tools); population dynamics (Pleasant Bay, Cape Cod; young of the year, Delaware Bay); spawning activity (male tactics, Florida); acoustic and radio tracking (Delaware Bay); by tagging, (Delaware Bay; Cape Cod, Mass.; Maine; a 10-year tagging study, Delaware Bay), and trawl survey design (for stock assessment).

The Horseshoe Crab Fisheries Management Plan (FMP) was recommended by the Horseshoe Crab Management Board, approved and adopted by the ASMFC in October 1998. It has four major goals: to conserve and protect the horseshoe crab resource, to maintain sustainable levels of spawning stock biomass, to ensure the continued role of *Limulus* in the ecology of coastal ecosystems, and to provide for continued use of horseshoe crabs over time. In Addendum I (April 2000), the HSC Management Board approved state plans to reduce horseshoe crab bait landings to at least 25 percent below 1995–1997 landing levels in the year 2000 in Massachusetts, Rhode Island, Connecticut, New York, New Jersey, Delaware, and Maryland. Notably, New Jersey and Maryland committed to continue existing harvest restrictions to reduce the 1995–1997 landings by 70 and 50 percent, respectively. The landings cap of 162,495 crabs for Virginia was also approved after negotiation in 2000.

In 2001, Addendum II was added to the FMP to regulate the transfer of harvest quotas from one state to another. It incorporated specific procedures to review each proposal for a state-to-state transfer, including considering: (1) whether discrete populations of horseshoe crabs exist along the coast (indicated by morphometric studies, see Shuster, 1955, 1982, and Riska 1981; by genetic studies, see Saunders et al., 1986, Selander et al., 1976, and Pierce et al., 2000), (2) whether transferring quotas from a region with a large horseshoe crab population to a region with a small population could threaten the smaller stock, and (3) impacts on competing uses, balancing the protection of migratory shorebird populations, the biomedical industry, and the fisheries (Berkson and Shuster, 1999, and Walls et al., 2002).

Establishment of a Fisheries Reserve

In April 2000, in response to evidence that young adult horseshoe crabs, especially the females on the continental shelf, were vulnerable to overexploitation, the ASMFC Board requested that the National Marine Fisheries Service investigate the feasibility of establishing a reserve off the mouth of Delaware Bay, within the exclusive economic zone (EEZ), from the edge of state-controlled waters at 3 mi to 200 naut mi offshore. Based on public review and fisheries considerations, an area encompassing a 30 naut mi radius was selected (U.S. Dept. Commerce, 2001). All harvesting of horseshoe crabs is banned from within the reserve except by special permit. Named the Carl N. Shuster, Jr., Horseshoe Crab Reserve, it occupies an area seaward from the mouth of Delaware Bay (Figure 15.6).

The extant horseshoe crabs are long-lived species (potentially reproductively active for up to 10 years) that produce stable, definable populations whose size is primarily governed by the availability of breeding sites, food, and harvest levels. Four factors govern their migrations, from hatching through adulthood: their numbers, their lack of gregariousness, the search for food, and the return to beaches to spawn. Fisheries managers can maintain stability by protecting the breeding areas and regulating the fisheries. The most destabilizing factor over the short term is overfishing, but long-term population stability will require preservation, and possibly replenishment, of spawning habitat. Maybe the time has come, if the management plan now in effect is maintained and no catastrophic event occurs, to let the horseshoe crabs do their thing in their own time and restore their populations.

This chapter outlines a management effort and related research, the results of which will probably be evident within the next decade or two. By then, fewer horseshoe crabs may be used for lysate production

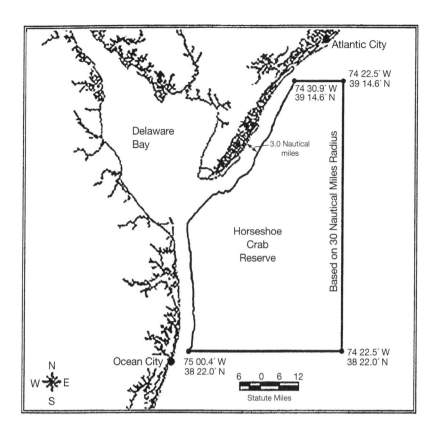

FIGURE 15.6 Three areas were considered for a horseshoe crab sanctuary (based on 15, 30, and 60 naut mi radii). The intermediate area was selected and named the Carl N. Shuster, Jr., Horseshoe Crab Reserve *(U.S. Department of Commerce, 2001).*

and bait and the larger horseshoe crab populations will increase. We may not see these same results in small, more local populations in which the size of the population is more easily impacted by harvest levels and environmental conditions. Because horseshoe crab larvae have very limited disposal potential (Botton and Loveland, 2003), repopulation of an area can occur only through adult or juvenile migration.

Meanwhile, in areas frequented by the migratory birds on their refueling stopovers, we need to know for certain how many available horseshoe crab eggs are needed to sustain migratory shorebird population. Perhaps we should also seek an answer to another question, from an evolutionary viewpoint: Is the shorebird–horseshoe crab connection, as in Delaware Bay, truly obligatory over the long term? Regardless, horseshoe crabs are worth conserving.

AUTHORS AND CONTRIBUTORS

REFERENCES

ACKNOWLEDGMENTS

INDEX

AUTHORS AND CONTRIBUTORS

Lyall I. Anderson
Curator of Invertebrate Palaeontology
Department of Geology and Zoology
National Museums of Scotland
Chamber Street
Edinburgh EH1 1JF, Scotland, UK

Peter B. Armstrong
Professor of Molecular and Cellular
  Biology
Division of Biological Sciences
University of California, Davis
Davis, CA 95616–8755

Robert B. Barlow
Professor of Ophthalmology
Director, Center for Vision Research
SUNY Upstate Medical University
Syracuse, NY 13210

Mark L. Botton
Professor of Biology
Department of Natural Sciences
Fordham University
New York, NY 10023

H. Jane Brockmann
Professor of Zoology
University of Florida
Gainesville, FL 32611

Brian A. Harrington
Manomet Center for Conservation
  Sciences
Box 1770
Manomet, MA 02345

Raymond P. Henry
Professor of Zoology
Auburn University
Auburn, AL 36849–5414

H. Donald Hochstein
Deputy Director, CBER (retired)
11313 Orleans Way
Kensington, MD 20895–1018

John A. Keinath
175 Herbal Lane
Midway, GA 31320

Louis Leibovitz (deceased, 1998)
Emeritus Professor
Department of Microbiology and
  Immunology
College of Veterinary Medicine
Cornell University
Ithaca, NY 14853

Jack Levin
Professor of Laboratory Medicine
Professor of Medicine
University of California School of
    Medicine
San Francisco, CA 94143

Gregory A. Lewbart
Associate Professor of Aquatic Medicine
Department of Clinical Sciences
College of Veterinary Medicine
North Carolina State University
Raleigh, NC 27606

Robert E. Loveland
Associate Professor of Ecology and
    Evolution
Cook College
Rutgers, The State University of New
    Jersey
New Brunswick, NJ 08903

David Mizrahi
Cape May Bird Observatory
Center for Research and Education
Cape May Court House, NJ 08210

Thomas J. Novitsky
President, Associates of Cape Cod, Inc.
Falmouth, MA 02540

Maureen K. Powers
Senior Scientist
Gemstone Foundation
Mission Viejo, CA 92692

Koichi Sekiguchi
354 Ushigasane
Kisai-machi
Saitama 347–01
Japan

Carl N. Shuster, Jr.
Adjunct Professor of Marine Science
Virginia Institute/School of Marine
    Science
The College of William and Mary
Gloucester Point, VA 23062

David W. Towle
Director, Marine DNA Sequencing
    Center
Mount Desert Island Biological
    Laboratory
P. O. Box 35, Old Bar Harbor Road
Salsbury Cove, ME 04672

Nellie Tsipoura
Natural Resources Defense Council
New York, NY 10011

REFERENCES

1 Synchronies in Migration

Alerstam, T., G. A. Gudmundsson, and K. Johannesson. 1992. Resources for long distance migration: intertidal exploitation of *Littorina* and *Mytilus* by knots *Calidris canutus* in Iceland. *Oikos* 65:179–189.

Baird, D., P. R. Evans, H. Milne, and M. Pienkowski. 1985. Utilization by shorebirds of benthic invertebrate production in intertidal areas. *Oceanogr. Mar. Biol. Ann. Rev.* 23:573–597.

Bent, A. C. 1962. *Life Histories of North American Shore Birds.* New York: Dover.

Botton, M. L. 1984. Effects of laughing gull and shorebird predation on the intertidal fauna at Cape May, New Jersey. *Est. Coast. Shelf Sci.* 18:209–220.

Botton, M. L., and R. E. Loveland. 1993. Predation by herring gulls and great black-backed gulls on horseshoe crabs. *Wilson Bull.* 105:518–521.

Botton, M. L., R. E. Loveland, and T. R. Jacobsen. 1988. Beach erosion and geochemical factors: influence on spawning success of horseshoe crabs *(Limulus polyphemus)* in Delaware Bay. *Mar. Biol.* 99:325–332.

———. 1994. Site selection by migratory shorebirds in Delaware Bay, and its relationship to beach characteristics and abundance of horseshoe crab *(Limulus polyphemus)* eggs. *Auk* 111:605–616.

Botton, M. L., and J. W. Ropes. 1987. The horseshoe crab, *Limulus polyphemus,* fishery and resource in the United States. *Mar. Fish. Rev.* 49:57–61.

Brewer, T. M. 1840. *Wilson's American Ornithology.* Boston: Otis, Broaders and Co.

Bryant, D. M. 1979. Effects of prey density and site character on estuary usage by overwintering waders (Charadrii). *Est. Coast. Mar. Sci.* 9:369–384.

Burger, J., L. Niles, and K. E. Clark. 1997. Importance of beach, mudflat and marsh habitats to migrant shorebirds on Delaware Bay. *Biol. Conserv.* 79:283–292.

Castro, G., and J. P. Myers. 1993. Shorebird predation on eggs of horseshoe crabs during spring stopover on Delaware Bay. *Auk* 110:927–930.

Castro, G., J. P. Myers, and A. R. Place. 1989. Assimilation efficiency of Sanderlings *(Calidris alba)* feeding on horseshoe crab *(Limulus polyphemus)* eggs. *Physiol. Zool.* 62:716–731.

Castro, G., J. P. Myers, and R. E. Ricklefs. 1992. Ecology and energetics of Sanderlings migrating to four latitudes. *Ecology* 73:833–844.

Clark, K. E., L. J. Niles, and J. Burger. 1993. Abundance and distribution of migrant shorebirds in Delaware Bay. *Condor* 95:694–705.

Fleischer, R. C. 1983. Relationships between tidal oscillations and ruddy turnstone flocking, foraging, and vigilance behavior. *Condor* 85:22–29.

Forbush, E. H. 1925. *Birds of Massachusetts and Other New England States. Part I. Water Birds, Marsh Birds, and Shore Birds.* Massachusetts Department of Agriculture.

Gonzalez, P. M., T. Piersma, and Y. Verkuil. 1996. Food, feeding, and refueling of red knots during northward migration at San Antonio Oeste, Rio Negro, Argentina. *J. Field Ornithol.* 67:575–591.

Goss-Custard, J. D. 1979. Effect of habitat loss on the numbers of overwintering shorebirds. In *Studies in Avian Biology,* no. 2, ed. F. A. Pitelka. Lawrence (Kansas): Allen Press. Pp. 167–177.

Grant, J. 1984. Sediment microtopography and shorebird foraging. *Mar. Ecol. Prog. Ser.* 19:293–296.

Gratto, G. W., M. L. H. Thomas, and C. L. Gratto. 1984. Some aspects of the foraging ecology of migrant juvenile sandpipers in the outer Bay of Fundy. *Can. J. Zool.* 62:1889–1892.

Haig, S. M., C. L. Gratto-Trevor, T. D. Mullins, and M. A. Colwell. 1997. Population identification of western hemisphere shorebirds throughout the annual cycle. *Mol. Ecol.* 6:413–427.

Harrington, B. A. 1982a. Morphometric variation and habitat use of semipalmated sandpipers during a migratory stopover. *J. Field Ornithol.* 53:258–262.

———. 1982b. Untying the enigma of the red knot. *Living Bird Quarterly* 1:4–7.

———. 1996. *The Flight of the Red Knot.* New York: W. W. Norton.

Harrington, B. A., J. M. Hagan, and L. E. Leddy. 1988. Site fidelity and survival differences between two groups of new world red knots *(Calidris canutus). Auk* 105:439–445.

Hicklin, P. W. 1987. The migration of shorebirds in the Bay of Fundy. *Wilson Bull.* 99:540–570.

Hicklin, P. W., and P. C. Smith. 1984. Selection of foraging sites and invertebrate prey by migrant semipalmated sandpipers, *Calidris pusilla* (Pallas), in Minas Basin, Bay of Fundy. *Can. J. Zool.* 62:2201–2210.

Holberton, R. L., J. D. Parrish, and J. C. Wingfield. 1996. Modulation of the adrenocortical stress response in neotropical migrants during autumn migration. *Auk* 113:558–564.

Howe, M. A., P. H. Geissler, and B. A. Harrington. 1989. Population trends of North American shorebirds based on the International Shorebird Survey. *Biol. Conserv.* 49:185–199.

Johnsgard, P. A. 1981. *The Plovers, Sandpipers, and Snipes of the World.* Lincoln, Nebraska: University of Nebraska Press.

Kelsey, M. G., and M. Hassall. 1989. Patch selection by Dunlin on a heterogeneous mudflat. *Ornis Scand.* 20:250–254.

Kersten, M., and T. Piersma. 1987. High levels of energy expenditure in shorebirds: metabolic adaptations to an energetically expensive way of life. *Ardea* 75: 175–187.

Knebel, H. J., C. H. Fletcher, III, and J. C. Kraft. 1988. Late Wisconsinian—Holocene paleogeography of Delaware Bay; a large coastal plain estuary. *Mar. Geol.* 83:115–133.

Mallory, E. P., and D. C. Schneider. 1979. Agonistic behavior in short-billed dowitchers feeding on a patchy resource. *Wilson Bull.* 91:271–278.

Mawhinnney, K., P. W. Hicklin, and J. S. Boates. 1993. A re-evaluation of the numbers of migrant semipalmated sandpipers, *Calidris pusilla,* in the Bay of Fundy during fall migration. *Can. Field-Nat.* 107:19–23.

McNeil, R., P. Drapeau, and J. D. Goss-Custard. 1992. The occurrence and adaptive significance of nocturnal habits in waterfowl. *Biol. Rev.* 67:381–419.

Morrison, R. I. G. 1984. Migration systems of some New World shorebirds. In *Shorebirds: Migration and Foraging Behavior,* eds. J. Burger and B. L. Olla. New York: Plenum Press.

Morrison, R. I. G., C. Downes, and B. Collins. 1994. Population trends of shorebirds on fall migration in eastern Canada, 1974–1991. *Wilson Bull.* 106:431–447.

Myers, J. P. 1986. Sex and gluttony on Delaware Bay. *Nat. Hist.* 95(5):68–77.

Myers, J. P., R. I. G. Morrison, P. Z. Antas, B. A. Harrington, T. E. Lovejoy, M. Sallaberry, S. E. Senner, and A. Tarak. 1987. Conservation strategy for migratory species. *Am. Sci.* 75:19–26.

Myers, J. P., M. Sallaberry, E. Ortiz, G. Castro, L. M. Gordon, J. L. Maron, C. T. Shick, E. Tabilo, P. Antas, and T. Below. 1990. Migration routes of New World sanderlings *(Calidris alba). Auk* 107:172–180.

Peterson, B. J., and B. Fry. 1987. Stable isotopes in ecosystem studies. *Ann. Rev. Ecol. Syst.* 181:293–320.

Piersma, T. 1987. Production by intertidal benthic animals and limits to their predation by shorebirds: a heuristic model. *Mar. Ecol. Prog. Ser.* 38:187–196.

Puttick, G. M. 1984. Foraging and activity patterns in wintering shorebirds. In *Shorebirds: Migration and Foraging Behavior,* eds. J. Burger and B. L. Olla. New York: Plenum Press.

Quammen, M. L. 1982. Influence of subtle substrate differences on feeding by shorebirds on intertidal mudflats. *Mar. Biol.* (Berl.) 71:339–343.

Recher, H. F., and J. A. Recher. 1969. Some aspects of the ecology of migrant shorebirds, part II: Aggression. *Wilson Bull.* 81:140–154.

Sewell, M. A. 1996. Detection of the impact of predation by migratory shorebirds: an experimental test in the Fraser River estuary, British Columbia (Canada). *Mar. Ecol. Prog. Ser.* 144:23–40.

Shuster, C. N., Jr. 1982. A pictorial review of the natural history and ecology of the horseshoe crab *Limulus polyphemus,* with reference to other Limulidae. In *Physiology and Biology of Horseshoe Crabs: Studies on Normal and Environmentally Stressed Animals,* eds. J. Bonaventura, C. Bonaventura, and S. Tesh. New York: Alan R. Liss.

Shuster, C. N., Jr., and M. L. Botton. 1985. A contribution to the population biology of horseshoe crabs, *Limulus polyphemus* (L.), in Delaware Bay. *Estuaries* 8:363–372.

Skagen, S. K., and H. D. Oman. 1996. Dietary flexibility of shorebirds in the western hemisphere. *Can. Field-Nat.* 110:419–444.

Stone, W. 1937. *Bird Studies at Old Cape May.* Philadelphia: Delaware Bay Ornithological Club.

Sullivan, K. A. 1986. Influence of prey distribution on aggression in ruddy turnstones. *Condor* 88:376–378.

Tsipoura, N., and J. Burger. 1999. Shorebird diet during spring migration stopover on Delaware Bay. *Condor* 101:635–644.

Tsipoura, N., C. G. Scanes, and J. Burger.

1999. Corticosterone and growth hormone levels in shorebirds during spring and fall migration stopover. *J. Exp. Zool.* 284:645–651.

Urner, C. A., and R. W. Storer. 1949. The distribution and abundance of shorebirds on the north and central New Jersey coast, 1928–1938. *Auk* 66:177–194.

Velasquez, C. R., and P. A. R. Hockey. 1992. The importance of supratidal foraging habitats for waders at a south temperate estuary. *Ardea* 80:243–253.

Wander, W., and P. Dunne. 1981. Species and numbers of shorebirds on the Delaware Bayshore of New Jersey—spring 1981. Occasional Paper 140, New Jersey Audubon Society. *Records of New Jersey Birds* 7:59–64.

Weber, L. M., and S. M. Haig. 1996. Shorebird use of South Carolina managed and natural coastal wetlands. *J. Wildl. Manage.* 80:73–82.

Weinstein, M. P., J. H. Balletto, J. M. Teal, and D. F. Ludwig. 1997. Success criteria and adaptive management for a large-scale wetland restoration project. *Wetlands Ecol. Manage.* 4:111–127.

Whitfield, D. P. 1990. Individual feeding specializations of wintering turnstone *Arenaria interpres. J. Anim. Ecol.* 59:193–211.

Wingfield, J. C., C. M. Vleck, and M. C. Moore. 1992. Seasonal changes of the adrenocortical response to stress in birds of the Sonoran desert. *J. Exp. Zool.* 264:419–428.

Wilson, W. H., Jr. 1989. Predation and the mediation of intraspecific competition in an infaunal community in the Bay of Fundy. *J. Exp. Mar. Biol. Ecol.* 132:221–245.

———. 1990. Relationship between prey abundance and foraging site selection by semipalmated sandpipers on a Bay of Fundy mudflat. *J. Field Ornithol.* 61:9–19.

———. 1991. The foraging ecology of migratory shorebirds in marine soft-sediment communities: The effects of episodic predation on prey populations. *Am. Zool.* 31:840–848.

2 Nesting Behavior

Ackerman, R. A. 1977. The respiratory gas exchange of sea turtle nest *(Chelonia, caretta). Respir. Physiol.* 31:19–38.

———. 1980. Physiological and ecological aspects of gas exchange by sea turtle eggs. *Am. Zool.* 20:575–583.

Alcock, J. 2001. *Animal Behavior.* Sunderland, MA: Sinauer Associates.

Badgerow, J. P., and M. A. Sydlik. 1989. Nest site selection in horseshoe crabs *(Limulus polyphemus)* using a Cape Cod Beach. *Am. Zool.* 29:36A.

Barash, L. 1993. Mass appeal. *Natl. Wildlife.* 31:14–19.

Barber, S. B. 1951. Contact chemoreception in *Limulus. Anat. Rec.* 111:561–562.

———. 1953. Action potential activity of *Limulus* chemoreceptor nerve fibers. *Anat. Rec.* 117:587.

Barber, S. B., and Itzkowitz, M. 1982. Crowding effects on hatching of *Limulus* embryos. *Am. Zool.* 22:879.

Barlow, R. B., M. K. Powers, H. Howard, H. and L. Kass. 1986. Migration of *Limulus* for mating: relation to lunar phase, tide height, and sunlight. *Biol. Bull.* 171:310–329.

———. 1987. Vision in *Limulus* mating and migration. In *Signposts in the Sea,* eds. W. F. Herrnkind and A. B. Thistle. Tallahassee, Fl.: Florida State University Press. Pp. 69–84.

Boaden, P. 1985. *An Introduction to Coastal Ecology.* New York: Chapman and Hall Publishers.

Botton, M. L. 1993. Predation by herring gulls and great black-backed gulls on horseshoe crabs. *Wilson Bull.* 105:518–521.

Botton, M. L., and R. E. Loveland. 1987. Orientation of the horseshoe crab, *Limulus polyphemus,* on a sandy beach. *Biol. Bull.* 173:289–298.

———. 1989. Reproductive risk: high mortality associated with spawning by horseshoe crabs *(Limulus polyphemus)* in Delaware Bay, USA. *Mar. Biol.* 101:143–151.

———. 1992. Body size, morphological constraints, and mated pair formation in four populations of horseshoe crabs *(Limulus polyphemus)* along a geographic cline. *Mar. Biol.* 112:409–415.

Botton, M. L., R. E. Loveland, and T. R. Jacobsen. 1988. Beach erosion and geochemical factors: influence on spawning success of horseshoe crabs *(Limulus polyphemus)* in Delaware Bay. *Mar. Biol.* 99:325–332.

———. 1992. Overwintering by trilobite larvae of the horseshoe crab *Limulus polyphemus* on a sandy beach of Delaware Bay (New Jersey, USA). *Mar. Ecol. Prog. Ser.* 88:289–292.

Brady, J. T., and E. Schrading. 1997. Habitat suitability index models: horseshoe crab (spawning beaches)—Delaware Bay, New Jersey and Delaware. U.S. Army Corps of Engineers.

Brafield, A. E. 1964. The oxygen content of interstitial water in sandy shores. *J. Anim. Ecol.* 33:97–116.

Brockmann, H. J. 1990. Mating behavior of horseshoe crabs, *Limulus polyphemus. Behaviour* 114:206–220.

———, in prep. A long-term study of spawning activity in a Florida Gulf Coast population of horseshoe crabs *(Limulus polyphemus).*

Brockmann, H. J., and D. Penn. 1992. Male mating tactics in the horseshoe crab, *Limulus polyphemus. Anim. Behav.* 44:653–665.

Burger, J., M. A. Howe, D. C. Hahn, and J. Chase. 1977. Effects of tide cycles on habitat selection and habitat partitioning by migrating shorebirds. *Auk* 94:743–758.

Burton, R. S. 1983. Protein polymorphisms and genetic differentiation of marine invertebrate populations. *Mar. Biol. Lett.* 4:193–206.

Cavanaugh, C. M. 1975. Observations on mating behavior in *Limulus polyphemus. Biol. Bull.* 149:422.

Cohen, J. A., and H. J. Brockmann. 1983. Breeding activity and mate selection in the horseshoe crab, *Limulus polyphemus. Bull. Mar. Sci.* 33:274–281.

deFur, P. L. 1988. Systemic respiratory adaptations to air exposure in intertidal decapod crustaceans. *Am. Zool.* 28:115–124.

Eagle, G. A. 1983. The chemistry of sandy beach ecosystems: a review. In *Sandy Beaches as Ecosystems,* ed. A. McLachlan. The Hague: Dr. W. Junk. Pp. 203–224.

Eagles, D. A. 1973a. Lateral spine mechanoreceptors in *Limulus polyphemus. Comp. Biochem. Physiol.* 44A:557–575.

———. 1973b. Tailspine movement and its motor control in *Limulus polyphemus. Comp. Biochem. Physiol.* 46A:391–407.

Ehlinger, G., and Bush, M. 2001. Identification of cause and extent of horseshoe crab population declines at Canaveral National Seashore. Titusville, FL: National Park Service, Canaveral National Seashore. Pp. 1–63.

Evans, P. R. 1988. Predation of intertidal fauna by shorebirds in relation to time of the day, tide and year. In *Behavioral Adaptation to Intertidal Life,* eds. G. Chelazzi, G. and M. Vannini. New York: Plenum Press. Pp. 65–78.

Gordon, M. S. 1960. Anaerobiosis in marine sandy beaches. *Science* 132:616–617.

Hasler, A. D. 1960. Guideposts of migrating fishes. *Science* 132:785–792.

Hayes, W. F. 1971. Fine structure of the chemoreceptor sensillum in *Limulus. J. Morphol.* 133:205–240.

Hayes, W. F., and S. B. Barber. 1967. Proprioceptor distribution and properties in *Limulus* walking legs. *J. Exp. Zool.* 165:195–210.

———. 1982. Peripheral synapses in *Limulus* chemoreceptors. *Comp. Biochem. Physiol.* 72A:287–293.

Hays, G. C., and J. R. Speakman. 1993. Nest placement by loggerhead turtles, *Caretta caretta. Anim. Behav.* 45:47–53.

Horrocks, J. A., and N. M. Scott. 1991. Nest site location and nest success in the hawksbill turtle *Eretmochelys imbricata* in Barbados, West Indies. *Mar. Ecol. Prog. Ser.* 69:1–8.

Howard, H. A., R. W. Fiordalice, M. D. Camara, L. Kass, M. K. Powers, and R. B. Barlow. 1984. Mating behavior of *Limulus:* relation to lunar phase, tide height, and sunlight. *Biol. Bull.* 167:527.

Laughlin, R. 1983. The effects of temperature and salinity on larval growth of the horseshoe crab *Limulus polyphemus. Biol. Bull.* 164:93–103.

Lockwood, S. 1870. The horse foot crab. *Am. Nat.* 4:257–274.

Maurmeyer, E. 1978. Geomorphology and evolution of transgressive estuarine washover barriers along the western shore of Delaware Bay. Ph.D. thesis, University of Delaware.

McLachlan, A., and I. Turner. 1994. The interstitial environment of sandy beaches. *Mar. Ecol.* 15:177–211.

Miyazaki, J., K. Sekiguchi, and T. Hirabayashi. 1987. Application of an improved method of two-dimensional electrophoresis to the systematic study of horseshoe crabs. *Biol. Bull.* 172:212–224.

Mrosovsky, N. 1983. Ecology and nest-site selection of leatherback turtles *Dermochelys coriacea. Biol. Conserv.* 26:47–56.

Palumbi, S. R. 1992. Marine speciation on a small planet. *Trends Ecol. Evol.* 7:114–118.

Palumbi, S. R., and B. A. Johnson. 1982. A note on the influence of life-history stage on metabolic adaptation: the responses of *Limulus* eggs and larvae to hypoxia. In *Physiology and Biology of Horseshoe Crabs: Studies on Normal and Environmentally Stressed Animals,* eds. J. Bonaventura, C. Bonaventura, and S. Tesh. New York: Alan R. Liss. Pp. 115–124.

Penn, D. 1992. The adaptive significance of nest-site selection and spawning synchronization in horseshoe crabs (*Limulus polyphemus*). M.S. thesis, University of Florida.

Penn, D., and H. J. Brockmann. 1994. Nest-site selection in the horseshoe crab, *Limulus polyphemus. Biol. Bull.* 187:373–384.

———. 1995. Age-biased stranding and righting in horseshoe crabs, *Limulus polyphemus. Anim. Behav.* 49:1531–1539.

Pierce, J. C., G. Tan, and P. M. Gaffney. 2000. Delaware Bay and Chesapeake Bay populations of the horseshoe crab *Limulus polyphemus* are genetically distinct. *Estuaries* 23:690–698.

Recher, H. F. 1966. Some aspects of the ecology of migrant shorebirds. *Ecology* 47:393–407.

Recher, H. F., and J. A. Recher. 1969. Some aspects of the ecology of migrant shorebirds. II: Aggression. *Wilson Bull.* 81:140–154.

Riska, B. 1981. Morphological variation in the horseshoe crab *Limulus polyphemus. Evolution* 35:647–658.

Rudloe, A. 1979. Locomotor and light responses of larvae of the horseshoe crab, *Limulus polyphemus* (L.). *Biol. Bull.* 157:494–505.

———. 1980. The breeding behavior and patterns of movement of horseshoe crabs, *Limulus polyphemus,* in the vicinity of breeding beaches in Apalachee Bay, Florida. *Estuaries* 3:177–183.

———. 1985. Variation in the expression of lunar and tidal behavioral rhythms in the horseshoe crab, *Limulus polyphemus. Bull. Mar. Sci.* 36:388–395.

Rudloe, A., and W. F. Herrnkind. 1976. Orientation of *Limulus polyphemus* in the vicinity of breeding beaches. *Mar. Behav. Physiol.* 4:75–89.

———. 1980. Orientation by horseshoe crabs, *Limulus polyphemus,* in a wave tank. *Mar. Behav. Physiol.* 7:199–211.

Saunders, N. C., L. G. Kessler, and J. C. Avise. 1986. Genetic variation and geographic differentiation in mitochondrial DNA of the horseshoe crab, *Limulus polyphemus. Genetics* 112:613–627.

Sekiguchi, K., ed. 1988. *Biology of Horseshoe Crabs.* Tokyo: Science House Co.

Sekiguchi, K., Y. Yamamichi, and J. D. Costlow. 1982. Horseshoe crab developmental studies I: Normal embryonic development of *Limulus polyphemus* compared with *Tachypleus tridentatus*. In *Physiology and Biology of Horseshoe Crabs: Studies on Normal and Environmentally Stressed Animals,* eds. J. Bonaventura, C. Bonaventura, and S. Tesh. New York: Alan R. Liss. Pp. 53–73.

Shuster, C. N., Jr. 1948. On the gross anatomy and histology of the alimentary tract in early developmental stages of *Xiphosura (Limulus) polyphemus* Linn. M.S. thesis, Rutgers University.

———. 1953. Odyssey of the horseshoe crab. *Audubon Magazine* 55:162–167.

———. 1955. On morphometric and serological relationships within the Limulidae, with particular reference to *Limulus polyphemus* (L.). Ph.D. thesis, New York University (1958 Diss. Abstr. 18:371–372).

———. 1960. Horseshoe "crabs." *Estuarine Bull.* 5:3–8.

———. 1979. Distribution of the American horseshoe "crab," *Limulus polyphemus* (L.). In *Biomedical Applications of the Horseshoe Crab (Limulidae),* ed. E. Cohen. New York: Alan R. Liss. Pp. 3–26.

———. 1982. A pictorial review of the natural history and ecology of the horseshoe crab *Limulus polyphemus,* with reference to other Limulidae. In *Physiology and Biology of Horseshoe Crabs: Studies on Normal and Environmentally Stressed Animals,* eds. J. Bonaventura, C. Bonaventura, and S. Tesh. New York: Alan R. Liss. Pp. 1–52.

———. 1993. Egg-laying activities of the American horseshoe crab *Limulus polyphemus.* unpub. ms.

———. 1995. On the evolution of mating in horseshoe crabs (Chelicerata: Merostomata: Xiphosurida). unpub. ms.

Shuster, C. N., Jr., and M. L. Botton.

1985. A contribution to the population biology of horseshoe crabs, *Limulus polyphemus* (L.) in Delaware Bay. *Estuaries* 8:363–572.

Sokoloff, A. 1978. Observations on populations of the horseshoe crab *Limulus* (=Xiphosura) *polyphemus. Res. Pop. Ecol.* 19:222–236.

Strahler, A. N. 1966. Tidal cycle of changes in an equilibrium beach, Sandy Hook, New Jersey. *J. Geol.* 74:247–268.

Swan, B. L. 1996. Annual survey of horseshoe crab spawning activity along the shores of Delaware Bay: 1990–1995 Summary. In *Horseshoe Crab Forum,* eds. J. Farrell and C. Martin. Lewes, DE: University of Delaware Sea Grant College Program. Pp. 35–39.

Teale, E. W. 1957. The oldest migration. *Natural Hist.* 66:364–369.

Tewksbury, H. T., II, and D. O. Conover. 1987. Adaptive significance of intertidal egg deposition in the Atlantic Silverside *Menidia menidia. Copeia* 1:76–83.

Thompson, C., and C. H. Page. 1975. Nervous control of respiration: oxygen-sensitive elements in the prosoma of *Limulus polyphemus. J. Exp. Biol.* 62:545–554.

Thompson, W. F. 1919. The spawning of the grunion. *Fish. Bull.* 3:2–29.

Truchot, J.-P. 1987. How do intertidal invertebrates breathe both water and air? In *Comparative Physiology: Life in Water and on Land,* eds. P. Dejours, L. Bolis, C. R. Taylor, and E. R. Weibel. Padova: Livian Press. Pp. 37–47.

Vosatka, E. D. 1970. Observations on the swimming, righting, and burrowing movements of young horseshoe crabs, *Limulus polyphemus. Ohio J. Sci.* 70:276–283.

Waterman, T. H., and D. F. Travis. 1953. Respiratory reflexes and the flabellum of *Limulus. J. Cell. Comp. Physiol.* 42:261–290.

Wyse, G. A. 1971. Receptor organization and function in *Limulus* chelae. *Z. Vergl. Physiol.* 73:249–273.

3 Male Competition and Satellite Behavior

Agresti, A. 1996. *An Introduction to Categorical Data Analysis.* New York: John Wiley & Sons.

Alcock, J. 2001. *Animal Behavior.* Sunderland, MA: Sinauer Associates.

Alcock, J., G. C. Eickwort, and K. R. Eickwort. 1977. The reproductive behavior of *Anthidium maculosum* (Hymenoptera: Megachilidae) and the evolutionary significance of multiple copulations by females. *Behav. Ecol. Sociobiol.* 2:385–396.

Armstrong, P. B. 1991. Cellular and humoral immunity in the horseshoe crab, *Limulus polyphemus.* In *Immunology of Insects and Other Arthropods,* ed. A. P. Gupta. Boca Raton, Fla.: CRC Press. Pp. 3–17.

Bang, F. B. 1979. Ontogeny and phylogeny of response to gram-negative endotoxins among the marine invertebrates. In *Biomedical Applications of the Horseshoe Crab (Limulidae),* ed. E. Cohen. New York: Alan R. Liss. Pp. 109–123.

Bannon, G. A., and Brown, G. G. 1980. Vesicle involvement in the egg cortical reaction of the horseshoe crab, *Limulus polyphemus. Dev. Biol.* 76:418–427.

Barlow, R. B., M. K. Powers, H. Howard, and L. Kass. 1986. Migration of *Limulus* for mating: relation to lunar phase, tide height, and sunlight. *Biol. Bull.* 171:310–329.

———. 1987. Vision in *Limulus* mating and migration. In *Signposts in the Sea,* eds. W. F. Herrnkind and A. B. Thistle. Tallahassee, Fla.: Florida State University Press. Pp. 69–84.

Barnard, C. J., and R. M. Sibly. 1981. Producers and scroungers: a general model and its application to captive flocks of house sparrows. *Anim. Behav.* 29:543–550.

Barthel, K. W. 1974. *Limulus:* a living fossil. Naturwissenschaften 61:428–433.

Botton, M. L. 1993. Predation by herring gulls and great black-backed gulls on horseshoe crabs. *Wilson Bull.* 105:518–521.

Botton, M. L., and R. E. Loveland. 1988. Sexual selection in crabs: testing the null hypothesis. *Am. Zool.* 28:52A.

———. 1989. Reproductive risk: high mortality associated with spawning by horseshoe crabs (*Limulus polyphemus*) in Delaware Bay, USA. *Mar. Biol.* 101:143–151.

———. 1992. Body size, morphological constraints, and mated pair formation in four populations of horseshoe crabs (*Limulus polyphemus*) along a geographic cline. *Mar. Biol.* 112:409–415.

Botton, M. L., and J. W. Ropes. 1988. An indirect method for estimating longevity of the horseshoe crab (*Limulus polyphemus*) based on epifaunal slipper shells (*Crepidula fornicata*). *J. Shellfish Res.* 7:407–412.

Brockmann, H. J. 1986. Decision making in a variable environment: lessons from insects. In *Behavioral Ecology and Population Biology,* ed. L. C. Drickamer. Toulouse, France: Privat, I.E.C. Pp. 95–111.

———. 1990. Mating behavior of horseshoe crabs, *Limulus polyphemus. Behaviour* 114:206–220.

———. 1996. Satellite male groups in horseshoe crabs, *Limulus polyphemus. Ethology* 102:1–21.

———. 2001. The evolution of alternative strategies and tactics. *Adv. Study Behav.* 30:1–51.

———. 2002. An experimental approach to altering mating tactics in male horseshoe crabs (*Limulus polyphemus*). *Behav. Ecol.* 13:232–238.

Brockmann, H. J., T. Colson, and W. Potts. 1994. Sperm competition in horseshoe crabs (*Limulus polyphemus*). *Behav. Ecol. Sociobiol.* 35:153–160.

Brockmann, H. J., and D. Penn. 1992. Male mating tactics in the horseshoe crab, *Limulus polyphemus. Anim. Behav.* 44:653–665.

Brockmann, H. T., C. Nguyen, and W.

Potts. 2000. Paternity in horseshoe crabs when spawning in multiple-male groups. *Anim. Behav.* 60:837–849.

Brooks, R. 1998. The importance of mate copying and cultural inheritance of mating preferences. *Trends Ecol. Evol.* 13:45–46.

Brown, G. G. 1976. Scanning electron-microscopical and other observations of sperm fertilization reactions in *Limulus polyphemus* L. (Merostomata: Xiphosura). *J. Cell Sci.* 22:547–562.

Brown, G. G., and S. R. Barnum. 1983. Postfertilization changes in the horseshoe crab *Limulus polyphemus* L. *Biol. Bull.* 164:163–175.

Brown, G. G., and D. L. Clapper. 1980. Cortical reaction in inseminated eggs of the horseshoe crab, *Limulus polyphemus. Dev. Biol.* 76:410–417.

———. 1981. Procedures for maintaining adults, collecting gametes, and culturing embryos and juveniles of the horseshoe crab, *Limulus polyphemus* L. In *Laboratory Animal Management, Marine Invertebrates,* eds. R. Hinegardner, J. Atz, R. Fay, M. Fingerman, R. Josephson, and N. Meinkoth, eds. Washington, D.C.: National Academy Press. Pp. 268–290.

Brown, G. G., and W. J. Humphreys. 1971. Sperm-egg interactions of *Limulus polyphemus* with scanning electron microscopy. *J. Cell Biol.* 51:904–907.

Brown, G. G., and J. R. Knouse. 1973. Effects of sperm concentration, sperm aging, and other variables on fertilization in the horseshoe crab, *Limulus polyphemus. Biol. Bull.* 144:462–470.

Carlsson, K., and G. Gade. 1986. Metabolic adaptation of the horseshoe crab, *Limulus polyphemus,* during exercise and environmental hypoxia and subsequent recovery. *Biol. Bull.* 171:217–235.

Cavanaugh, C. M. 1975. Observations on mating behavior in *Limulus polyphemus. Biol. Bull.* 149:422.

Charnov, E. 1982. *The Theory of Sex Allocation.* Princeton, N.J.: Princeton University Press.

Chevalier, R. L., and H. B. Steinbach. 1969. A chemical signal attracting the flatworm *Bdelloura candida* to its host, *Limulus polyphemus. Biol. Bull.* 137:394.

Clapper, D. L., and G. G. Brown. 1980. Sperm motility in the horseshoe crab, *Limulus polyphemus* L. *Dev. Biol.* 76:341–349.

Clapper, D. L., and D. Epel. 1985. The *Limulus* sperm motility-initiating peptide initiates acrosome reactions in sea water lacking potassium. *J. Exp. Zool.* 236:211–217.

Clark, C. 1994. Antipredator behavior and the asset-protection principle. *Behav. Ecol.* 5:159–170.

Clark, C., and M. Mangel. 1984. Foraging and flocking strategies: information in an uncertain environment. *Am. Nat.* 123:626–641.

Cohen, J. A., and H. J. Brockmann. 1983. Breeding activity and mate selection in the horseshoe crab, *Limulus polyphemus. Bull. Mar. Sci.* 33:274–281.

Curtis, L. 1987. Vertical distribution of an estuarine snail altered by a parasite. *Science* 235:1509–1511.

Davies, N. B., and T. R. Halliday. 1979. Competitive mate searching in male common toads, *Bufo bufo. Anim. Behav.* 27:1253–1267.

Dominey, W. J. 1981. Maintenance of female mimicry as a reproductive strategy in bluegill sunfish (*Lepomis macrochirus*). *Environ. Biol. Fish.* 6:59–64.

Emlen, S. T., and L. W. Oring. 1977. Ecology, sexual selection, and the evolution of mating systems. *Science* 197:215–223.

Fahrenbach, W. H. 1973. Spermiogenesis in the horseshoe crab, *Limulus polyphemus. J. Morphol.* 140:31–52.

Georges, M., A. S. Laquarre, M. Castelli, R. Hanset, and G. Vassart. 1988. DNA fingerprinting in domestic animals using four different minisatellite probes. *Cytogenet. Cell Genet.* 47:127–131.

Groff, J. M., and L. Leibovitz. 1982. A gill disease of *Limulus polyphemus* associated with triclad turbellarid worm infection. *Biol. Bull.* 163:92.

Gross, M. R. 1984. Sunfish, salmon, and the evolution of alternative reproductive strategies and tactics in fishes. In *Fish Reproduction: Strategies and Tactics,* eds. G. W. Potts and R. J. Wootton. New York: Academic Press. Pp. 55–75.

———. 1991a. Salmon breeding behavior and life history evolution in changing environments. *Ecology* 72:1180–1186.

———. 1991b. Evolution of alternative reproductive strategies: frequency-dependent sexual selection in male bluegill sunfish. *Phil. Trans. R. Soc. Lond. B* 332:59–66.

Halliday, T. 1983. Do frogs and toads choose their mates? *Nature* 306:226–227.

Halliday, T., and S. J. Arnold. 1987. Multiple mating by females: a perspective from quantitative genetics. *Anim. Behav.* 35:939–941.

Halliday, T. R., and P. A. Verrell. 1984. Sperm competition in amphibians. In *Sperm Competition and the Evolution of Animal Mating Systems,* ed. R. L. Smith. New York: Academic Press. Pp. 487–508.

Hanstrom, B. 1926. Das Nervensystem und die Sinnesorgane von *Limulus polyphemus. Lunds Universitets Arsskrift* 22:1–79.

Harrington, J. M., and P. B. Armstrong. 1999. A cuticular secretion of the horseshoe crab, *Limulus polyphemus:* a potential anti-fouling agent. *Biol. Bull.* 197:274–275.

———. 2000. Initial characteristics of a potential anti-fouling system in the American horseshoe crab, *Limulus polyphemus. Biol. Bull.* 199:189–190.

Hassler, C. L. 1999. Satellite male groups in horseshoe crabs *(Limulus polyphe-mus):* how and why are males choosing females. M.S. thesis, University of Florida.

Hassler, C. L., and H. J. Brockmann. 2001. Evidence for use of chemical cues by male horseshoe crabs when locating nesting females *(Limulus polyphemus). J. Chem. Ecol.* 27:2319–2335.

Hock, C. W. 1940. Decomposition of chitin by marine bacteria. *Biol. Bull.* 79:199–206.

Höglund, J. 1989. Pairing and spawning patterns in the common toad, *Bufo bufo:* the effects of sex ratios and the time available for male-male competition. *Anim. Behav.* 38:423–429.

Höglund, J., and J. G. M. Robertson. 1987. Random mating by size in a population of common toads *(Bufo bufo). Amphibia-Reptilia* 8:321–330.

Höglund, J., and S. Säterberg. 1989. Sexual selection in common toads: correlates with age and body size. *J. Evol. Biol.* 2:367–372.

Howard, R. 1978. The evolution of mating strategies in bullfrogs, *Rana catesbeiana. Evolution* 32:850–871.

———. 1980. Mating behavior and mating success in woodfrogs, *Rana sylvatica. Anim. Behav.* 28:705–716.

Huggins, L. G., and H. Waite. 1993. Eggshell formation in *Bdelloura candida,* an ectoparasitic Turbellarian of the horseshoe crab *Limulus polyphemus. J. Exp. Zool.* 265:549–557.

Hutchings, J. A., and R. A. Myers. 1988. Mating success of alternative maturation phenotypes in male Atlantic salmon, *Salmo salar. Oecologia* 75:169–174.

Jamieson, B. G. M. 1987. *The Ultrastructure and Phylogeny of Insect Spermatozoa.* New York: Cambridge University Press.

Jeffreys, A. J., V. Wilson, and S. L. Thein. 1985. Hypervariable "minisatellite" regions in human DNA. *Nature* 314:67–73.

Jordan, W. C., and A. F. Youngson. 1992. The use of genetic marking to assess the reproductive success of mature male Atlantic salmon parr *(Salmo salar,* L.) under natural spawning conditions. *J. Fish. Biol.* 41:613–618.

Kirkpatrick, M. 1987. Sexual selection by female choice in polygynous animals. *Ann. Rev. Ecol. Syst.* 18:43–70.

Knowlton, N., and S. R. Greenwell. 1984. Male sperm competition avoidance mechanisms: the influence of female interests. In *Sperm Competition and the Evolution of Animal Mating Systems,* ed. R. L. Smith. New York: Academic Press. Pp. 61–84.

Koons, B. F. 1883. Sexual characters of *Limulus. Am. Nat.* 17:1297–1299.

Leibovitz, L. 1986. Cyanobacterial diseases of the horseshoe crab *(Limulus polyphemus). Biol. Bull.* 171:482–483.

Leibovitz, L., and G. A. Lewbart. 1987. A green algal (chlorophycophytal) infection of the dorsal surface of the exoskeleton and associated organ structures, in the horseshoe crab, *Limulus polyphemus. Biol. Bull.* 173:430.

Lindberg, R. B., W. W. Inge, Jr., A. D. Mason, Jr., and B. A. Pruitt, Jr. 1972. Natural variation in sensitivity of *Limulus* amoebocyte lysate to endotoxin. *Fed. Proc.* 31:791Abs.

Loveland, R. E., and M. L. Botton. 1992. Size dimorphism and the mating system in horseshoe crabs *(Limulus polyphemus* L.). *Anim. Behav.* 44:907–916.

MacKenzie, C. L., Jr. 1962. Transportation of oyster drills by horseshoe "crabs." *Science* 137:36–37.

Maekawa, K., and H. Onozato. 1986. Reproductive tactics and fertilization success of mature male Miyabe charr, *Salvelinus malma miyabei. Environ. Biol. Fishes* 15:119–129.

Manning, J. T. 1985. Choosy females and correlates of male age. *J. Theoret. Biol.* 116:349–354.

Maynard Smith, J. 1978. *The Evolution of Sex.* Cambridge, England: Cambridge University Press.

McLachlan, A., and I. Turner. 1994. The interstitial environment of sandy beaches. *Mar. Ecol.* 15:177–211.

Milinski, M. 1981. Games fish play:

making decisions as a social forager. *Trends Ecol. Evol.* 3:325–330.

Milinski, M., and G. A. Parker. 1991. Competition for resources. In *Behavioral Ecology: An Evolutionary Approach,* eds. J. R. Krebs and N. B. Davies. Oxford: Blackwell Scientific. Pp. 122–147.

Moore, J. 2002. *Parasites and the Behavior of Animals.* Oxford: Oxford University Press.

Parker, G. 1990. Sperm competition games: raffles and roles. *Proc. R. Soc. Lond. B* 242:120–126.

Parker, G. A., and W. J. Sutherland. 1986. Ideal free distributions when individuals differ in competitive ability: phenotype-limited ideal free models. *Anim. Behav.* 34:1222–1242.

Patil, J. S., and A. C. Anil. 2000. Epibiotic community of the horseshoe crab *Tachypleus gigas. Mar. Biol.* 136:699–713.

Patten, W. 1912. *Dermal Skeleton of Limulus: The Evolution of the Vertebrates and Their Kin.* Philadelphia: P. Blakiston's Son and Co. Pp. 295–302.

Pearson, F. C., III, and M. Weary. 1980. The *Limulus* amebocyte lysate test for endotoxin. *Bioscience* 30:461–464.

Penn, D., and H. J. Brockmann. 1994. Nest-site selection in the horseshoe crab, *Limulus polyphemus. Biol. Bull.* 187:373–384.

———. 1995. Age-biased stranding and righting in horseshoe crabs, *Limulus polyphemus. Anim. Behav.* 49:1531–1539.

Petrie, M., M. Hall, T. Halliday, H. Budgey, and C. Pierpoint. 1992. Multiple mating in a lekking bird: why do peahens mate with more than one male and with the same male more than once? *Behav. Ecol. Sociobiol.* 31:349–358.

Pistole, T. G., and S. A. Graf. 1986. Antibacterial activity in *Limulus.* In *Hemocytic and Humoral Immunity in Arthropods,* ed. A. P. Gupta. New York: John Wiley and Sons. Pp. 331–344.

Pomerat, C. M. 1933. Mating in *Limulus polyphemus. Biol. Bull.* 12:243–252.

Pulliam, H. R., and T. Caraco. 1984. Living in groups: is there an optimal group size? In *Behavioral Ecology: An Evolutionary Approach,* eds. J. R. Krebs and N. B. Davies. Oxford: Blackwell Scientific. Pp. 122–147.

Queller, D. C., J. E. Strassmann, and C. R. Hughes. 1993. Microsatellites and kinship. *Trends Ecol. Evol.* 8:285–288.

Quigley, J. P. 1997. A hemolytic activity secreted by the endotoxin challenged horseshoe crab: a novel immune system operating at the surface of the carapace. *Biol. Bull.* 193:273.

Ridley, M. 1988. Mating frequency and fecundity in insects. *Biol. Rev.* 63:509–549.

Riedl, R. J., and R. Machan. 1972. Hydrodynamic patterns in lotic intertidal sands and their bioclimatological implications. *Mar. Biol.* 13:179–209.

Rudloe, A. 1980. The breeding behavior and patterns of movement of horseshoe crabs, *Limulus polyphemus,* in the vicinity of breeding beaches in Apalachee Bay, Florida. *Estuaries* 3:177–183.

Sargent, W. 1988. *The Year of the Crab.* New York: W. W. Norton and Co.

Sekiguchi, K., ed. 1988. *Biology of Horseshoe Crabs.* Tokyo: Science House Co.

Sekiguchi, K., H. Seshimo, and H. Sugita. 1988. Post-embryonic development of the horseshoe crab. *Biol. Bull.* 174:337–345.

Shine, R. 1988. The evolution of large body size in females: a critique of Darwin's "fecundity advantage" model. *Am. Nat.* 131:124–131.

Shoger, R. L., and D. W. Bishop. 1967. Sperm activation and fertilization in *Limulus polyphemus. Biol. Bull.* 133:485.

Shoger, R. L., and G. G. Brown. 1970. Ultrastructural study of sperm-egg interactions of the horseshoe crab, *Limulus polyphemus* L. (Merostomata: Xiphosura). *J. Submicr. Cytol.* 2:167–179.

Shuster, C. N., Jr. 1950. Observations on the natural history of the American horseshoe crab, *Limulus polyphemus.* Woods Hole Oceanogr. Inst. Contr. No. 564:18–23.

———. 1953. Odyssey of the horseshoe crab. *Audubon* 55:162–167.

———. 1955. On morphometric and serological relationships within the Limulidae, with particular reference to *Limulus polyphemus* (L.). Ph.D. thesis, New York University (1958 Diss. Abstr. 18:371–372).

———. 1982. A pictorial review of the natural history and ecology of the horseshoe crab *Limulus polyphemus,* with reference to other Limulidae. In *Physiology and Biology of Horseshoe Crabs: Studies on Normal and Environmentally Stressed Animals,* eds. J. Bonaventura, C. Bonaventura, and S. Tesh. New York: Alan R. Liss. Pp. 1–52.

———. 1993. Amplexus in the American horseshoe crab *Limulus polyphemus.* Unpub. ms.

Shuster, C. N., Jr., and M. L. Botton. 1985. A contribution to the population biology of horseshoe crabs, *Limulus polyphemus* (L.) in Delaware Bay. *Estuaries* 8:363–572.

Shuster, C. N., Jr., C. P. Randall, and Robert Shevock. 1961. Biometry of *Limulus* mating. Unpub. ms.: 100 pairs at Slaughter Beach, Del., on May 31, 1961.

Stagner, J. I., and J. R. Redmond. 1975. The immunological mechanisms of the horseshoe crab, *Limulus polyphemus. Mar. Fish. Rev.* 37:11–19.

Stunkard, H. W. 1951. Observations on the morphology and life-history of *Microphallus limuli* n. sp. (Trematoda: Microphallidae). *Biol. Bull.* 101:307–318.

———. 1953. Natural hosts of *Microphallus limuli.* Stunkard, 1951. *J. Parasitol.* 39:225.

Svard, L., and C. Wiklund. 1986. Different ejaculate delivery strategies in first versus subsequent matings in the swallowtail butterfly *Papilio macchaon* L. *Behav. Ecol. Sociobiol.* 18:325–330.

Sydlik, M. A., and L. L. Turner. 1989. Temporal variation in the composition of epibiotic assemblages found on horseshoe crab carapaces. *Am. Zool.* 29:37A.

———. 1990. Behavioral observations of horseshoe crabs moving along shore. *Mich. Acad. Sci.* 22:115–124.

Teale, E. W. 1957. The oldest migration. *Nat. Hist.* 66:364–369.

Thompson, W. F. 1919. The spawning of the grunion. *Fish. Bull.* 3:2–29.

Tilney, L. G. 1975. Actin filaments in the acrosomal reaction of *Limulus* sperm. *J. Cell Biol.* 64:289–310.

Turner, L. L., C. Kammire, and M. A. Sydlik. 1988. Preliminary report: composition of communities resident on *Limulus* carapaces. *Biol. Bull.* 175:312.

Vickery, W. L., L. Giraldeau, J. J. Templeton, D. L. Kramer, and C. A. Chapman. 1991. Producers, scroungers and group foraging. *Am. Nat.* 137:847–863.

Waltz, E. C. 1982. Alternative tactics and the law of diminishing returns: the satellite threshold model. *Behav. Ecol. Sociobiol.* 10:75–83.

Wasserman, G. S., and Z. Cheng. 1996. Electroretinographic measures of vision in horseshoe crabs with uniform versus variegated carapaces. *Biol. Signals* 5:247–262.

Watson, P. J. 1993. Foraging advantage of polyandry for female sierra dome spiders (*Linyphia litigiosa:* Linyphiidae) and assessment of alternative direct benefit hypothesis. *Am. Nat.* 141:440–465.

Watson, W. H., III. 1980a. *Limulus* gill cleaning behavior. *J. Comp. Physiol.* 141:67–75.

———. 1980b. Long-term patterns of gill cleaning, ventilation and swimming in *Limulus. J. Comp. Physiol.* 141:77–85.

Weatherhead, P. J. 1984. Mate choice in avian polygyny: why do females prefer older males? *Am. Nat.* 123:873–875.

Wells, K. D. 1977. The social behaviour of anuran amphibians. *Anim. Behav.* 25:666–693.

———. 1979. Reproductive behavior and male mating success in a neotropical toad, *Bufo typhonius. Biotropica* 11:301–307.

Westneat, D. F., P. W. Sherman, and M. L. Morton. 1990. The ecology and evolution of extra-pair copulations in birds. In: Power, D. M. (ed.), *Current Ornithology* 7:331–369.

Wheeler, W. M. 1894. *Syncoelidium pellucidum,* a new marine triclad. *J. Morphol.* 9:167–195.

Williams, D. P. 1955. The role of microorganisms in the settlement of *Ophelia bicornis* Savigny. *J. Mar. Biol. Assoc. U.K.* 34:513–543.

Zuk, M. 1988. Parasite load, body size and age of wild-caught male field crickets (Orthoptera: Gryllidae): effect on sexual selection. *Evolution* 42:969–976.

4 Seeing at Night and Finding Mates

Atherton, J. L., M. A. Krutky, J. M. Hitt, F. A. Dodge, and R. B. Barlow. 2000. Optic nerve responses of *Limulus* in its natural habitat at night. *Biol. Bull.* 199:176–178.

Barlow, R. B. 1986. From string galvanometer to digital computer: Haldane Keffer Hartline (1903–1983). *Trends Neurosci.* 6:552–555.

———. 1990. What the brain tells the eye. *Sci. Am.* April: 90–95.

Barlow, R. B., R. R. Birge, E. Kaplan, and J. Tallent. 1993. On the molecular origin of photoreceptor noise. *Nature* 366:64–66.

Barlow, R. B., S. J. Bolanowski, and M. L. Brachman. 1977. Efferent optic nerve fibers mediate circadian rhythms in the *Limulus* eye. *Science* 19:86–89.

Barlow, R. B., S. C. Chamberlain, and L. Kass. 1984. Circadian rhythms in retinal function. In *The Molecular and Cellular Basis of Visual Acuity,* eds. S. R. Hilfer and J. B. Sheffield. New York: Springer-Verlag. Pp. 31–35.

Barlow, R. B., S. C. Chamberlain, and H. K. Lehman. 1989. Circadian rhythms in the invertebrate retina. In *Facets of Vision,* eds. D. C. Stavenga and R. C. Hardie. New York: Springer-Verlag. Pp. 257–280.

Barlow, R. B., S. C. Chamberlain, and J. Z. Levinson. 1980. *Limulus* brain modulates the structure and function of the lateral eyes. *Science* 210:1037–1039.

Barlow, R. B., J. M. Hitt, and F. A. Dodge. 2001. *Limulus* vision in the marine environment. *Biol. Bull.* 200:169–176.

Barlow, R. B., L. C. Ireland, and L. Kass. 1982. Vision has a role in *Limulus* mating behavior. *Nature* 296:65–66.

Barlow, R. B., and E. Kaplan. 1971. *Limulus* lateral eye: properties of receptor units in the unexcised eye. *Science* 174:1027–1029.

———. 1977. Properties of visual cells in the lateral eye of *Limulus* in situ: intracellular recordings. *J. Gen. Physiol.* 69:2–3.

Barlow, R. B., E. Kaplan, G. H. Renninger, and T. Saito. 1987. Circadian rhythms in *Limulus* photoreceptors. I: Intracellular recordings. *J. Gen. Physiol.* 89:353–378.

Barlow, R. B., M. K. Powers, H. Howard, and L. Kass. 1986. Migration of *Limulus* for mating: relation to lunar phase, tide height, and sunlight. *Biol. Bull.* 171:310–329.

Batra, R., and R. B. Barlow. 1990. Efferent control of temporal response properties of the *Limulus* lateral eye. *J. Gen. Physiol.* 95:229–244.

Battelle, B-A. 2002. Circadian efferent input to Limulus eyes: anatomy, circuitry, and impact. *Microsc. Res. Tech.* 15:58(4):345–355.

Battelle, B-A., A. W. Andrews, B. G. Calman, J. R. Sellers, R. M. Greenberg, and W. C. Smith. 1998. A myosin III from Limulus eyes is a clock-regulated phosphoprotein. *J. Neurosci.* 18:4548–4559.

Battelle, B-A., and J. A. Evans. 1984.

Octopamine release from centrifugal fibers of the Limulus peripheral visual system. *Vis. Neurosci.* 17(2):71–79.

Battelle, B-A., J. A. Evans, and S. C. Chamberlain. 1982. Efferent fibers to *Limulus* eyes synthesize and release octopamine. *Science* 216:1250–1252.

Battelle, B-A., C. D. Williams, J. L. Schremser-Berlin, and C. Cacciatore. 2000. Circadian rhythms in *Limulus* photoreceptors. *Vis Neurosci.* 17(2):217–227.

Behrens, M. E., and J. L. Fahey. 1981. Slow potentials in nonspiking optic nerve fibers in the peripheral visual system of *Limulus. J. Comp. Physiol.* 141:239–247.

Calman, B. G., and B-A. Battelle. 1991. Central origin of the efferent neurons projecting to the eyes of Limulus polyphemus. *Vis. Neurosci.* 6:481–495.

Chamberlain, S. C., and R. B. Barlow. 1979. Light and efferent activity control rhabdom turnover in *Limulus* photoreceptors. *Science* 206:361–363.

Dodge, F. A., Jr., B. W. Knight, and J. Toyoda. 1968. Voltage noise in *Limulus* visual cells. *Science* 160:88–90.

Dowling, J. E. 1968. Discrete potentials in the dark-adapted eye of *Limulus. Nature* 217(5123):28–31.

Edwards, S. C., and B-A. Battelle. 1987. Octopamine—and cyclic AMP-stimulated phosphorylation of a protein in *Limulus* ventral and lateral eyes. *J. Neurosci.* 7(9):2811–2820.

Fahrenbach, W. H. 1973. The morphology of the *Limulus* visual system. V: Protocerebral neurosecretion and ocular innervation. *Z. Zellforsch.* 144:153–166.

Fuortes, M. G. F., and S. Yeandle. 1964. Probability of occurrence of discrete potential waves in the eye of *Limulus. J. Gen. Physiol.* 47:433–463.

Graham, C. H., and H. K. Hartline. 1935. The response of single visual sense cells to lights of different wave lengths. *J. Gen. Physiol.* 18:917.

Hanna, W. J., J. A. Horne, and G. H. Renninger. 1988. Circadian photoreceptor organs in *Limulus.* II:

The telson. *J. Comp. Physiol. A* 162:133–140.

Hartline, H. K. 1928. A quantitative and descriptive study of the electric response to illumination of the arthropod eye. *Am. J. Physiol.* 83:466–483.

———. 1934. Intensity and duration in the excitation of single photoreceptor units. *J. Cell. Comp. Physiol.* 5:229.

———. 1972. Visual receptors and retinal interaction. In *Nobel Lectures: Physiology or Medicine, 1963–1970.* New York: Elsevier. Pp. 269–288.

Hartline, H. K., and C. H. Graham. 1932. Nerve impulses from single receptors in the eye of *Limulus. Proc. Soc. Exp. Biol. Med.* 20.

Hartline, H. K., and P. R. McDonald. 1947. Light and dark adaptation of single photoreceptor elements in the eye of *Limulus. J. Cell. Comp. Physiol.* 30:225–254.

Hartline, H. K., and F. Ratliff. 1957. Inhibitory interaction of receptor units in the eye of *Limulus. J. Gen. Physiol.* 40:357.

———. 1958. Spatial summation of inhibitory influences in the eye of *Limulus,* and the mutual interaction of receptor units. *J. Gen. Physiol.* 41:1049.

Hartline, H. K., H. G. Wagner, and E. F. MacNichol, Jr. 1952. The peripheral origin of nervous activity in the visual system. *Cold Spring Harbor Symposia Quant. Biol.* 17:125–141.

Hartline, H. K., H. G. Wagner, and F. Ratliff. 1956. Inhibition in the eye of *Limulus. J. Gen. Physiol.* 39:651–673.

Herzog, E., and R. B. Barlow. 1991. Ultraviolet light from the nighttime sky enhances retinal sensitivity of *Limulus. Biol. Bull.* 181:321–322.

Herzog, E. D., M. K. Powers, and R. B. Barlow. 1996. *Limulus* vision in the ocean day and night: effects of image size and contrast. *Vis. Neurosci.* 13:31–41.

Hubbard, R., and G. Wald. 1960. Visual pigment of the horseshoe crab, *Limulus polyphemus. Nature* 186:212–215.

Kaplan, E., and R. B. Barlow. 1980. Cir-

cadian clock in *Limulus* brain increases response and decreases noise of retinal photoreceptors. *Nature* 286:393–395.

Kaplan, E., R. B. Barlow, G. Renninger, and K. Purpura. 1990. Circadian rhythms in *Limulus* photoreceptors. II: Quantum bumps. *J. Gen. Physiol.* 96:665–685.

Kass, L., and R. B. Barlow. 1984. Efferent neurotransmission of circadian rhythms in Limulus lateral eye. I: Octopamine-induced increases in retinal sensitivity. *J. Neurosci.* 4:908–917.

MacNichol, E. F., Jr. 1956. Visual receptors as biological transducers, in molecular structure and functional activity of nerve cells. *Am. Inst. Biol. Scis. Publ.* 1:34.

Manglapus, M. K., H. Uchiyama, N. F. Buelow, and R. B. Barlow. 1998. Circadian rhythms of rod-cone dominance in the Japanese quail retina. *J. Neurosci.* 18:4775–4784.

Millechia, R., and A. Mauro. 1969. The ventral photoreceptor cells of *Limulus.* II: The basic photoresponse. *J. Gen. Physiol.* 54:1199–1201.

Passaglia, C., F. A. Dodge, and R. B. Barlow. 1998. Cell based model of the *Limulus* lateral eye. *Am. Physiol. Soc.* 1800–1815.

Passaglia, C., F. Dodge, E. Herzog, S. Jackson, and R. B. Barlow. 1997. Deciphering a neural code for vision. *Proc. Natl. Acad. Sci.* 94:12649–12654.

Powers, M. K., and R. B. Barlow. 1985. Behavioral correlates of circadian rhythms in the *Limulus* visual system. *Biol. Bull.* 169:578–591.

Powers, M. K., R. B. Barlow, and L. Kass. 1991. Visual performance of horseshoe crabs day and night. *Vis. Neurosci.* 7:179–189.

Ratliff, F. 1965. *Mach Bands: Quantitative Studies on Neural Networks in the Retina.* San Francisco: Holden Day.

Ratliff, F., and H. K. Hartline. 1974. *Studies on Excitation and Inhibition in the Retina: A Collection of Papers from the Laboratories of H. Keffer Hartline.*

New York: Rockefeller University Press.

Ridings, C., D. Borst, K. Smith, F. Dodge, and R. Barlow. 2002. Visual behavior of juvenile Limulus in their natural habitat and in captivity. *Biol. Bull.* 203:224–225.

Tomita, T. 1956. The nature of action potentials in the lateral eye of the horseshoe crab as revealed by simultaneous intra- and extracellular recording. *Jap. J. Physiol.* 6:27–340.

5 Growing Up Takes about Ten Years and Eighteen Stages

Baptist, J. P. 1953. Record of a hermaphrodite horseshoe crab. *Breviora* No. 14:1–4.

Barthel, K. W. 1974. *Limulus:* a living fossil. Horseshoe crabs aid interpretation of an Upper Jurassic environment (Solnhofen). *Naturwissenschaften* 61:428–433.

Battey, T. F. 1883. The freshly hatched young of the horseshoe crab. *Am. Nat.* 17:91.

Bennett, J. 1979. The cytochemistry of *Limulus* eggs. *Biol. Bull.* 156: 141–156.

Botton, M. L., R. E. Loveland, and T. R. Jacobsen. 1988. Beach erosion and geochemical factors: influence on spawning success of horseshoe crabs *(Limulus polyphemus)* in Delaware Bay. *Mar. Biol.* 99:325–332.

———. 1992. Overwintering by trilobite larvae of the horseshoe crab *Limulus polyphemus* on a sandy beach of Delaware Bay (New Jersey, USA). *Mar. Ecol. Prog. Ser.* 88:289–292.

Botton, M. L., and J. W. Ropes. 1988. An indirect method for estimating longevity of the horseshoe crab *(Limulus polyphemus)* based on epifaunal slipper shells *(Crepidula fornicata)*. *J. Shellfish Res.* 7:407–412.

Brown, G. G., and D. L. Clapper. 1981. Procedures for maintaining adults, collecting gametes, and culturing embryos and juveniles of the horseshoe crab, *Limulus polyphemus* L. In *Laboratory Animal Management: Marine Invertebrates,* eds. Committee on Marine Invertebrates. Washington, D.C.: National Academy Press. Pp. 268–290.

Carmichael, R. H., D. Rutecki, and I. Valiela. 2003. Abundance and population structure of the Atlantic horseshoe crab, *Limulus polyphemus,* in Pleasant Bay, Cape Cod. *Mar. Ecol. Prog. Ser.* 246:225–239.

Dumont, J. N., and E. Anderson. 1967. Vitellogenesis in the horseshoe crab, *Limulus polyphemus. J. Microscopic* 6:791–806.

Eldredge, N. 1970. Observations on burrowing behavior in *Limulus polyphemus* (Chelicerata, Merostomata); with implications on the functional anatomy of trilobites. *Am. Mus. Novitates* 2436:1–17.

French, K. A. 1979. Laboratory culture of embryonic and juvenile *Limulus.* In *Progress in Clinical and Biological Research,* vol. 29: *Biomedical Applications of the Horseshoe Crab (Limulidae),* eds. E. Cohen, F. B. Bang, J. Levin, J. J. Marchalonis, T. G. Pistole, R. A. Prendergast, C. N. Shuster, Jr., and S. W. Watson. New York: Alan R. Liss. Pp. 61–71.

Gauvry, G. D., A. Hata, A. Wesche, S. F. Michels, and C. N. Shuster, Jr. 2002. Do adult female horseshoe crabs, *Limulus polyphemus,* molt? (unpublished manuscript).

Goto, S., and O. Hatori. 1927. Notes on the spawning habits and growth stages of the Japanese king crab. *Congres Internat. Xe. Ecol. A Budapest,* Ft. 2, Sect. IV:1147–1155.

Grady, S., P. D. Rutecki, R. Carmichael, and I. Valiela. 2001. Age structure of the Pleasant Bay population of *Crepidula fornicata:* a possible tool for estimating horseshoe crab age. *Biol. Bull.* 201:296–297.

Haefner, P. A., Jr., G. Gauvry, and C. N. Shuster, Jr. 2002. Poster. Reading horseshoe crab shells: the condition of the *Limulus* exoskeleton reveals much about the natural history of its possessor—its age, health, and activities. Am. Fish. Soc. annual meeting, Baltimore, Md.

Itow, T. 1993. Crisis in the Seto Inland Sea: the decimation of the horseshoe crab. *Ecol. Management Enclosed Coastal Seas (EMECS) Newsletter* No. 3:10–11.

———. 1997. My journey to U.S.A. for horseshoe crabs. Dept. Biology, Faculty of Education, Shizuoka Univ., pp. 7–11. [In Japanese]

Iwasaki, Y., T. Iwami, and K. Sekiguchi. 1988. Karyology. In *The Biology of Horseshoe Crabs,* ed. K. Sekiguchi. Tokyo: Science House Co. Pp. 309–314.

Jegla, T. C. 1982. A review of the molting physiology of the trilobite larvae of *Limulus.* In *Physiology and Biology of Horseshoe Crabs: Studies on Normal and Environmentally Stressed Animals,* eds. J. Bonaventura, C. Bonaventura, and S. Tesh. New York: Alan R. Liss. Pp. 83–101.

Jegla, T. C., and J. D. Costlow. 1982. Temperature and salinity effects on development and early posthatch stages of *Limulus.* In *Physiology and Biology of Horseshoe Crabs: Studies on Normal and Environmentally Stressed Animals,* eds. J. Bonaventura, C. Bonaventura, and S. Tesh. New York: Alan R. Liss. Pp. 103–113.

Kingsley, J. S. 1892. The embryology of *Limulus. J. Morphol.* 7:35–66.

———. 1893. The embryology of *Limulus. J. Morphol.* 8:195–268.

Krishnakum, A., and H. A. Schneiderman. 1970. Control of molting in mandibulate and chelicerate arthropods by ecdysones. *Biol. Bull.* 139:520–538.

Laverock, W. S. 1927. On the casting of the shell in *Limulus. Proc. & Trans. Liverpool Biol. Soc.* 41:13–16.

Lockwood, S. 1870. The horse foot crab. *Am. Nat.* 4:257–274.

———. 1884. Moulting of *Limulus. Am. Nat.* 18:200–201.

Makioka, T. 1988. Internal morphology. In *Biology of Horseshoe Crabs,* ed. K. Sekiguchi. Tokyo: Science House Co. Pp. 118–132.

Natur Cine Pro Co. (producer). 1993. Horseshoe crabs—the guardian spirit of the Earth. Planned by Kasaoka Municipal Museum of Horseshoe Crab, Okayama, Japan. Scientific advisor: K. Sekiguchi.

Oka, H. 1943. Recherches sur l'embryologie du Limule. II. Sci. Rept. Tokyo Bunrika Daigaku, Sect. B6:87–127.

Osborne, H. L. 1885. Metamorphosis of *Limulus polyphemus. Johns Hopkins Univ. Circular* 5 (No. 43):2.

Packard, A. S. 1870. The embryology of *Limulus polyphemus. Am. Nat.* 4:498–502.

———. 1872. On the embryology of *Limulus polyphemus. Memoir Boston Soc. Nat. Hist.* 2, Part II, No. 1:155–202.

———. 1883. Moulting of the shell of *Limulus. Am. Nat.* 176:1075–1076.

———. 1885. On the embryology of *Limulus polyphemus.* III. *Am. Nat.* 19:722–727.

Palumbi, S. R., and B. A. Johnson. 1982. A note on the influence of life history stage on metabolic adaptation: responses of *Limulus* eggs and larvae to hypoxia. In *Physiology and Biology of Horseshoe Crabs: Studies on Normal and Environmentally Stressed Animals,* eds. J. Bonaventura, C. Bonaventura, and S. Tesh. New York: Alan R. Liss. Pp. 115–124.

Patten, W. 1896. Variations on the development of *Limulus polyphemus. J. Morphol.* 12:17–149.

———. 1912. *The Evolution of the Vertebrates and Their Kin.* Philadelphia: P. Blakiston's Son & Co.

Riska, B. 1981. Morphological variation in the horseshoe crab *Limulus polyphemus. Evolution* 35:647–658.

Rudloe, A. 1979. Locomotor and light responses of larvae of the horseshoe crab, *Limulus polyphemus* (L.). *Biol. Bull.* 157:494–505.

Schaller, S. 2002. Horseshoe crab *(Limulus polyphemus)* spawning surveys in Maine, 2001. Poster. Am. Fish. Soc. Annual Meeting, Baltimore, Md.

Sekiguchi, K. 1988a. Artificial insemination. In *Biology of Horseshoe Crabs,* ed.

K. Sekiguchi. Tokyo: Science House Co.

——— 1988b. Embryonic development. In *Biology of Horseshoe Crabs,* ed. K. Sekiguchi. Tokyo: Science House Co.

Sekiguchi, K., and K. Nakamura. 1979. Ecology of the extant horseshoe crabs. In *Progress in Clinical and Biological Research,* vol. 29: *Biomedical Applications of the Horseshoe Crab (Limulidae),* eds. E. Cohen, F. B. Bang, J. Levin, J. J. Marchalonis, T. G. Pistole, R. A. Prendergast, C. N. Shuster, Jr., and S. W. Watson. New York: Alan R. Liss. Pp. 37–45.

Sekiguchi, K., H. Seshimo, and H. Sugita. 1988. Post-embryonic development. In *Biology of Horseshoe Crabs,* ed. K. Sekiguchi. Tokyo: Science House Co. Pp. 181–195.

Sekiguchi, K. and Y. Yamamichi. 1988. Eggs and sperm. In *Biology of Horseshoe Crabs,* ed. K. Sekiguchi. Tokyo: Science House Co. Pp. 133–144.

Sekiguchi, K., Y. Yamamichi and J. D. Costlow. 1982. Horseshoe crab developmental studies: normal embryonic development of *Limulus polyphemus* compared with *Tachypleus tridentatus.* In *Physiology and Biology of Horseshoe Crabs: Studies on Normal and Environmentally Stressed Animals,* eds. J. Bonaventura, C. Bonaventura, and S. Tesh. New York: Alan R. Liss. Pp. 53–73.

Shuster, C. N., Jr. 1948. On the gross anatomy and histology of the alimentary tract in the early developmental stages of *Xiphosura* (= *Limulus*) *polyphemus* Linn. MSc thesis, Zoology, Rutgers University.

———. 1950. Observations on the natural history of the American horseshoe crab, *Limulus polyphemus.* In *Third Rept.: Investigations of Methods of Improving the Shellfish Resources of Massachusetts.* Woods Hole Oceanogr. Inst. Contrib. No. 564:18–21.

———. 1954. A horseshoe "crab" grows up. *Ward's Natural Sci. Bull.* 28(1)1–6.

———. 1955. On morphometric and serological relationships within the Limulidae, with particular reference to Limulus polyphemus (L.). Ph.D. thesis, New York University (1958 Diss. Abstr. 18:371–372).

———. 1960. "Horseshoe crabs." In former years, during the month of May, these animals dominated Delaware Bay shores. *Estuarine Bull.* 5(2):1–8.

———. 1982. A pictorial review of the natural history and ecology of the horseshoe crab *Limulus polyphemus* with reference to other Limulidae. In *Physiology and Biology of Horseshoe Crabs: Studies on Normal and Environmentally Stressed Animals,* eds. J. Bonaventura, C. Bonaventura, and S. Tesh. New York: Alan R. Liss. Pp. 1–52.

Shuster, C. N., Jr., C. P. Randall, and R. Shevock. 1961. Biometry of *Limulus* mating (unpublished ms.: 100 pairs at Slaughter Beach, DE, on 31 May 1961).

Sugita, H., Y. Yamamichi, and K. Sekiguchi. 1988. Experimental hybridization. In *Biology of Horseshoe Crabs,* ed. K. Sekiguchi. Tokyo: Science House Co. Pp. 288–308.

6 Horseshoe Crabs in a Food Web

Allee, W. C. 1922. Studies in marine biology. II: An annotated catalog of the distribution of common invertebrates in the Woods Hole littoral. Deposited in the Marine Biological Laboratory Library, Woods Hole, Massachusetts.

Baptist, J. P., O. R. Smith, and J. W. Ropes. 1957. Migrations of the horseshoe crab, *Limulus polyphemus,* in Plum Island Sound, Massachusetts. *U.S. Dept. Interior Spec. Sci. Rept. Fisheries* 220:1–15.

Barber, S. B. 1956. Chemoreception and proprioreception in *Limulus. J. Exp. Zool.* 131:51–73.

Botton, M. L. 1981. The gill book of the horseshoe crab *(Limulus polyphemus)* as a substrate for the blue mussel (*Mytilus edulis*). *Bull. N. J. Acad. Sci.* 26:26–28.

———. 1984a. Diet and food preferences of the adult horseshoe crab *Limulus polyphemus* in Delaware Bay, New Jersey, USA. *Mar. Biol.* 81:199–207.

———. 1984b. The importance of predation by horseshoe crabs, *Limulus polyphemus,* to an intertidal sand flat community. *J. Mar. Res.* 42:139–161.

Botton, M. L., and H. H. Haskin. 1984. Distribution and feeding of the horseshoe crab, *Limulus polyphemus,* on the continental shelf off New Jersey. *Fish. Bull.* 82:383–389.

Botton, M. L., and R. E. Loveland. 1993. Predation by herring gulls and great black-backed gulls on horseshoe crabs. *Wilson Bull.* 105:518–521.

Botton, M. L., R. E. Loveland, and T. R. Jacobsen. 1992. Overwintering by trilobite larvae of the horseshoe crab *Limulus polyphemus* on a sandy beach of Delaware Bay (New Jersey, USA). *Mar. Ecol. Prog. Ser.* 88:289–292.

Botton, M. L., and J. W. Ropes. 1988. An indirect method for estimating longevity of the horseshoe crab *(Limulus polyphemus)* based on epifaunal slipper shells *(Crepidula fornicata). J. Shellfish Res.* 7:407–412.

———. 1989. Feeding ecology of horseshoe crabs on the continental shelf, New Jersey to North Carolina. *Bull. Mar. Sci.* 45:637–647.

Byles, R. A. 1988. Behavior and ecology of sea turtles in the Chesapeake Bay, Virginia. Ph.D. thesis, College of William and Mary, Virginia Institute of Marine Science.

Chatterji, A., J. K. Mishra, and A. H. Parulekar. 1992. Feeding behaviour and food selection in the horseshoe crab, *Tachypleus gigas* (Muller). *Hydrobiologia* 246:41–48.

Comitto, J. A., C. A. Currier, L. R. Kane, K. A. Reinsel, and I. M. Ulm. 1995. Dispersal dynamics of the bivalve *Gemma gemma* in a patchy environment. *Ecol. Monogr.* 65:1–20.

Deaton, L. E., and K. D. Kempler. 1989. Occurrence of the ribbed mussel, *Beukensia demissa,* on the book gills of a horseshoe crab, *Limulus polyphemus. Nautilus* 103:42.

Debnath, R., and A. Choudhury. 1988. Predation of Indian horseshoe crab *Tachypleus gigas* by *Corvus splendens. Tropical Ecol.* 29:86–89.

Debnath, R., S. K. Nag, A. Choudhury, R. Dasgupta, and R. K. Sur. 1989. Feeding habit and digestive physiology of the Indian horseshoe crab, *Tachypleus gigas* (Muller). *Ind. J. Physiol. Allied Sci.* 43:44–49.

deSylva, D. P., F. A. Kalber, Jr., and C. N. Shuster, Jr. 1962. *Fishes and Ecological Conditions in the Shore Zone of the Delaware River Estuary, with Notes on Other Species Collected in Deeper Water.* University of Delaware Marine Laboratories, Information Series Publication 5:1–170.

Dietl, J., C. Nascimento, and R. Alexander. 2000. Influence of ambient flow around the horseshoe crab Limulus polyphemus on the distribution and orientation of selected epizoans. *Estuaries* 23:509–520.

Eldredge, N. 1970. Observations on burrowing behavior in *Limulus polyphemus* (Chelicerata, Merostomata), with implications on the functional anatomy of trilobites. *Am. Mus. Novitates* 2436:1–17.

Fowler, H. 1907. *Annual Report of the New Jersey State Museum, 1907.* Part 3: The king crab fisheries in Delaware Bay (pp. 111–119 and plates 59–65).

French, K. A. 1979. Laboratory culture of embryonic and juvenile *Limulus.* In *Biomedical Applications of the Horseshoe Crab (Limulidae),* ed. E. Cohen. New York: Alan R. Liss. Pp. 61–71.

Grant, D. 1998. Living on *Limulus. Underwater Naturalist* 24(2):13–21.

Harrington, J. M., and P. B. Armstrong. 2000. Initial characterization of a potential anti-fouling system in the American horseshoe crab, *Limulus polyphemus. Biol. Bull.* 199:189–190.

Hedeen, R. A. 1986. *The Oyster: The Life and Lore of the Celebrated Bivalve.* Centreville, Md.: Tidewater Publishers. P. 237. (Odostomia on *Limulus,* p. 147.)

Hummon, W. D., J. W. Fleeger, and M. R. Hummon. 1976. Meiofauna-macrofauna interactions. I: Sand beach meiofauna affected by maturing *Limulus* eggs. *Ches. Sci.* 17:297–299.

Keinath, J. A. 1993. Movements and behavior of wild and head-started sea turtles *(Caretta caretta, Lepidochelys kempii).* Ph.D. thesis, College of William and Mary, Virginia Institute of Marine Science.

Keinath, J. A., J. A. Musick, and R. A. Byles. 1987. Aspects of the biology of Virginia's sea turtles: 1979–1986. *Virginia J. Sci.* 38:329–336.

Key, M. M., W. B. Jeffries, H. K. Voris, and C. M. Yang. 1996. Epizoic bryozoans, horseshoe crabs, and other mobile benthic substrates. *Bull. Mar. Sci.* 58:368–384.

Kraeuter, J. N., and S. R. Fegley. 1994. Vertical disturbance of sediments by horseshoe crabs *(Limulus polyphemus)* during their spawning season. *Estuaries* 17:288–294.

Kropach, C. 1979. Observations on the potential of *Limulus* aquaculture in Israel. In *Biomedical Applications of the Horseshoe Crab (Limulidae),* ed. E. Cohen. New York: Alan R. Liss. Pp. 103–106.

Lauer, D., and B. Fried. 1977. Observations on nutrition of *Bdelloura candida* (Turbellaria), an ectocommensal of *Limulus polyphemus* (Xiphosura). *Am. Midl. Nat.* 97:240–247.

Lockhead, J. H. 1950. *Xiphosura polyphemus.* In *Selected Invertebrate Types,* ed. F. A. Brown. New York: John Wiley and Sons. Pp. 360–381.

Lockwood, S. 1870. The horse-foot crab. *Am. Nat.* 4:257–274.

Loveland, R. E., M. L. Botton, and T. R. Jacobsen. 1989. Seasonal aspects of organic carbon levels on sandy intertidal beaches of Delaware Bay, NJ. Abstracts, 10th Biennial International Estuarine Research Conference, p. 47.

Lutcavage, M. E. 1981. The status of marine turtles in Chesapeake Bay and Virginia coastal waters. M. S. thesis, Virginia Institute of Marine Science, The College of William and Mary.

Lutcavage, M. E., and J. A. Musick. 1985. Aspects of the biology of sea turtles in Virginia. *Copeia* 1985(2):449–456.

MacKenzie, C. L., Jr. 1962. Transportation of oyster drills by horseshoe "crabs." *Science* 137:36–37.

Makioka, T. 1988. Internal morphology. In *Biology of Horseshoe Crabs*, ed. K. Sekiguchi. Tokyo: Science House. Pp. 104–132.

Manton, S. M. 1977. *The Arthropoda: Habits, Functional Morphology, and Evolution.* Oxford: Clarendon Press.

Maurer, D. 1977. Estuarine benthic invertebrates of Indian River and Rehoboth Bays, Delaware. *Int. Rev. Ges. Hydrobiol.* 62:591–629.

Maurer, D., L. Watling, P. Kinne, W. Leathem, and C. Wethe. 1978. Benthic invertebrate assemblages of Delaware Bay. *Mar. Biol.* 45:65–78.

Musick, J. A. 1979. *The Marine Turtles of Virginia (Families Chelonidae and Dermochelyidae) with Notes on Identification and Natural History.* Virginia Institute of Marine Science, Sea Grant Program Educational Series No. 24.

Patil, J. S., and A. C. Anil. 2000. Epibiotic community of the horseshoe crab *Tachypleus gigas. Mar. Biol.* 136:699–713.

Patten, W., and W. A. Redenbaugh. 1899. Studies on *Limulus* II. The nervous system of *Limulus polyphemus,* with observations upon the general anatomy. *J. Morphol.* 16:91–200.

Pearse, A. S. 1947. On the occurrence of ectoconsortes on marine animals at Beaufort, NC. *J. Parasitol.* 33:453–458.

Perry, L. M. 1931. Catfish feeding on the eggs of the horseshoe crab, *Limulus polyphemus. Science* 74:312.

Price, K. 1962. Biology of the sand shrimp, *Crangon septemspinosa,* in the shore zone of the Delaware Bay region. *Ches. Sci.* 3:244–255.

Rhoads, D. C. 1967. Biogenic reworking of intertidal and subtidal sediments in Barnstable Harbor and Buzzards Bay, Massachusetts. *J. Geol.* 75:461–476.

Rudloe, A. 1979. Locomotor and light responses of larvae of the horseshoe crab, *Limulus polyphemus* (L.). *Biol. Bull.* 157:494–505.

———. 1981. Aspects of the biology of juvenile horseshoe crabs, *Limulus polyphemus. Bull. Mar. Sci.* 31:125–133.

Schlottke, E. 1934. Histologische Beobachtungen am Darmkanal von *Limulus* und Vergleich seines Zwischengewebes mit dem von Spinnen. *Jahrb. Morph. v. Mikrosk. Anat., Abt. II, Zeitsch. Mikrosk.-Anat. Forsch.* 35:57–70.

Sekiguchi, K., H. Seshimo, and H. Sugita. 1988. Post-embryonic development of the horseshoe crab. *Biol. Bull.* 174:337–345.

Shuster, C. N., Jr. 1948. On the gross anatomy and histology of the alimentary tract in the early developmental stages of *Xiphosura* (= *Limulus*) *polyphemus* Linn. M.Sc. thesis, Zoology, Rutgers University.

———. 1950. Observations on the natural history of the American horseshoe crab, *Limulus polyphemus. Woods Hole Oceanogr. Inst. Contr.* 564:18–23.

———. 1955. *On Morphometric and Serological Relationships within the Limulidae, with Particular Reference to Limulus polyphemus.* Ph.D. thesis, New York University. (1958 Diss. Abstr. 18:371–372).

———. 1982. A pictorial review of the natural history and ecology of the horseshoe crab *Limulus polyphemus,* with reference to other Limulidae. In *Physiology and Biology of Horseshoe Crabs: Studies on Normal and Environmentally Stressed Animals,* eds. J. Bonaventura, C. Bonaventura, and S. Tesh. New York: Alan R. Liss. Pp. 1–52.

Sluys, R. 1989. A monograph of the marine triclads. Rotterdam: A. A. Balkema.

Smith, O. R. 1953. Notes on the ability of the horseshoe crab, *Limulus polyphemus,* to locate soft-shell clams, *Mya arenaria. Ecology* 34:636–637.

Smith, O. R., J. P. Baptist, and E. Chin. 1955. Experimental farming of the soft-shell clam, *Mya arenaria,* in Massachusetts, 1949–1953. *Comm. Fish. Rev.* 17:1–16.

Smith, O. R., and E. Chin. 1951. The effects of predation on soft clams, *Mya arenaria. Nat. Shellfish. Assoc. Conv. Addr.* 1951:37–44.

Spraker, H., and H. M. Austin. 1997. Diel feeding periodicity of Atlantic Silverside, *Menidia menidia,* in the York River, Chesapeake Bay, Virginia. *J. Elisha Mitchell Sci. Soc.* 113:171–182.

Turner, H. J., Jr. 1949. Report on investigations of methods of improving the shellfish resources of Massachusetts. *Woods Hole Oceanogr. Inst. Contr.* 510:3–22.

Vosatka, E. D. 1970. Observations on the swimming, righting, and burrowing movements of young horseshoe crabs, *Limulus polyphemus. Ohio J. Sci.* 70:276–283.

Wahl, M. 1989. Marine epibiosis. I: Fouling and antifouling: some basic aspects. *Mar. Ecol. Prog. Ser.* 58:175–189.

Wahl, M., and O. Mark. 1999. The predominantly facultative nature of epibiosis: experimental and observational evidence. *Mar. Ecol. Prog. Ser.* 187:59–66.

Warwell, H. C. 1897. Eels feeding on the eggs of *Limulus. Am. Nat.* 31:347–348.

Wasserman, G. S., and Z. Cheng. 1996. Electroretinographic measures of vision in horseshoe crabs with uniform versus variegated carapaces. *Biol. Signals* 5:247–262.

Woodin, S. A. 1978. Refuges, disturbance, and community structure: a marine soft-bottom example. *Ecology* 59:274–284.

Wyse, G. A. 1971. Receptor organization and function in *Limulus* chelae. *Z. Vergl. Physiol.* 73:249–273.

Wyse, G. A., and N. K. Dwyer. 1973. The neuromuscular basis of coxal feeding and locomotor movements in *Limulus. Biol. Bull.* 144:567–579.

Yonge, C. M. 1937. Evolution and adaptation in the digestive system of the Metazoa. *Biol. Rev.* 12:87–115.

7 A History of Skeletal Structure

Anderson, L. I. 1994. Xiphosurans from the Westphalian D of the Radstock Basin, Somerset Coalfield, the South Wales Coalfield and Mazon Creek, Illinois. *Proc. Geol. Assn.* 105:265–275.

———. 1999. A new specimen of the Silurian synziphosurine arthropod *Cyamocephalus. Proc. Geol. Assoc.* 110:211–216.

Anderson, L. I., J. A. Dunlop, R. M. C. Eagar, C. A. Horrocks, and H. M. Wilson. 1999. Soft-bodied fossils from the roof shales of the Wigan Four Foot coal seam, Westhoughton, Lancashire, UK. *Geol. Mag.* 135:321–329.

Anderson, L. I., and P. A. Selden. 1997. Opishosomal fusion and phylogeny of Palaeozoic Xiphosura. *Lethaia* 30:19–31.

Austin, P. R. 1979. Letter of 10 May.

Austin, P. R., C. J. Brine, J. E. Castle, and J. P. Zikakis. 1981. Chitin: new facets of research. *Science* 212: 749–753.

Babcock, L. E. 1992. Lectotype of *Phacops rana milleri* Stewart, 1927 (Trilobita, Devonian of Ohio). *J. Paleontol.* 66:692–693.

Babcock, L. E., and W. T. Chang. 1997.Comparative taphonomy of two nonmineralized arthropods: *Naraoia* (Nektaspida, early Cambrian, Chengjland, China) and *Limulus polyphemus* (Xiphosura, Miocene, Atlantic Ocean). *Bull. Nat. Mus. Sci.* 10:233–250.

Babcock, L. E., D. F. Merriam, and R. R. West. 2000. *Paleolimulus,* an early limuine (Xiphosurida), from Pennsylvanian-Permian Lagerstatten of Kansas and taphonomic comparison with modern Limulus. *Lethaia* 33:129–141.

Babcock, L. E., M. D. Wegweiser, A. E. Wegweiser, T. M. Stanley, and S. C. McKenzie, 1995. Horseshoe crabs and their trace fossils from the Devonian of Pennsylvania. *Penn. Geol.* 26:2–7.

Baird, G. C. 1997a. Paleoenvironmental setting of the Mazon Creek biota. In *Richardson's Guide to the Fossil Fauna of Mazon Creek,* eds. C. W. Shabica and A. A. Hay. Chicago: Northeastern Illinois University Press. Pp. 35–51.

———. 1997b. Geologic setting of the Mazon Creek area fossil deposit. In: C. W. Shabica, and A. A. Hay (eds.). *Richardson's Guide to the Fossil Fauna of Mason Creek.* Northeastern Illinois University Press. Pp. 16–20.

Baird, G. C., and J. L. Anderson. 1997. Relative abundance of different Mazon Creek organisms. In: Shabica, C. W. and A. A. Hay (eds.). *Richardson's Guide to the Fossil Fauna of Mason Creek.* Chicago: Northeastern Illinois University Press. Pp. 27–29.

Barthel, K. W. 1974. *Limulus:* a living fossil. *Naturwissenschaften* 61:428–433.

Bergström, J. 1975. Functional morphology and evolution of xiphosurids. *Fossils Strata* 4:291–305.

Bleicher, M. 1897. Sur la découverte d'une espèce de limule dans les marnes irisées de Lorraine. *Bulletin des séances de la Societe des Sciences de Nancy* 14:116–126.

Botton, M. L. 1984. The importance of predation by horseshoe crabs, *Limulus polyphemus,* in an intertidal and flat community. *J. Mar. Res.* 42:139–161.

Botton, M. L. 1999. Ecology of planktonic horseshoe crabs in Delaware Bay; patterns of abundance and the potential for dispersal. *Am. Zool.* 29:51A–52A.

Botton, M. L., and R. E. Loveland. 1987. Reproductive risk: high mortality associated with spawning in horseshoe crabs *Limulus polyphemus* in Delaware Bay, U.S.A. *Mar. Biol.* 101:143–151.

Botton, M. L., C. N. Shuster, Jr., K. Sekiguchi, and H. Sugita. 1996. Amplexus and mating behavior in the Japanese horseshoe crab, *Tachypleus tridentatus. Zool. Sci.* 13:151–159.

Briggs, D. E. G., and R. A. Fortey. 1989. The early radiation and relationships of the major arthropod groups. *Science* 246:241–243.

Buisonjé, P. H. de. 1985. Climatological conditions during deposition of the Solnhofen limestones. In *The Beginnings of Birds. Proceedings of the International Archaeopterx Conference Eichstätt 1984,* eds. M. K. Hecht, J. H. Ostrom, G. Viohl, and P. Wellnhofer. Eichstätt, Germany: Freunde des Jura-Museums Eichstätt, Willibaldsburg. Pp. 45–65.

Caster, K. E. 1938. A restudy of the tracks of *Paramphibius. J. Paleontol.* 12:3–60.

Chisholm, J. I. 1983. Xiphosurid traces, *Kouphichniun*aff. *Variabilis* (Linck), from the Namuian Upper Haslingden Flags of Whitworth, Lancashire. *Rep. Inst. Geol. Sci.* No. 83/10:37–44.

Cisne, J. L. 1975. Anatomy of *Triarthrus* and the relationships of the Trilobita. *Fossils and Strata* No. 4 (Univ. Oslo): 45–63.

Dunlop, J. A., and P. A. Selden. 1997. The early history and phylogeny of the chelicerates. In *Arthropod Relationships,* eds. R. A. Fortey and R. H. Thomas. Special Volume, Series 55:221–235. London: Chapman & Hall.

Dunbar, C. O. 1923. Kansas Permian insects. Part 2: *Paleolimulus,* a new genus Paleozoic Xiphosura, with notes on other genera. *Am. J. Sci.* 5:443–454.

Eldredge, N. 1970. Observations on the burrowing behavior in *Limulus polyphemus* (Chelicerata, Merostomasta), with implications on the functional morphology of trilobites. *Am. Mus. Novitates.* No. 2436:17 pp.

———. 1974. A revision of the suborder Synziphosurina (Chelicerata, Merostomata), with remarks on merostome phylogeny. *Am. Mus. Novitates.* No. 2543: 1–41.

———. 1991. *Fossils. The Evolution and Extinction of Species.* New York: Harry N. Abrams.

Fisher, D. C. 1975a. Evolution and functional morphology of the Xiphosurida. Ph.D. diss., Department of Geological Sciences, Harvard University.

———. 1975b. Swimming and burrowing in *Limulus* and *Mesolimulus. Fossils and Strata.* No. 4:281–290.

———. 1977a. Mechanisms and significance of enrollment in xiphosurans (Chelicerata, Merostomata). *Geol. Soc. Am. Abst. Prog.* 9:264–265.

———. 1977b. Functional significance of spines in the Pennsylvanian horseshoe crab *Euproops danae. Paleobiol.* 3:175–195.

———. 1981. The role of functional analysis in phylogenetic inference: examples from the history of the Xiphosura. *Am. Zool.* 21:47–62.

———. 1984. The Xiphosurida: archetypes of bradytely? In *Living Fossils,* eds. N. Eldredge and S. M. Stanley. New York: Springer-Verlag. Pp. 196–213.

———. 1990. Rates of evolution—living fossils. In *Paleobiology: A Synthesis,* eds. D. E. G. Briggs and P. Crothers. Oxford: Blackwell Scientific Publishers. Pp. 152–159.

Fortey, R. 1999. *Trilobite! Eyewitness to Evolution.* New York: Alfred A. Knopf. P. 137.

Fortey, R. A., and R. M. Owens. 1999. Feeding habits in trilobites. *Palaeontol.* 42:429–465.

Fourtner, C. R., C. D. Drewes, and R. A. Pax. 1971. Rhythmic motor outputs co-ordinating the respiratory movement of the gill plates of *Limulus polyphemus. Comp. Biochem. Physiol.* 38A:751–762.

Goldring, R., and A. Seilacher. 1971. Limuloid undertracks and their sedimentological implications. *N. Jb. Geol. Paläont., Abh.* 137:19–20.

Guanzhong, W. 1993. Xiphosurid trace fossils for the Westbury Formation (Rhaetien) of Southwest Britain. *Palaeontol.* 36:111–122.

Hock, C. W. 1940. Decomposition of chitin by marine bacteria. *Biol. Bull.* 79: 199–299.

Kraeuter, J. N., and S. R. Fegley 1994. Vertical disturbance of sediments by horseshoe crabs *(Limulus polyphemus)* during their spawning season. *Estuaries* 17:288–294.

Kraus, O. 1976. Zur phylogenetischen stellung und evolution der Chelicerata. *Entomologica Germanica* 3:1–12.

Lankester, E. R. 1881. *Limulus* an arachnid. *Q. J. Microsc. Sci.* (new series) 21:504–548, 609–649.

Levine, R. J. C., P. D. Chantler, and R. W. Kensler. 1988. Arrangement of myosin heads on *Limulus* thick filaments. *J. Cell Biol.* 107:1739–1747.

Levine, R. J. C., P. D. Chantler, R. W. Kensler, and J. L. Woodhead. 1991a. Effects of phosphorylation by myosin light chain kinase on the structure of *Limulus* thick filaments. *J. Cell Biol.* 113:563–572.

Levine, R. J. C., S. Davidheiser, A. M. Kelly, R. W. Kensler, J. Leferovich, and R. E. Davies. 1989. Fibre types in *Limulus* telson muscles: morphology and histochemistry. *J. Muscle Res. Cell Motil.* 10:53–66.

Levine, R. J. C., and R. W. Kensler. 1985. Structure of short thick filaments from *Limulus* muscle. *J. Mol. Biol.* 182:347–352.

Levine, R. J. C., R. W. Kensler, M. Stewart, and J. C. Haselgrove. 1982. Molecular organization of *Limulus* thick filaments. In *Basic Biology of Muscles: A Comparative Approach,* eds. B. M. Twarog, R. J. C. Levine, and M. M. Dewey. New York: Raven Press. Pp. 37–52.

Levine, R. J. C., J. L. Woodhead, and H. A. King. 1991b. The effect of calcium activation of skinned fiber bundles on the structure of *Limulus* thick filaments. *J. Cell Biol.* 113:573–583.

Levi-Setti, R. 1993. *Trilobites,* 2nd ed. Chicago: University of Chicago Press.

Luckenbach, M. W., and C. N. Shuster, Jr. 1997. Behavior of juvenile horseshoe crabs (*Limulus polyphemus* [L.]) in flowing water. Unpublished manuscript.

Malz, H. 1964. *Kouphichnium walchi,* die Geshichte einer Fährte und ihres Tieres. *Natur und Museum* 94:81–97.

Manton, S. M. 1977. *The Arthropoda: Habits, Functional Morphology, and Evolution.* Oxford: Clarendon Press.

Meek, F. B., and A. H. Worthen. 1865. Notice of some new types of organic remains, from the coal measures of Illinois. *Proc. Acad. Nat. Sci.* 17:41–53.

Packard, A. S. 1886. On the Carboniferous xiphosurous fauna of North America. *Natl. Acad. Sci.* (Washington), Memoir 3, pt. 2:143–157.

Patten, W. 1912. *The Evolution of the Vertebrates and Their Kin.* Philadelphia: P. Blakison's Son & Co.

Penn, D., and H. J. Brockmann. 1995. Age-biased stranding and righting in horseshoe crabs, *Limulus polyphemus. Anim. Behav.* 49:1531–1539.

Reeside, J. B., Jr., and D. V. Harris. 1952. A Cretaceous horseshoe crab from Colorado. *J. Wash. Acad. Sci.* 42:174–187.

Richter, R., and E. Richter. 1929. *Weinbergina opitzi,* n.g., n.sp., ein Schwertträger (Merost., Xiphos.) aus dem Devon (Rheinland). *Senckenbergiana* 11:193–209.

Romano, M., and M. A. Whyte. 1987. A limulid trace fossil from the Scarborough Formation (Jurassic) of Yorkshire; its occurrence, taxonomy and interpretation. *Proc. Yorkshire Geol. Soc.* 46:85–95.

Rudloe, A., and J. Rudloe. 1981. The changeless horseshoe crab. *Natl. Geogr. Mag.* 159: 562–572.

Scholl, G. 1977. Beitrage zur Embryonalenentwicklung von *Limulus polyphemus* L. (Chelicerata, Xiphosura). *Zoomorphologie* 86:99–154.

Sekiguchi, K., ed. 1988. *Biology of Horseshoe Crabs.* Tokyo: Science House Co.

Sekiguchi, K., and K. Nakamura. 1979. Ecology of the extant horseshoe crabs. In *Biomedical Applications of the*

Horseshoe Crab (Limulidae), eds. E. Cohen, F. B. Bang, J. Levin, J. J. Marchalonis, T. G. Pistole, R. A. Prendergast, C. N. Shuster, Jr., and S. W. Watson. New York: Alan R. Liss. Pp. 37–45.

Sekiguchi, K., and T. Yamasaki. 1988. Evolution of horseshoe crabs. In *Biology of Horseshoe Crabs,* ed. K. Sekiguchi. Tokyo: Science House Co. Pp. 408–421.

Selden, P. A. 1993. Arthropods (Aglaspidida, Pycnogonida and Chelicerata). In *The Fossil Record 2,* ed. M. J. Benton. London: Chapman & Hall. Pp. 297–320.

Selden, P. A., and D. J. Siveter. 1987. The origin of the limuloids. *Lethaia* 20:383–392.

Sewell, R. B. 1912. Capture of *Limulus* on the surface. *Rec. Indian Museum (Calcutta)* 7:87–88.

Shultz, J. W. 2001. Gross muscular anatomy of *Limulus polyphemus* (Xiphosura, Chelicerata) and its bearing on evolution in the arachnida. *J. Arachnol.* 29:283–303.

Shuster, C. N., Jr. 1948. On the gross anatomy and histology of the alimentary tract in early developmental stages of *Xiphosura (Limulus) polyphemus* Linn., M.S. thesis. Rutgers University.

———. 1955. On morphometric and serological relationships within the Limulidae, with particular reference to *Limulus polyphemus* (L.). Ph.D. thesis, New York University (1958 Diss. Abstr. 18:371–372).

———. 1979. Distribution of the American horseshoe "crab," *Limulus polyphemus* (L.). In *Biomedical Applications of the Horseshoe Crab (Limulidae),* eds. E. Cohen, F. B. Bang, J. Levin, J. J. Marchalonis, T. G. Pistole, R. A. Prendergast, C. N. Shuster, Jr., and S. W. Watson. New York: Alan R. Liss. Pp. 3–26.

Spraker, H., and H. M. Austin. 1997. Diel feeding periodicity of Atlantic Silverside, *Menidia menidia,* in the York River, Chesapeake Bay, Virginia. *J. Elisha Mitchell Sci. Soc.* 113:171–182.

Stewart, M., R. W. Kensler, and R. J. C.

Levine. 1985. Three-dimensional reconstruction of thick filaments from *Limulus* and scorpion muscle. *J. Cell Biol.* 1101:402–411.

Størmer, L. 1952. Phylogeny and taxonomy of fossil horseshoe crabs. *J. Paleontol.* 26:630–639.

———. 1955. Merostomata. In *Treatise on Invertebrate Paleontology, Part P, Arthropoda 2,* ed. R. C. Moore. Geological Society of America: University of Kansas Press. Pp. P4–P41.

Sugita, H. 1988. *Hybridization and phylogeny.* In *Biology of Horseshoe Crabs,* ed. K. Sekiguchi. Tokyo: Science House Co. Pp. 288–302.

Townes, H. K., Jr. 1938. *Ecological Studies on the Long Island Marine Invertebrates of Importance as Fish Food or Bait. A Biological Survey of the Salt Waters of Long Island,* Part II. State of New York Conserv. Dept. Salt-Water Survey No. 14.

Vosatka, E. D. 1970. Observations on the swimming, righting, and burrowing movements of young horseshoe crabs, *Limulus polyphemus. Ohio J. Sci.* 70:276–283.

Waterman, T. H. 1958. On the doubtful validity of *Tachypleus hoeveni* Pocock, an Indonesian horseshoe crab (Xiphosura). *Postilla Yale Peabody Mus. Nat. Hist.* No. 36:1–17.

Waterston, C. D. 1975. Gill structure in the Lower Devonian eurypterid *Tarsopterella scotia. Fossils and Strata.* No. 4:241–254.

———. 1979. Problems of functional morphology and classification in stylonuroid euryterids (chelicerata, merostomata), with observations on the Scottish Silurian Stylonuroidea. *Trans. Royal Soc. Edinburgh* 70:251–322.

Yamasaki, T. 1988. Taxonomy of horseshoe crabs. In *Biology of Horseshoe Crabs,* ed. K. Sekiguchi. Tokyo: Science House Co. Pp. 419–421.

8 Throughout Geologic Time

Anderson, L. I. 1994. Xiphosurans from the Westphalian D of the Radstock

Basin, Somerset Coalfield, the South Wales Coalfield and Mazon Creek, Illinois. *Proc. Geologist's Assoc.* 105:265–275.

———. 1996. Taphonomy and taxonomy of Palaeozoic Xiphosura. Ph.D. thesis. University of Manchester, England.

Anderson, L. I., and P. A. Selden. 1997. Opisthosomal fusion and phylogeny of Palaeozoic Xiphosura. *Lethaia* 30:19–31.

Annandale, N. 1909. The habits of the Indian king crabs. *Rec. Indian Mus.* 3:294–295.

Babcock, L. E., M. D. Wegweiser, A. E. Wegweiser, S. C. McKensie, and A. Ostrander. 1998. Marginal-marine lithofacies, biofacies, and ichnofacies, Chadrokoin and Vendango Formation (Upper Devonian), Unior City dam, Erie County, Pennsylvania. In *Guidebook, 63rd Ann. Field Conf. of Penna. Geologists, Geotectonic Environment of the Lake Erie Crustal Block,* ed. J. A. Harper. Pp. 26–52.

Baird, G. C. 1997a. Geologic setting of the Mazon Creek area fossil deposit. In *Richardson's Guide to the Fossil Fauna of Mazon Creek,* eds. C. W. Shabica and A. A. Hay. Chicago: Northeastern Illinois University Press. Pp. 16–20.

———. 1997b. Fossil distribution and fossil associations. In *Richardson's Guide to the Fossil Fauna of Mazon Creek,* eds. C. W. Shabica and A. A. Hay. Chicago: Northeastern Illinois University Press. Pp. 21–26.

———. 1997c. Paleoenvironmental setting of the Mazon Creek biota. In *Richardson's Guide to the Fossil Fauna of Mazon Creek,* eds. C. W. Shabica and A. A. Hay. Chicago: Northeastern Illinois University Press. Pp. 35–51.

Baptist, J. P., O. R. Smith, and J. W. Ropes. 1957. Migrations of the horseshoe crab, *Limulus polyphemus,* in Plum Island Sound, Mass. *Fish and Wildlife Service, USDOI, Special Sci. Rept.—Fisheries* No. 220:1–12.

Barthel, K. W. 1974. *Limulus:* a living fossil. Horseshoe crabs aid in an interpretation of an Upper Jurassic environment (Solnhofen). *Die Naturwissenschaften* 61:428–433.

Barthel, K. W., N. H. M. Swinburne, and S. C. Morris. 1990. *Solnhofen: A Study in Mesozoic Palaeontology.* New York: Cambridge University Press.

Belknap, D. F., and J. C. Kraft. 1981. Preservation potential of transgressive coastal lithosomes on the U.S. Atlantic shelf. *Mar. Geol.* 42:429–442.

Botton, M. L., and H. H. Haskin. 1984. Distribution and feeding of the horseshoe crab, *Limulus polyphemus,* on the continental shelf of New Jersey. *Fish. Bull.* 82(2):383–389.

Botton, M. L., R. E. Loveland, and T. R. Jacobsen. 1988. Beach erosion and geochemical factors: influence on spawning success of horseshoe crabs (*Limulus polyphemus*) in Delaware Bay. *Mar. Biol.* 99:325–332.

———. 1994. Site selection by migratory shorebirds in Delaware Bay, and its relationship to beach characteristics and abundance of horseshoe crab, *Limulus polyphemus,* eggs. *Auk* 111:605–615.

Botton, M. L., and J. W. Ropes. 1987. Populations of horseshoe crabs, *Limulus polyphemus,* on the northwestern Atlantic continental shelf. *Fish. Bull.* 85(4):805–812.

———. 1989. Feeding ecology of horseshoe crabs on the continental shelf, New Jersey to North Carolina. *Bull. Mar. Sci.* 45(3):537–647.

Botton, M. L., C. N. Shuster, Jr., K. Sekiguchi, and H. Sugita. 1996. Amplexus and mating behavior in the Japanese horseshoe crab, *Tachypleus tridentatus. Zool. Sci.* 13:151–159.

Braddy, S. J. 2001. Eurypterid palaeoecology: palaeobiological, ichnological and comparative evidence for a "mass-moult-mate" hypothesis. *Palaeogeogr. Palaeoclim. Palaeoecol.* 172:115–132.

Brady, J. T., and E. P. Schrading. 1997. *Habitat Suitability Models: Horseshoe Crab (Spawning) Delaware Bay, New Jersey and Delaware* (developed for the Cape May Villas and Reeds Beach Habitat Evaluation Procedures). Philadelphia: U.S. Army Corps of Engineers.

Broadhurst, F. M. 1964. Some aspects of the palaeoecology of non-marine faunas and rates of sedimentation in the Lancashire Coal Measures. *Am. J. Sci.* 262:858–869.

Brockmann, H. J. 1990. Mating behavior of horseshoe crabs, *Limulus polyphemus. Behaviour* 114:206–220.

Chatterjee, S. 1997. *225 Million Years of Evolution: The Rise of Birds.* Baltimore: The Johns Hopkins University Press.

Chatterji, A. 1994. *The Horseshoe Crab— A Living Fossil.* Project Swaarajya, Dona Paula, Goa.

Chatterji, A., J. K., Mishra, and A. H. Parulekar. 1992. Feeding behavior and food selection in the horseshoe crab, *Tachypleus gigas* (Müller). *Hydrobiologica* 246: 41–48.

Chatterji, A., and A. H. Parulekar. 1992. Fecundity of the Indian horse-shoe crab (Latreille). *Trop. Ecol.* 33: 97–102.

Chrzastowski, M. J. 1986. Stratigraphy and geologic history of a Holocene lagoon: Rehoboth Bay and Indian River Bay, Delaware. Ph.D. thesis, Department of Geology, University of Delaware.

Cochrane, J. D. 1966. The Yucatan current. In an unpublished report, Department of Oceanography, Texas A&M University, Ref. 66–23T:14–25.

Cook, G. H. 1857. *King Crabs, or Horse-Feet. Geology of the County of Cape May, State of New Jersey.* Trenton: Office of the True American. Pp. 105–112.

Daiber, F. C. 1960. Mangroves—the tidal marshes of the tropics. *Estuarine Bull.* (Univ. Delaware) 5(2):10–15.

Debnath, R. 1991. Studies on species morphology and sexual dimorphism of Indian horseshoe crabs. *Geobios. New Reports* 10:128–131.

———. 1992. Studies on Indian horse-shoe crabs (Merostomata: Xiphosura) with special reference to its feeding behavior. Ph.D. thesis, Department of Marine Sciences, University of Calcutta (India).

Debnath, R., and A. Choudhury. 1988a. Predation of Indian horseshoe crab *Tachypleus gigas* by *Corvus splendens. Trop. Ecol.* 29(2):86–89.

———. 1988b. Population estimation of horseshoe crab *Tachypleus gigas* by capture-recapture method at Chandipur sea shore (Orissa), India. *J. Mar. Biol. Assn. India* 30(1&2):8–12.

Debnath, R., S. K. Nag, A. Choudhury, R. Dasgupta, and R. K. Sur. 1989. Feeding habit and digestive physiology of the Indian horseshoe crab, *Tachypleus gigas* (Muller). *Ind. J. Physiol. & Allied Sci.* 43:44–49.

Dexter, R. W. 1947. The marine communities of a tidal inlet at Cape Ann, Massachusetts: A study in bio-ecology. *Ecol. Monogr.* 17:263–294.

Dunbar, C. O. 1923. Kansas Permian insects. Part 2: *Paleolimulus,* a new genus of Paleozoic Xiphosura, with notes on other genera. *Am. J. Sci.* 5th Ser. 30:443–454.

Eager, R. M. C., J. G. Baines, J. D. Collinson, P. G. Hardy, S. A. Okolo, and J. E. Pollard. 1985. Trace fossil assemblages and their occurrence in Silesian (Mid-Carboniferous) deltaic sediments of the Central Pennine Basin, England. In *Biogenic Structures: Their Use in Interpreting Depositional Environments,* ed. H. A. Curran. Soc. Economic Paleontologists and Mineralogists, Special Publication No. 35. Pp. 99–149.

Finn, J. J., C. N. Shuster, Jr., and B. L. Swan. 1991. *Limulus Spawning Activity on Delaware Bay Shores, 1990.* Cape May Court House, N.J.: Finn-Tech Industries brochure.

Fisher, D. C. 1975. Evolution and functional morphology of the Xiphosuridae. Ph.D. thesis, Harvard University. Pp. 134–141.

———. 1979. Evidence for subaerial activity of *Euproops danae* (Merostomata, Xiphosurida). In *Mazon Creek Fossils,* ed. M. H. Nitecki. New York: Academic Press. Pp. 379–447.

Fletcher, C. H., III, H. J. Knebel, and J. C. Kraft. 1990. Holocene evolution of an estuarine coast and tidal wetlands. *Geol. Soc. Amer. Bull.* 102:283–297.

Fowler, H. W. 1908. The king crab fisheries in Delaware Bay, and further notes on New Jersey fishes, amphibians, and reptiles. In *Annual Report, New Jersey State Museum 1907*, Part III: 111–119 + Pls. 59–65.

Fraenkel, G. 1960. Lethal high temperatures for three marine invertebrates: *Limulus polyphemus, Littorina littorea* and *Pagurus longicarpus. Oikos* 11:171–182.

Gómez-Aquirre, S. 1979. Notas para estudios de población de Limulus polyphemus L. (Xiphosura: Xiphosuridae) en la isla del Carmen, Campeche (1964–1978). *Ann. Inst. Biol. Univ. Nal. Auton, Mexico, Ser. Zoologica* 50:769–772.

———. 1987. Resultados preliminarios de la iniciativa de alicmar para la protección de *Limulus polyphemus* L. (Arthopoda Merostomata). *Congreso Latinoamericano sobre Ciencias del Mar* 2:193–198.

———. 1993. Cacerolita de Mar (*Limulus polyphemus* L.) en la Península de Yucatán. In *Biodiversidad Marina y Costera de México*, eds. S. I. Salazar-Vallejo and N. E. Gonzalez. Com. Natl. Biodiversidad y CIQRO, Mexico, D.F.

Hata, D., and J. Berkson, 2002. Abundance of horseshoe crabs, *Limulus polyphemus*, in the Delaware Bay area (submitted).

Hecht, M. K. 1985. The biological significance of *Archaeopteryx*. In *The Beginnings of Birds. Proceedings of the International Archaeopteryx Conference, Eichstätt 1984,* eds. M. K. Hecht, J. H. Ostrom, G. Viohl, and P. Wellnhofer. Willibaldsburg, Germany: Freunde des Jura-Museum Eichstätt. Pp. 149–160.

Hoyt, W. H., J. C. Kraft, and M. J. Chrzastowski. 1990. Prospecting for submerged archaeological sites on the continental shelf: Southern mid-Atlantic Bight of North America. In *Archaeological Geology of North America,* eds. N. P. Lasca and J. Donahue. Geol. Soc. Amer., Centennial Special Vol. 4:147–160.

Hutchins, L. W. 1947. The basis for temperature zonation in geographical distribution. *Ecol. Monogr.* 17:325–335.

Ives, J. E. 1891. Crustacea from the north coast of Yucatan, the harbor of Vera Cruz, the west coast of Florida and the Bermuda Islands. *Proc. Acad. Natl. Sci. (Philadelphia)* 43:176–200.

Jackson, N. L., K. F. Nordstrom, and D. R. Smith. 2002. Geomorphic-biotic interactions on beach foreshores in estuaries. *J. Coastal Res.* Special Issue: 414.

Knebel, H. J., C. H. Fletcher III, and J. C. Kraft. 1988. Late Wisconsin—Holocene paleogeography of Delaware Bay: a large coastal plain estuary. *Mar. Geol.* 83:115–133.

Kraft, J. C. 1971. Sedimentary facies patterns and geologic history of a Holocene marine transgression. *Geol. Soc. Am. Bull.* 82:2131–2158.

Kropach, C. 1979. Observations on the potential of *Limulus* aquaculture in Israel. In *Progress in Clinical and Biological Research,* vol. 29: *Biomedical Applications of the Horseshoe Crab (Limulidae),* eds. E. Cohen, F. B. Bang, J. Levin, J. J. Marchalonis, T. G. Pistole, R. A. Prendergast, C. N. Shuster, Jr., and S. W. Watson. New York: Alan R. Liss. Pp. 103–106.

Luckenbach, M., and C. N. Shuster, Jr. 1997. Unpublished manuscript. Preliminary study on the behavior of juvenile *Limulus* in a flume tank.

Marmer, H. A. 1954. Tides and sea level in the Gulf of Mexico. In *Gulf of Mexico: Its Origin, Waters, and Marine Life,* coordinator P. S. Galtsoff. U.S. Department of the Interior, Fish and Wildlife Service, *Fishery Bull.* 89:101–118.

Mayer, A. G. 1914. The effects of temperature upon tropical marine animals. *Pap. Tortugas Lab. Carnegie Inst. Publ.* 183(6): 1–24.

Michels, S. F. 1997. Summary of trends in horseshoe crab abundance in Delaware. In *Proceedings of the Horseshoe Crab Forum: Status of the Resource,* eds. J. Farrell and C. Martin. University of Delaware Sea Grant College Program (DEL-SG-05-97). Pp. 26–29.

Mikkelsen, T. 1988. *The Secret in the Blue Blood.* Beijing, China: Science Press.

Mikulic, D. G. 1997. Xiphosura. In *Richardson's Guide to the Fossil Fauna of Mazon Creek,* eds. C. W. Shabica and A. A. Hay. Chicago: Northeastern Illinois University Press. Pp. 134–139.

National Geographic Maps. 1988–1999. Ocean maps. Washington, D.C.: National Geographic Society. On CD-ROM.

Nowlin, W. D., Jr. 1972. Winter circulation patterns and property distributions. In *Gulf of Mexico*, vol. 2, eds. L. R. A. Capurro and J. L. Reid. Contr. Physical Oceanogr. Texas A&M Univ. *Oceanogr. Stud.* Pp. 3–51.

Palmer, D. 1999. *Atlas of the Prehistoric World.* New York: Discovery Channel (Discovery Books).

Penn, D., and H. J. Brockmann. 1994. Nest site selection in the horseshoe crab, *Limulus polyphemus. Biol. Bull.* 187:373–384.

Petuch, E. J. 1992. *The Edge of the Fossil Sea. Life along the Shores of Prehistoric Florida.* Sanibel Island, Fl.: Bailey-Matthews Shell Museum.

Pierce, J. C., G. Tan, and P. M. Gaffney. 2000. Delaware and Chesapeake Bay populations of the horseshoe crab *Limulus polyphemus* are genetically distinct. *Estuaries* 23:690–698.

Powers, M. K., and R. B. Barlow. 1985. Behavioral correlates of circadian rhythms in the *Limulus* visual system. *Biol. Bull.* 169:578–591.

Price, W. A. 1954. Shorelines and coasts of the Gulf of Mexico. In *U.S. Dept. Interior, Fish and Wildl. Ser., Fish. Bull. Ser.* (coordinator P. S. Galtsoff). 55:38–65.

Riska, B. 1981. Morphological variation in the horseshoe crab *Limulus polyphemus* (L.). *Evolution* 35:647–658.

Ropes, J. W., C. N. Shuster, Jr., L. O'Brien, and R. Mayo. 1982. Data on the occurrence of horseshoe crabs, *Limulus polyphemus* (L.), in NMFS-NEFC survey samples. Woods Hole Lab. Ref. Doc. No. 82–23.

Rudloe, A. E., and W. F. Herrnkind. 1976. Orientation of *Limulus polyphemus* in the vicinity of breeding beaches. *Mar. Behav. Physiol.* 4:75–89.

Saha, D. 1989. Status survey, breeding biology and some aspects of morphology by SEM study of Xiphosurid arthropods in Indian coastal waters. Ph.D. thesis, University of Calcutta (India).

Saunders, N. C., L. G. Kessler, and J. C. Avise. 1986. Genetic variation and geographic differentiation in mitochondrial DNA of the horseshoe crab, *Limulus polyphemus. Genetics* 112:613–627.

Say, T. 1818. An account of the Crustacea of the United States. *J. Acad. Natl. Sci. (Philadelphia)*, Pt. II, 1(5):433–436.

Schaller, S. Y., P. E. Thayer, and S. Hanson. 2001. *Survey of Maine Horseshoe Crab (Limulus polyphemus) Spawning Populations 2001.* Boothbay Harbor: Maine Dept. Mar. Res.

Schaller, S. Y., P. E. Thayer, and S. Hanson. 2002. *Survey of Maine Horseshoe Crab (Limulus polyphemus) Spawning Populations, 2002.* Buxton, Maine: Bar Mills Ecological.

Schram, F. R. 1979. Limulines of the Mississippian Bear Gultch limestone of central Montana. *Trans. San Diego Soc. Nat. Hist.* 19:67–73.

Scotese, C. R. et al. 1979. Paleozoic base maps, *J. Geol.* 87:217–277.

Seilacher, A. 1990. Taphonomy of Fossil-Lagerstätten. In *Palaeobiology: A Synthesis,* eds. D. E. G. Briggs and P. R. Crowther. Blackwell Scientific Publications. Pp.266–270.

Sekiguchi, K., ed. 1988a. Biogeography. In *Biology of Horseshoe Crabs,* ed. K. Sekiguchi. Tokyo: Science House. Pp. 22–49.

———. 1988b. Ecology. In *Biology of Horseshoe Crabs.* Tokyo: Science House. Pp. 50–68.

———. 1988c. Distribution pattern of horseshoe crabs. In *Biology of Horseshoe Crabs.* ed. K. Sekiguchi. Tokyo: Science House. Pp. 410–414.

Sekiguchi, K., and K. Nakamura. 1979. Ecology of the extant horseshoe crabs. In *Progress in Clinical and Biological Research,* vol. 29: *Biomedical Applications of the Horseshoe Crab (Limulidae),* eds. E. Cohen, F. B. Bang, J. Levin, J. J. Marchalonis, T. G. Pistole, R. A. Prendergast, C. N. Shuster, Jr., and S. W. Watson. New York: Alan R. Liss. Pp. 37–45.

———. 1980. Sympatric distribution pattern of three species of Asian horseshoe crabs. *Proc. Jap. Soc. Syst. Zool.* 18:1–4.

Sekiguchi, K., K. Nakamura, T. K. Sen, and H. Sugita. 1976. Morphological variation and distribution of a horseshoe crab, *Tachypleus gigas,* from the Bay of Bengal and the Gulf Coast of Siam. *Proc. Jap. Soc. Syst. Zool.* 12:13–20.

Sekiguchi, K., K. Nakamura, and H. Seshimo. 1978. Morphological variation of a horseshoe crab, *Carcinoscorpius rotundicauda,* from the Bay of Bengal and the Gulf of Siam. *Proc. Jap. Soc. Syst. Zool.* 15:24–30.

Sekiguchi, K., S. Nishiwaki, T. Makioka, S. Srithunya, S. Machjajib, K. Nakamura, and T. Yamasaki. 1977. A study of the egg-laying habits of the horseshoe crabs, *Tachypleus gigas* and *Carcinoscorpius rotundicauda,* in Chonburi area of Thailand. *Proc. Jap. Soc. Syst. Zool.* 13:39–45.

Sewell, R. B. S. 1912. Capture of *Limulus* on the surface. *Rec. Indian Mus. (Calcutta)* 7:87–88.

Shabica, C. W., and A. A. Hay, eds. 1997. *Richardson's Guide to the Fossil Fauna of Mazon Creek.* Chicago: Northeastern Illinois University Press.

Shipman, P. 1998. *Taking Wing: Archaeopteryx and the Evolution of Bird Flight.* New York: Simon & Schuster.

Shuster, C. N., Jr. 1950. Observations on the natural history of the American horseshoe crab, *Limulus polyphemus. Third Rept. Investigations of Methods of Improving the Shellfish Resources of Mas-*

sachusetts. Woods Hole Oceanogr. Inst., Contr. 564:18–23.

———. 1955. On morphometric and serological relationships within the Limulidae, with particular reference to *Limulus polyphemus* (L.). Ph.D. thesis, New York University (1958 Diss. Abstr. 18:371–372).

———. 1958. "Study These." *The Staff Reporter* (Wilmington Public Schools, Delaware) 10, no. 8:4–5.

———. 1959. Biological evaluation of the Delaware River estuary. In *State of Delaware Intrastate Water Resources Survey,* eds. A. J. Kaplovsky and C. O. Simpson. Wilmington, Del.: William N. Cann. Pp. 1–73.

———. 1979. Distribution of the American horseshoe crab, *Limulus polyphemus* (L.). In *Progress in Clinical and Biological Research,* vol. 29: *Biomedical Applications of the Horseshoe Crab (Limulidae),* eds. E. Cohen, F. B. Bang, J. Levin, J. J. Marchalonis, T. G. Pistole, R. A. Prendergast, C. N. Shuster, Jr., and S. W. Watson. New York: Alan R. Liss. Pp. 3–26.

———. 1982. A pictorial review of the natural history and ecology of the horseshoe crab *Limulus polyphemus,* with references to other Limulidae. In *Progress in Clinical and Biological Research,* vol. 81: *Physiology and Biology of Horseshoe Crabs: Studies on Normal and Environmentally Stressed Animals,* eds. J. Bonaventura, C. Bonaventura, and S. Tesh. New York: Alan R. Liss. Pp. 1–52.

———. 1997. Abundance of adult horseshoe crabs, *Limulus polyphemus,* in Delaware Bay, 1850–1990. In *Proceedings of the Horseshoe Crab Forum: Status of the Resource.* University of Delaware Sea Grant College Program (DEL-SG-05-97). Pp. 5–14.

Shuster, C. N., Jr., and M. L. Botton. 1985. A contribution to the population biology of horseshoe crabs, *Limulus polyphemus* (L.), in Delaware Bay. *Estuaries* 8:363–372.

Shuster, C. N., Jr., C. H. P. Langrall, and R. C. Shevock. 1961. Unpublished manuscript. Biometric analysis of mating of the horseshoe crab, *Limulus polyphemus.*

Siveter, D. J., and P. A. Selden. 1987. A new, giant xiphosurid from the lower Namurian of Weardale, County Durham. *Proc. Yorkshire Geol. Soc.* 45 (Pt. 2):153–168.

Smith, D. R., P. S. Pooler, R. E. Loveland, M. L. Botton, S. F. Michels, R. G. Weber, and D. B. Carter. 2002. Horseshoe crab *(Limulus polyphemus)* reproductive activity on Delaware Bay beaches: interactions with beach characteristics. *J. Coastal Res.* 18:730–740.

Smith, O. R. 1953. Notes on the ability of the horseshoe crab, *Limulus polyphemus,* to locate soft-shell clams, *Mya arenaria. Ecology* 34:636–637.

Sokoloff, A. 1978. Observations on populations of the horseshoe crab *Limulus (Xiphosura) polyphemus. Res. Popul. Ecol.* 19:222–236.

Størmer, L. 1955. Merostomata. In *Treatise on Invertebrate Paleontology Pt. P Arthropoda 2,* ed. R. C. Moore. Lawrence: Geol. Soc. Amer. and Univ. Kansas Press. Pp. 4–31.

Sumner, F. B., R. C. Osburn, and L. J. Cole. 1911. A biological survey of the waters of Woods Hole and vicinity. *Bull. Bur. Fish.* 31:1–794.

Swan, B. L., W. R. Hall, Jr., and C. N. Shuster, Jr. 1992. *Limulus* spawning activity on Delaware Bay shores, 25 May 1991. Wilmington: Delaware Estuary Program.

———. 1999. *Ten Years and Still Counting. 1999 Horseshoe Crab Spawning Survey along the Delaware Bay Shore.* File report; limited distribution.

———. 2000. *Limulus Spawning Activity on Delaware Bay Shores.* File report; limited distribution.

Verrill, A. E., and S. I. Smith. 1873. Report upon the invertebrate animals of Vineyard Sound and the adjacent waters with an account of the physical character of the region. Rept., U.S. Fish Comm. 1871–72: 295–778 + 38pls.

Viohl, G. 1985. Geology of the Sohlhofen lithographic limestone and

the habitat of *Archaeopteryx.* In *The Beginning of Birds, Proc. Internatl. Archaeopteryx Conference 1994,* eds. M. H. Hecht, J. H. Ostrom, G. Viohl, and P. Wellnofer. Willbaldsburg: Freunde des Jura-Museums Eichstätt. Pp. 31–44.

Ward, P. D. 1992. Timeless design: the horseshoe crabs. In *Methuselah's Trail. Living Fossils and the Great Extinctions.* New York: W. H. Freeman and Company. Pp. 135–150.

Waterston, C. D. 1985. Chelicerata from the Dinantian of Foulden, Berwickshire, Scotland. *Trans. Roy. Soc. Edinburgh: Earth Scis.* 76:25–33.

Weil, C. B. 1977. *Sediments, Structural Framework, and Evolution of Delaware Bay: A Transgressive Estuarine Delta.* Delaware Sea Grant College Program, University of Delaware (DEL-SG-05–97).

Wigley, R. L., and R. B. Theroux. 1981. Atlantic Continental Shelf and Slope of the United States—Macrobenthic Invertebrate Fauna of the Middle Atlantic Region—Faunal Composition and Quantitative Distribution. U.S.D.I., Geological Survey Professional Paper 929-N.

Wolff, T. 1977. The horseshoe crab *(Limulus polyphemus)* in North European waters. *Vidensk. Meddr. Dansk. naturh. Foren* 140:39–52.

Yamasaki, T. 1988 Taxonomy of horseshoe crabs. In *Biology of Horseshoe Crabs,* ed. K. Sekiguchi. Tokyo Science House Co. Pp. 419–421.

9 Coping with Environmental Changes

Bang, F. B. 1979. Ontogeny and phylogeny of response to gram-negative endotoxins among the marine invertebrates. In *Biomedical Applications of the Horseshoe Crab (Limulidae),* eds. E. Cohen, F. B. Bang, J. Levin, J. J. Marchalonis, T. G. Pistole, R. A. Prendergast, C. N. Shuster, Jr., and S. W. Watson. New York: Alan R. Liss. Pp. 109–123.

Briggs, R. T., and B. L. Moss. 1997. Ultrastructure of the coxal gland of the horseshoe crab *Limulus polyphemus:* evidence for ultrafiltration and osmoregulation. *J. Morphol.* 234:233–252.

Burnett, L. E. 1988. Physiological responses to air exposure: Acid-base balance and the role of branchial water stores. *Am. Zool.* 28:125–136.

Burnett, L. E., D. A. Scholnick, and C. P. Mangum. 1988. Temperature sensitivity of molluscan and arthropod hemocyanins. *Biol. Bull.* 174:153–162.

Cameron, J. N. 1978. NaCl balance in blue crabs, *Callinectes sapidus,* in fresh water. *J. Comp. Physiol.* 123:127–135.

Cameron, J. N., and C. V. Batterton. 1978. Antennal gland function in the fresh water blue crab *Callinectes sapidus:* water, electrolyte, acid base, and ammonia excretion. *J. Comp. Physiol. B* 123:143–148.

Crabtree, R. L., and C. H. Page. 1974. Oxygen-sensitive elements in the book gills of *Limulus polyphemus. J. Exp. Biol.* 60:631–639.

Dragolovich, J., and S. K. Pierce. 1992. Comparative time courses of inorganic and organic osmolyte accumulation as horseshoe crabs *Limulus polyphemus* adapt to high salinity. *Comp. Biochem. Physiol. A* 102:79–84.

———. 1994. Characterization of partially purified betaine aldehyde dehydrogenase from horseshoe crab *(Limulus polyphemus)* cardiac mitochondria. *J. Exp. Zool.* 270:417–425.

Dunson, W. A. 1984. Permeability of the integument of the horseshoe crab *Limulus polyphemus* to water, sodium and bromide. *J. Exp. Zool.* 230:495–499.

Dykens, J. A., R. W. Wiseman, and C. D. Hardin. 1996. Preservation of phosphagen kinase function during transient hypoxia via enzyme abundance or resistance to oxidative inactivation. *J. Comp. Physiol. B* 166:359–368.

Edwards, S. C., and S. K. Pierce. 1986. Octopamine potentiates intracellular Na^+ and Cl^- reductions during cell volume regulation in *Limulus* exposed to hypoosmotic stress. *J. Comp. Physiol. B* 156:481–489.

Falkowski, P. G. 1974. Facultative anaerobiosis in *Limulus polyphemus:* phosphoenolpyruvate carboxykinase and heart activities. *Comp. Biochem. Physiol. B* 49:749–759.

Fourtner, C. R., C. D. Drewes, and R. A. Pax. 1971. Rhythmic motor outputs co-ordinating the respiratory movement of the gill plates of *Limulus polyphemus. Comp. Biochem. Physiol. A* 38:751–762.

Fraenkel, G. 1960. Lethal high temperatures for three marine invertebrates: *Limulus polyphemus, Littorina littorea* and *Pagarus logicarpus. Oikos* 11:171–182.

Freadman, M. A., and W. H. Watson III. 1989. Gills as possible accessory circulatory pumps in *Limulus polyphemus. Biol. Bull.* 177:386–395.

Gaede, G., R. A. Graham, and W. R. Ellington. 1986. Metabolic disposition of lactate in the horseshoe crab *Limulus polyphemus* and the stone crab *Mennippe mercenaria. Mar. Biol.* 91:473–480.

Griffin, A. J., and W. H. Fahrenbach. 1977. Gill receptor arrays in the horseshoe crab *Limulus polyphemus. Tiss. Cell* 9:745–750.

Hannan, J. V., and D. H. Evans. 1973. Water permeability in some euryhaline decapods and *Limulus polyphemus. Comp. Biochem. Physiol. A* 44:1199–1213.

Henry, R. P. 1984. The role of carbonic anhydrase in blood ion and acid-base regulation. *Am. Zool.* 24:241–251.

———. 1988. Multiple functions of carbonic anhydrase in the crustacean gill. *J. Exp. Zool.* 248:19–24.

———. 1994. Morphological, behavioral, and physiological characterization of bimodal breathing crustaceans. *Am. Zool.* 34:205–215.

———. 1995. Nitrogen metabolism and excretion for cell volume regulation in invertebrates. In *Nitrogen Metabolism and Excretion,* eds. P. J. Walsh and P. Wright. Boca Raton: CRC Press. Pp. 63–74.

———. 1996. Multiple roles of carbonic anhydrase in cellular transport and metabolism. *Ann. Rev. Physiol.* 58:523–538.

Henry, R. P., S. A. Jackson, and C. P. Mangum. 1996. Ultrastructure and transport-related enzymes of the gills and coxal gland of the horseshoe crab *Limulus polyphemus. Biol. Bull.* 191:241–250.

Johansen, K., and J. A. Petersen. 1975. Respiratory adaptation in *Limulus polyphemus* (L.). In *Physiological Ecology of Estuarine Organisms,* ed. F. J. Vernberg. Columbia: University of South Carolina Press. Pp. 129–145.

Laughlin, R. B., Jr. 1981. Sodium, chloride and water exchange in selected developmental stages of the horseshoe crab, *Limulus polyphemus (Linnaeus). J. Exp. Mar. Biol. Ecol.* 52:135–146.

Mangum, C. P., C. E. Booth, P. L. De Fur, N. A. Heckel, R. P. Henry, L. C. Oglesby, and G. Polites. 1976. The ionic environment of hemocyanin in *Limulus polyphemus. Biol. Bull.* 150:453–467.

Mangum, C. P., M. A. Freadman, and K. Johansen. 1975. The quantitative role of hemocyanin in aerobic respiration of *Limulus polyphemus. J. Exp. Zool.* 191:279–285.

Mangum, C. P., and J. Ricci. 1989. The influence of temperature on oxygen uptake and transport in the horseshoe crab *Limulus polyphemus* L. *J. Exp. Mar. Biol. Ecol.* 129:243–250.

McMannus, J. J. 1969. Osmotic relations in the horseshoe crab, *Limulus polyphemus. Am. Mid. Natur.* 81:569–573.

Meglitsch, P. A. 1967. *Invertebrate Zoology.* New York: Oxford University Press.

Page, C. H. 1973. Localization of *Limulus polyphemus* oxygen sensitivity. *Biol. Bull.* 144:383–390.

Perry, S. F. 1997. The chloride cell: structure and function in the gills of freshwater fishes. *Ann. Rev. Physiol.* 59:325–347.

Robertson, J. D. 1970. Osmotic and ionic regulation in the horseshoe crab *Limulus polyphemus* (Linnaeus). *Biol. Bull.* 138:157–183.

Shuster, C. N., Jr. 1982. A pictorial review of the natural history and ecology of the horseshoe crab *Limulus polyphemus,* with reference to other Limulidae. In *Physiology and Biology of Horseshoe Crabs: Studies on Normal and Environmentally Stressed Animals,* eds. J. Bonaventura, C. Bonaventura, and S. Tesh. New York: Alan R. Liss. Pp. 11–52.

Sullivan, B., J. Bonaventura, and C. Bonaventura. 1974. Functional differences in the multiple hemocyanins of the horseshoe crab *Limulus polyphemus* L. *Proc. Natl. Acad. Sci.* 71:2558–2562.

Taylor, H. H., and E. W. Taylor. 1992. Gills and lungs: the exchange of gases and ions. In *Microscopic Anatomy of Invertebrates.* Vol. 10: *Decapod Crustacea,* eds. F. W. Harrison and A. G. Humes. New York: Wiley-Liss (John Wiley & Sons). Pp. 203–293.

Thompson, C., and C. H. Page. 1975. Nervous control of respiration: oxygen-sensitive elements in the prosoma of *Limulus polyphemus. J. Exp. Biol.* 62:545–554.

Towle, D. W. 1990. Sodium transport systems in gills. In *Comparative Aspects of Sodium Cotransport Systems,* ed. R. K. H. Kinne. Basel: Karger Publishing. Pp. 241–263.

———. 1997. Molecular approaches to understanding salinity adaptation of estuarine animals. *Am. Zool.* 37:575–584.

Towle, D. W., and J. Litteral. 1999. Preliminary identification of $Na+/H+$ exchanger mRNA in gill and coxal gland of the horseshoe crab *Limulus polyphemus. Bull. Mt. Des. Isl. Biol. Lab.* 38:24.

Towle, D. W., C. P. Mangum, B. A. Johnson, and N. A. Mauro. 1982. The role of the coxal gland in ionic, osmotic, and pH regulation in the horseshoe crab *Limulus polyphemus.* In *Physiology and Biology of Horseshoe Crabs: Studies on Normal and Environmentally Stressed Animals,* eds. J. Bonaventura, C. Bonaventura, and S. Tesh. New York: Alan R. Liss. Pp. 147–172.

Warren, M. K., and S. K. Pierce. 1982. Two cell volume regulatory systems in the *Limulus* myocardium: an interaction of ions and quaternary ammonium compounds. *Biol. Bull.* 163:504–516.

Waterman, T. H., and D. F. Travis. 1953. Respiratory reflexes and the flabellum of *Limulus. J. Cell. Comp. Physiol.* 41:261–289.

Wheatly, M. G. 1985. The role of the antennal gland in ion and acid-base regulation during hyposaline exposure of the Dungeness crab *Cancer magister* (Dana). *J. Comp. Physiol. B* 155:445–454.

Wyse, G. A., and C. H. Page. 1976. Sensory and central nervous control of gill ventilation in *Limulus. Fed. Proc.* 35:2007–2012.

Yancey, P. H., M. E. Clark, S. C. Hand, R. D. Bowlus, and G. N. Somero. 1982. Living with water stress: evolution of osmolyte systems. *Science* 217:1214–1222.

10 Diseases and Symbionts

Allee, W. C. 1923. An annotated catalog of the distribution of common invertebrates of the Woods Hole littoral. Unpublished manuscript in Marine Biological Laboratory Library.

Andrews, E. A., and T. C. Nelson. 1942. A folliculinid carried by *Limulus. Anat. Rec.* 84:495.

Ball, I. R. 1977. On the phylogenetic classification of aquatic planarians. *Acta Zoo. Fenn.* 154: 21–35.

Bang, F. B. 1953. The toxic effect of a marine bacterium on *Limulus* and the formation of blood clots. *Biol. Bull.* 105:361–362.

———. 1956. A bacterial disease of *Limulus polyphemus. Bull. Johns Hopkins Hosp.* 98:325–351.

———. 1979. Ontogeny and phylogeny of response to gram-negative endotoxins among the marine invertebrates. In *Biomedical Applications of the Horseshoe Crab (Limulidae),* eds. E. Cohen, F. B. Bang, J. Levin, J. J.

Marchalonis, T. G. Pistole, R. A. Prendergast, C. N. Shuster, Jr., and S. W. Watson. New York: Alan R. Liss. Pp. 109–123.

Barnes, R. D. 1986. *Invertebrate Zoology.* Philadelphia: Saunders College (W. B. Saunders).

Baumann, P., L. Baumann, M. J. Woolkslid, and S. Bang. 1983. Evolutionary relationships in *Vibrio* and *Photobacterium:* a basis for a natural classification. *Ann. Rev. Microbiol.* 37:369–398.

Belov, A. P., J. D. Giles, and R. J. Wiltshire. 1999. Toxicity in a water column following the stratification of a cyanobacterial population development in a calm lake. *J. Math. Appl. Med. Biol.* 16(1):93–110.

Bonaventura, J., C. Bonaventura, and S. Tesh, eds. 1982. *Progress in Clinical and Biological Research,* vol. 81: *Physiology and Biology of Horseshoe Crabs: Studies on Normal and Environmentally Stressed Animals.* New York: Alan R. Liss. P. 316.

Botton, M. L. 1981. The gill books of the horseshoe crab *(Limulus polyphemus)* as a substrate for the blue mussel *(Mytilus edulis). Bull. N.J. Acad. Sci.* 26:26–28.

Brandin, E. R., and T. G. Pistole. 1985. Presence of microorganisms in hemolymph of the horseshoe crab *Limulus polyphemus. Appl. Environ. Microbiol.* 49(3):718–720.

Braten, T. 1975. Observations on mechanisms of attachment in green alga *Ulva mutabilis* Foyn. *Protoplasma* 84:161–173.

Braten, T., and A. Lovlie. 1968. On the ultrastructure of vegetative and sporulating cells of the multicellular green alga *(Ulva mutabilis) Foyn., Nytt Mag. Botany* 15:209–219.

Brown, G. G., and D. L. Clapper. 1981. Procedures for maintaining adults, collecting gametes, and culturing embryo and juveniles of the horseshoe crab, *Limulus polyphemus* L. In *Laboratory Animal Management: Marine Invertebrates,* eds. E. Cohen et al. Washington, D.C.: National Academy Press. Pp. 268–290 and references.

Bursey, C. R. 1977. Histological response to injury in the horseshoe crab, *Limulus polyphemus. Can. J. Zool.* 55:1158–1165.

Cohen, E., F. B. Bang, J. Levin, J. J. Marchalonis, T. G. Pistole, R. A. Prendergast, C. N. Shuster, Jr., and S. W. Watson, eds. 1979. *Progress in Clinical and Biological Research,* vol. 29: *Biomedical Applications of the Horseshoe Crab (Limulidae).* New York: Alan R. Liss. P. 683.

Dekhuysen, M. C. 1901. Ueber die Tshrombocyten (Blutplatchen). *Anat. Anzeiger* 19:529–540.

Fahrenbach, W. H. 1970. The cyanoblast: hemocyanin formation in *Limulus polyphemus. J. Cell Biol.* 44:445–453.

French, K. A. 1979. Laboratory culture of embryonic and juvenile *Limulus.* In *Progress in Clinical and Biological Research,* vol. 29: *Biomedical Applications of the Horseshoe Crab (Limulidae),* eds. E. Cohen, F. B. Bang, J. Levin, J. J. Marchalonis, T. G. Pistole, R. A. Prendergast, C. N. Shuster, Jr., and S. W. Watson. New York: Alan R. Liss. Pp. 61–71.

Furman, R. M., and T. G. Pistole. 1976. Bacterial activity of hemolymph from the horseshoe crab, *Limulus polyphemus. J. Invertebr. Pathol.* 28:239–244.

Gould, J. S. 1989. *Wonderful Life. The Burgess Shale and the Nature of History.* New York: W. W. Norton & Co. P. 43.

Groff, J. F., and L. Leibovitz. 1982. A gill disease of *Limulus polyphemus* associated with triclad turbellarid worm infections. *Biol. Bull.* 163:392.

Hanstrom, B. 1926. Ueber einen Fall von pathologischer Chitinbildung in Inneren des Körpers von *Limulus polyphemus. Zool. Anz.* 66:213–219.

Ijima, I., and T. Kaburaki. 1916. Preliminary descriptions of some Japanese triclads. *Annot. Zool. Japon.* 9:153–171.

Isberg, O. 1917. Ein regeneriertes Trilobitenauge. *Geol. Foren. Stockholm, Forhandl.* 39:593–596.

Johannsen, R., R. S. Anderson, R. A. Good, and N. K. Day. 1973. A comparative study of the bactericidal activity of horseshoe crab *(Limulus polyphemus)* hemolymph and vertebrate serum. *J. Vertebr. Pathol.* 22:372–376.

Kawakatsu, M., L. Oki, S. Tamura, and K. Sekiguchi. 1988. Karyological and taxonomical studies of ectoparasitic marine triclads collected from the four extant species of horseshoe crabs. In *Fortschritte der Zoologic (Progress in Zoology),* Vol. 36: *Free-living and Symbiotic Platyhelminthes,* Ax/Ehlers/Sopot-Ehlers, eds. Stuttgart and New York: Gustav Fischer Verlag.

Kawakatsu, M., and K. Sekiguchi. 1988. Redescription of *Ectoplana limuli* (Ijima et Kaburaki, 1916) and *Ectoplana undata,* 1983 (Turbellaria, Tricladida, Maricola), collected from three extant species of Asian horseshoe crabs. *Jobu Journal of Management and Information Service* 5:57–94.

———. 1989. Redescription of an ectoparasitic marine triclad, *Bdelloura candida* (Girard, 1850) (Burbellaria; Tricladida; Maricola), collected from the American horseshoe crab, *Limulus polyphemus. Bull. Biogeograph. Soc. Japan* 44:183–198.

Kawakatsu, M., K. Sekiguchi, K. Miyazaki, and T. Makioka. 1989. The egg-capsules of ectoparasitic marine triclads collected from the four extant species of horseshoe crabs. *Bull. Dept. Management Inform. Sci. Jobu University* 1:15–29.

Keleti, G., and J. L. Sykora. 1982. Production and properties of cyanobacterial endotoxins. *Appl. Environ. Microbiol.* 43(1):104–109.

Kerr, L. A., C. P. McCoy, and D. Eaves. 1987. Blue-green algae toxicosis in five dairy cows. *Journal of the American Veterinary Medical Association* 191(7):829–830.

Kropch, C. 1979. Observations on the potential of *Limulus* aquaculture in Israel. In *Progress in Clinical and Biological Research,* vol. 29: *Biomedical Applications of the Horseshoe Crab (Limulidae),* E. Cohen, F. B. Bang, J. Levin, J. J. Marchalonis, T. G. Pistole, R. A. Prendergast, C. N. Shuster, Jr., and

S. W. Watson. Alan R. Liss: New York. Pp. 103–106.

Landy, R. B., and L. Leibovitz. 1983. A preliminary study of the toxicity and therapeutic efficacy of formalin in the treatment of Triclad turbellarid worm infestations in *Limulus polyphemus. Proc. Ann. Mtg. Soc. Invert. Pathol.,* Ithaca, N.Y.

Leibovitz, L. 1986. Cyanobacterial diseases of the horseshoe crab *(Limulus polyphemus). Biol. Bull.* 171:482.

Leibovitz, L., and T. R. Capo. 1982. A phytomastigophorean infection of embryonating sea hares *(Aplysia californica). Biol. Bull.* 163:393.

Leibovitz, L., and G. A. Lewbart. 1987. A green algal (chlorophycophytal) infection of the dorsal surface of the exoskeleton and associated organ structures in the horseshoe crab *(Limulus polyphemus). Biol. Bull.* 173:430.

Leibovitz, L., J. A. Paige, and J. Bidwell. 1984. Amoebiasis of larval sea hares *(Aplysia californica). Biol. Bull.* 167: 537.

Levin, J., and F. B. Bang. 1964. Clottable protein in *Limulus:* its localization and kinetics of its coagulation by endotoxin. *Throm. Diathes. Hemorrh. (Stugg)* 19:186–197.

Liddell, J. A. 1912. *Nitoceramira bdellourae,* nov. gen. et sp., a copepod of the family Canthocamptridae, parasitic in the egg-cases of *Bdelloura. J. Linn. Soc. London* 32(213):87–94.

Loeb, L. 1902. The blood lymph cells and inflammatory processes of *Limulus. J. Med. Res.* 2:145–158.

Ludvigsen, R. 1977. Rapid repair of traumatic injury by an Ordovician trilobite. *Lethaia* 10:205–207.

MacKenzie, C. L. 1979. Management for increasing clam abundance. *Mar. Fisheries Review* 41:10–22.

Nachum, R., S. E. Siegel, J. D. Sullivan, and S. W. Watson. 1978. Inactivation of endotoxin by *Limulus* amoebocyte lysate. *J. Invertebr. Pathol.* 32:51.

Paerl, H. W., and B. M. Bebout. 1988. Direct measurement of O_2-depleted microzones in marine Oscillatoria: re-

lation to N_2 fixation. *Science* 241:442–445.

Perry, L. M. 1940. Marine shells of the southwest coast of Florida. *Bull. Am. Paleontol.* 16:1–260.

Pistole, T. G. 1978. Bacterial agglutinins from *Limulus polyphemus:* an overview. In *Progress in Clinical and Biological Research,* vol. 29: *Biomedical Applications of the Horseshoe Crab (Limulidae),* eds. E. Cohen, F. B. Bang, J. Levin, J. J. Marchalonis, T. G. Pistole, R. A. Prendergast, C. N. Shuster, Jr., and S. W. Watson. New York: Alan R. Liss. Pp. 547–553.

Rudloe, A. 1983. The effect of heavy bleeding on mortality of the horseshoe crab, *Limulus polyphemus,* in the natural environment. *J. Invertebr. Pathol.* 42:167–176.

Ryder, J. A. 1882. Observations on the species of planarians parasitic on *Limulus. Am. Nat.* 16:48–51.

Seed, R. 1976. Ecology. In *Marine Mussels: Their Ecology and Physiology,* ed. B. L. Bayne. Cambridge: Cambridge University Press.

Shuster, C. N., Jr. 1982. A pictorial review of the natural history and ecology of the horseshoe crab *Limulus polyphemus,* with reference to other Limulidae. In *Progress in Clinical and Biological Research,* vol. 81: *Physiology and Biology of Horseshoe Crabs: Studies on Normal and Environmentally Stressed Animals,* eds. J. Bonaventura, C. Bonaventura, and S. Tesh. New York: Alan R. Liss. Pp. 1–52.

Sluys, R. 1983. A new marine triclad ectoparasitic on Malaysian and Indonesian horseshoe crabs (Platyhelminthes, Turbellaria, Tricladida). *Bijdragen Tot de Dierkunde* 53:218–226.

Smith, S. A., J. M. Berkson, and R. A. Barratt. 2002. Horseshoe crab *(Limulus polyphemus)* hemolymph, biochemical and immunological parameters. *Proc. IAAM* 33:101–102.

Smith, W. R. 1964. Interactions between *Limulus polyphemus* and two species of marine bacteria. *Biol. Bull.* 127:390.

Snyder, G. K., and C. Mangum. 1982. The relationship between the capacity for oxygen transport, size, shape and aggregation state of an extracellular oxygen carrier. In *Progress in Clinical and Biological Research,* vol. 81: *Physiology and Biology of Horseshoe Crabs: Studies on Normal and Environmentally Stressed Animals,* eds. J. Bonaventura, C. Bonaventura, and S. Tesh. New York: Alan R. Liss. Pp. 173–188.

Stedman's Medical Dictionary, 24th ed. 1982. Baltimore: Williams and Wilkins.

Stunkard, H. W. 1950. Microphallid metacercariae encysted in *Limulus. Biol. Bull.* 99:347.

———. 1951. Observation on the morphology and life-history of *Microphallus* n. sp. (Trematoda: Microphallidae). *Biol. Bull.* 101:307–318.

———. 1953. Natural hosts of *Microphallus limuli. J. Parasitol.* 39:225.

———. 1968. The sexual generation, life cycle, and systemic relations of *Microphallus limuli* (Trematoda: Digenea). *Biol. Bull.* 134:332–343.

Sullivan, J. D., Jr., and S. W. Watson. 1975. Purification and properties of the clotting enzyme from *Limulus* lysate. *Biochem. Biophys. Res. Commun.* 66:648–655.

Vasconcelos, V. M., and E. Pereira. 2001. Cyanobacterial diversity and toxicity in a wastewater treatment plant (Portugal). *Water Res.* 35(5):1354–1357.

Verrill, A. E. 1893. Marine planarians of New England. *Trans. Ct. Acad.* 8:459–520.

———. 1895. Supplement to the marine nemerteans and planarians of New England. *Trans. Conn. Acad.* 9:141–152.

Waterman, T. H. 1950. A light polarization analyzer in the compound eye of *Limulus. Science* 111:252–254.

Wheeler, W. M. 1894. *Syncoelidium pellucidum,* a new marine triclad. *J. Morphol.* 9:167–194.

11   The Circulatory System

Brouwer, M., C. Bonaventura, and J. Bonaventura. 1982. Chloride and pH dependence of cooperative interactions in *Limulus polyphemus* hemocyanin. In *Progress in Clinical and Biological Research,* vol. 81: *Physiology and Biology of Horseshoe Crabs: Studies on Normal and Environmentally Stressed Animals,* eds. J. Bonaventura, C. Bonaventura, and S. Tesh. New York: Alan R. Liss. Pp. 231–256.

Carlson, A. J. 1904. The nervous origin of the heartbeat in *Limulus,* and the nervous nature of co-ordination or conduction in the heart. *Am. J. Physiol.* 12:67–74.

———. 1906a. On the mechanism of co-ordination and conduction in the heart, with special reference to the heart of *Limulus. Am. J. Physiol.* 15:99–120.

———. 1906b. Osmotic pressure and heart activity. *Am. J. Physiol.* 15:357–370.

———. 1907. The relation of the normal heart rhythm to the artificial rhythm produced by sodium chloride. *Am. J. Physiol.* 17:478–486.

Carlson, A. J., and W. J. Meek. 1908. On the mechanism of the embryonic heart rhythm in *Limulus. Am. J. Physiol.* 21:1–10.

Chao, I. 1933. Action of electrolytes on the dorsal median nerve cord of the *Limulus* heart. *Biol. Bull.* 64:358–382.

———. 1935. Hydrogen ion concentration and the rhythmic activity of the nerve cells in the ganglion of the *Limulus* heart. *Biol. Bull.* 68:69–73.

Cole, W. H., and J. B. Allison. 1940. The nitrogen, copper, and hemocyanin content of the sera of several arthropods. *J. Biol. Chem.* 136:259–265.

Engel, G. L., and I. Chao. 1935. Comparative distribution of organic phosphates in the skeletal and cardiac muscles of *Limulus polyphemus. J. Biol. Chem.* 108:389–393.

Fahrenbach, W. H. 1999. Merostomata. In *Microscopic Anatomy of Invertebrates,* Vol. 8A: *Chelicerae Arthropoda,* eds. F. W. Harrison and R. F. Foelix. New York: Wiley-Liss (John Wiley & Sons). Pp. 21–115.

Fourtner, C. R., C. D. Drewes, and R. A. Pax. 1971. Rhythmic outputs coordinating the respiratory movement of the gill plates of *Limulus polyphemus. Comp. Biochem. Physiol.* 28A:751–762.

Freadman, M. A., and W. H. Watson, III. 1989. Gills as possible accessory circulatory pumps. *Biol. Bull.* 177:386–395.

Fredericq, H. 1947. Les nerfs cardioregulateurs de invertebres et la theorie des mediateurs chimiques. *Biol. Rev.* 22:298–314.

Garrey, W. E. 1912. Compression of the cardiac nerves of *Limulus,* and some analogies which apply to the mechanism of heart block. *Am. J. Physiol.* 30:283–302.

———. 1930. The pace maker of the cardiac ganglion of *Limulus polyphemus. Am. J. Physiol.* 93:178–185.

Groome, J. R., M. A. Townlet, and W. H. Watson III. 1994. Excitatory actions of FMRFamide-related peptides (FaRPs) on the neurogenic *Limulus* heart. *Biol. Bull.* 186:309–318.

Heinbecker, P. 1933. The heart and median cardiac nerve of *Limulus polyphemus. Am. J. Physiol.* 103:104–120.

———. 1936. The potential analysis of a pacemaker mechanism in *Limulus polyphemus. Am. J. Physiol.* 117:686–700.

Johnson, M. L., and D. A. Yphantis. 1978. Subunit association and heterogeneity of *Limulus polyphemus* hemocyanin. *Biochem.* 17:1448–1455.

Jordan, H. E. 1917. The microscopic structure of striped muscle of *Limulus.* Department of Marine Biology, Carnegie Institution of Washington. 9:273–290.

Kingsley, J. S. 1885. Notes on the embryology of *Limulus. Quart. J. Microsc. Sci.* 25:521–576.

———. 1892. The embryology of *Limulus. J. Morphol.* 7:25–66.

———. 1893. The embryology of *Limulus,* part II. *J. Morphol.* 8:195–268.

Milne-Edwards, A. 1873. Recherches sur l'anatomie des *Limulus. Ann. Sci. Nat.* 17:1–67.

Packard, A. S., Jr. 1872. The development of *Limulus polyphemus*. *Mem. Boston Soc. Nat. Hist.* 2:155–202.

Patten, W. 1912. *The Evolution of the Vertebrates and Their Kin*. Philadelphia: P. Blakiston's Son & Co. Pp. 195–209.

Patten. W., and W. A. Redenbaugh. 1899. Studies on *Limulus,* part II: the nervous system of *Limulus polyphemus,* with observations upon the general anatomy. *J. Morphol.* 16:91–200.

Prosser, C. L. 1942. An analysis of the action of acetylcholine on hearts, particularly in arthropods. *Biol. Bull.* 83:145–164.

———. 1943. Single unit analysis of the heart ganglion discharge in *Limulus polyphemus. J. Cell. Comp. Physiol.* 21:295–305.

Redfield, A. C. 1934. The haemocyanins. *Biol. Rev.* 9:175–212.

Redmond, J. R., D. D. Jorgensen, and G. B. Bourne. 1982. Circulatory physiology of *Limulus*. In *Progress in Clinical and Biological Research,* vol. 81: *Physiology and Biology of Horseshoe Crabs: Studies on Normal and Environmentally Stressed Animals,* eds. J. Bonaventura, C. Bonaventura, and S. Tesh. New York: Alan R. Liss. Pp. 133–148.

Roche, J. 1930. Recherches sur quelques propriétés physico-chimiques des hémocyanins du Poulpe et de la Limule. *Arch. Physiol. Biol.* 7:207–220.

Samojloff, A. 1930. The extra systolic impulse of the ganglion of *Limulus* heart. *Am. J. Physiol.* 93:186–189.

Shuster, C. N., Jr. 1955. On morphometric and serological relationships within the Limulidae, with particular reference to *Limulus polyphemus* (L.). Ph.D thesis, New York University (1958 Diss. Abstr. 18: 371–372).

———. 1982. A pictorial review of the natural history and ecology of the horseshoe crab *Limulus polyphemus,* with reference to other Limulidae. In *Progress in Clinical and Biological Research,* vol. 81: *Physiology and Biology of Horseshoe Crabs: Studies on Normal and Environmentally Stressed Animals,* eds. J.

Bonaventura, C. Bonaventura, and S. Tesh. New York: Alan R. Liss. Pp. 1–52.

Watson, W. H., III, and J. R. Groome. 1989. Modulation of the *Limulus* heart. *Am. Zool.* 29:1287–1303.

Watson, W. H., III, and G. A. Wyse. 1978. Coordination of heart and fill rhythm in *Limulus. J. Comp. Physiol.* 124A:267–275.

Wyse, G. A., and C. H. Page. 1976. Coordination of the heart and gill rhythms in *Limulus. J. Comp. Physiol.* 124:265–275.

Wyse, G. A., D. H. Sanes, and W. H. Watson III. 1980. Central neural motor programs underlying short- and long-term patterns of *Limulus* respiratory activity. *J. Comp. Physiol.* 141: 87–92.

Yeager, J. F., and O. E. Tauber. 1935. On the hemolymph cell counts of some marine invertebrates. *Biol. Bull.* 69:66–70.

12 Internal Defense against Pathogenic Invasion

Agarwala, K. L., S.-i. Kawabata, Y. Miura, Y. Kuroki, and S. Iwanaga. 1996. *Limulus* intracellular coagulation inhibitor type 3. Purification, characterization, cDNA cloning, and tissue localization. *J. Biol. Chem.* 271:23768–23774.

Agarwala, K. L., S. Kawabata, M. Hirata, M. Miyagi, S. Tsunasawa, and S. Iwanaga. 1996. A cysteine protease inhibitor stored in the large granules of horseshoe crab hemocytes: purification, characterization, cDNA cloning and tissue localization. *J. Biochem. (Tokyo)* 119:85–94.

Armstrong, P. B. 1977. Interaction of the motile blood cells of the horseshoe crab, *Limulus*. Studies on contact paralysis of pseudopodial activity and cellular overlapping in vitro. *Exp. Cell Res.* 107:127–138.

———. 1979. Motility of the *Limulus* blood cell. *J. Cell Sci.* 37:169–180.

———. 1980. Adhesion and spreading of *Limulus* blood cells on artificial surfaces. *J. Cell Sci.* 44:243–262.

———. 1985. Adhesion and motility of the blood cells of *Limulus*. In *Blood Cells of Marine Invertebrates,* ed. W. D. Cohen. New York: Alan R. Liss. Pp. 77–124.

———. 1991. Cellular and humoral immunity in the horseshoe crab, *Limulus polyphemus*. In *Immunology of Insects and Other Arthropods,* ed. A. P. Gupta. Boca Raton: CRC Press. Pp. 3–17.

———. 2001. The contribution of proteinase inhibitors to immune defense. *Trends Immunol.* 22:47–52.

Armstrong, P. B., M. T. Armstrong, and J. P. Quigley. 1993. Involvement of α_2-macroglobulin and C-reactive protein in a complement-like hemolytic system in the arthropod, *Limulus polyphemus. Mol. Immunol.* 30:929–934.

Armstrong, P. B., and J. Levin. 1979. In vitro phagocytosis by *Limulus* blood cells. *J. Invert. Pathol.* 34:145–151.

Armstrong, P. B., R. Melchior, and J. P. Quigley. 1996a. Humoral immunity in long-lived arthropods. *J. Insect Physiol.* 42:53–64.

Armstrong, P. B., and J. P. Quigley. 1992. Humoral immunity: α_2-macroglobulin activity in the plasma of mollusks. *Veliger* 35:161–164.

Armstrong, P. B., J. P. Quigley, and F. R. Rickles. 1990. The *Limulus* blood cell secretes α_2-macroglobulin when activated. *Biol. Bull. (Woods Hole)* 178: 137–143.

Armstrong, P. B., and F. R. Rickles. 1982. Endotoxin-induced degranulation of the *Limulus* amebocyte. *Exp. Cell Res.* 140:15–24.

Armstrong, P. B., M. T. Rossner, and J. P. Quigley. 1985. An α_2-macroglobulin-like activity in the blood of chelicerate and mandibulate arthropods. *J. Exp. Zool.* 236:1–9.

Armstrong, P. B., S. Swarnakar, S. Srimal, S. Misquith, E. A. Hahn, R. T. Aimes, and J. P. Quigley. 1996. A cytolytic function for a sialic acid-binding

lectin that is a member of the
pentraxin family of proteins. *J. Biol.
Chem.* 271:14717–14721.

Bang, F. B. 1979. Ontogeny and phylog-
eny of response to gram-negative
endotoxins among the marine inver-
tebrates. In *Progress in Clinical and
Biological Research*, vol. 29: *Biomedical
Applications of the Horseshoe Crab
(Limulidae)*, eds. E. Cohen, F. B. Bang,
J. Levin, J. J. Marchalonis, T. G.
Pistole, R. A. Prendergast, C. N.
Shuster, Jr., and S. W. Watson. New
York: Alan R. Liss. Pp. 109–123.

Bender, R. C., S. E. Fryer, and C. J.
Bayne. 1992. Proteinase inhibitory
activity in the plasma of a mollusc—
Evidence for the presence of α_2-
macroglobulin in *Biomphalaria
glabrata*. *Comp. Biochem. Physiol. B.*
102:821–824.

Bohn, H. 1986. Hemolymph clotting in
insects. In *Immunity in Invertebrates.
Cells, Molecules, and Defense Reactions*,
ed. M. Brehélin. Berlin: Springer-
Verlag. Pp. 188–207.

Cohen, E., F. B. Bang, J. Levin, J. J.
Marchalonis, T. G. Pistole, R. A.
Prendergast, C. Shuster, and S. W.
Watson. 1979. *Progress in Clinical and
Biological Research*, vol. 29: *Biomedical
Applications of the Horseshoe Crab
(Limulidae)*. New York: Alan R.
Liss.

Dimarcq, J. L., D. Zachary, J. A.
Hoffmann, D. Hoffmann, and J. M.
Reichhart. 1990. Insect immunity: expression of the two major inducible
antibacterial peptides, defensin and
diptericin, in *Phormia terranovae*.
EMBO J. 9:2507–2515.

Dunn, P. E., T. J. Bohnert, and V. Russell.
1994. Regulation of antibacterial protein synthesis following infection and
during metamorphosis of *Manduca
sexta*. *Ann. N. Y. Acad. Sci.* 712:117–
130.

Finch, C. E. 1990. *Longevity, Senescence
and the Genome*. Chicago: University
of Chicago Press.

Furie, B., and B. C. Furie. 1992. Molecular and cellular biology of blood co-

agulation. *N. Engl. J. Med.* 326:800–
806.

Gabay, J. E. 1994. Ubiquitous natural antibiotics. *Science* 264:373–374.

Gabay, J. E., and R. P. Almeida. 1993.
Antibiotic peptides and serine protease homologs in human poly-
morphonuclear leukocytes: defensins
and azurocidin. *Curr. Opin. Immunol.*
5:97–102.

Hoffmann, D., and J. A. Hoffmann.
1990. Cellular and molecular aspects
of insect immunity. *Res. Immunol.*
141:895–896.

Holme, R., and N. O. Solum. 1973.
Electron microscopy of the gel protein formed by clotting of *Limulus
polyphemus* hemocyte extracts. *J.
Ultrastruct. Res.* 44:329–338.

Ikawi, D., S.-I. Kawabata, Y. Miura, A.
Kato, P. B. Armstrong, J. P. Quigley,
K. L. Nielsen, and K. Dolmer. 1996.
Molecular cloning of *Limulus* α_2-
macroglobulin. *Eur. J. Biochem.*
242:822–831.

Iwaki, D., T. Osaki, Y. Mizunoe, S. N.
Wai, S. Iwanaga, and S. Kawabata.
1999. Functional and structural diver-
sities of C-reactive proteins present in
horseshoe crab hemolymph plasma.
Eur. J. Biochem. 264:314–326.

Iwanaga, S. 2002. The molecular basis of
innate immunity in the horseshoe
crab. *Curr. Opin. Immunol.* 14:87–95.

Iwanaga, S., and S. Kawabata. 1998. Evolution and phylogeny of defense molecules associated with innate immunity in horseshoe crab. *Frontiers Biosci.*
3:D973-D984.

Iwanaga, S., T. Miyata, F. Tokunaga, and
T. Muta. 1992. Molecular mechanism
of hemolymph clotting system in
Limulus. Thromb. Res. 68:1–32.

Iwanaga, S., T. Muta, T. Shigenaga, N.
Seki, K. Kawano, T. Katsu, and S.
Kawabata. 1994. Structure-function
relationships of tachyplesins and their
analogues. *Ciba Foundation Symposium*
186:160–174; discussion, 174–175.

Janeway, C. A., and R. Medzhitov. 2002.
Innate immune recognition. *Annual
Rev. Immunol.* 20:197–216.

Kawabata, S., H. G. Beisel, R. Huber, W.
Bode, S. Gokudan, T. Muta, R. Tsuda,
K. Koori, T. Kawahara, N. Seki, Y.
Mizunoe, S. N. Wai, and S. Iwanaga.
2001. Role of tachylectins in host de-
fense of the Japanese horseshoe crab
*Tachypleus tridentatus. Adv. Exp. Med.
Biol.* 484:195–202.

Kawabata, S., and S. Iwanaga. 1999. Role
of lectins in the innate immunity of
horseshoe crab. *Dev. Comp. Immunol.*
23:391–400.

Kawabata, S., R. Nagayama, M. Hirata, T.
Shigenaga, K. L. Agarwala, T. Saito, J.
Cho, H. Nakajima, T. Takagi, and S.
Iwanaga. 1996. Tachycitin, a small
granular component in horseshoe
crab hemocytes, is an antimicrobial
protein with chitin-binding activity. *J.
Biochem. (Tokyo)* 120:1253–1260.

Kawabata, S., T. Saito, K. Saeki, N.
Okino, A. Mizutani, Y. Toh, and S.
Iwanaga. 1997. CDNA cloning, tissue
distribution, and subcellular localiza-
tion of horseshoe crab big defensin. *J.
Biol. Chem.* 378:289–292.

Kloczewiak, M., K. M. Black, P. Loiselle,
J. M. Cavaillon, N. Wainwright, and
H. S. Warren. 1994. Synthetic peptides that mimic the binding site of
horseshoe crab antilipopolysaccha-
ride factor. *J. Infect. Dis.* 170:1490–
1497.

Kristensen, T., S. K. Moestrup, J.
Gliemann, L. Bendtsen, O. Sand, and
L. Sottrup-Jensen. 1990. Evidence
that the newly cloned low-density-li-
poprotein receptor related protein
(LRP) is the α_2-macroglobulin receptor. *FEBS Let.* 276:151–155.

Lantz, M. S. 1997. Are bacterial proteases important virulence factors? *J.
Periodontal Research* 32:126–132.

Law, S. K., and K. B. M. Reid. 1995.
Complement, 2nd ed. Oxford: IRL
Press.

Levashina, E. A., L. F. Moita, S. Blandin,
G. Vriend, M. Lagueux, and F. C.
Kafatos. 2001. Conserved role of a
complement-like protein in
phagocytosis revealed by dsRNA
knockout in cultured cells of the

mosquito, *Anopheles gambiae. Cell* 104:709–718.

Loeb, L. 1920. The movements of the amoebocytes and the experimental production of amoebocyte (cell fibria) tissue. *Washington University Studies* 8:3–79.

Matsuzaki, K., M. Nakayama, M. Fukui, A. Otaka, S. Funakoshi, N. Fujii, K. Bessho, and K. Miyajima. 1993. Role of disulfide linkages in tachyplesin-lipid interactions. *Biochem.* 32:11704–11710.

Mekalanos, J. J. 1992. Environmental signals controlling expression of virulence determinants in bacteria. *J. Bacteriol.* 174:1–7.

Melchior, R., J. P. Quigley, and P. B. Armstrong. 1995. α_2-macroglobulin-mediated clearance of proteases from the plasma of the American horseshoe crab, *Limulus polyphemus. J. Biol. Chem.* 270:13496–13502.

Miura, Y., S. Kawabata, and S. Iwanaga. 1994. A *Limulus* intracellular coagulation inhibitor with characteristics of the serpin superfamily. Purification, characterization, and cDNA cloning. *J. Biol. Chem.* 269:542–547.

Miura, Y., S. Kawabata, Y. Wakamiya, T. Nakamura, and S. Iwanaga. 1995. A *Limulus* intracellular coagulation inhibitor type 2. Purification, characterization, cDNA cloning, and tissue localization. *J. Biol. Chem.* 270:558–565.

Miyata, T., F. Tokunaga, T. Yoneya, K. Yoshikawa, S. Iwanaga, M. Niwa, T. Takao, and Y. Shimonishi. 1989. Antimicrobial peptides, isolated from horseshoe crab hemocytes, tachyplesin II, and polyphemusins I and II: chemical structures and biological activity. *J. Biol. Chem. (Tokyo)* 106:663–668.

Murer, E. H., J. Levin, and R. Holme. 1975. Isolation and studies of the granules of the amebocytes of *Limulus polyphemus,* the horseshoe crab. *J. Cell Physiol.* 86:533–542.

Muta, T., R. Hashimoto, T. Miyata, H. Nishimura, Y. Toh, and S. Iwanaga.

1990. Proclotting enzyme from horseshoe crab hemocytes. cDNA cloning, disulfide locations, and subcellular localization. *J. Biol. Chem.* 265:22426–22433.

Muta, T., T. Miyata, Y. Misumi, F. Tokunaga, T. Nakamura, Y. Toh, Y. Ikehara, and S. Iwanaga. 1991. *Limulus* factor C. An endotoxin-sensitive serine protease zymogen with a mosaic structure of complement-like, epidermal growth factor-like, and lectin-like domains. *J. Biol. Chem.* 266:6554–6561.

Muta, T., T. Miyata, F. Tokunaga, T. Nakamura, and S. Iwanaga. 1987. Primary structure of anti-lipopolysaccharide factor from American horseshoe crab, *Limulus polyphemus. J. Biochem. (Tokyo)* 101:1321–1330.

Muta, T., T. Oda, and S. Iwanaga. 1993. Horseshoe crab coagulation factor B. A unique serine protease zymogen activated by cleavage of an Ile-Ile bond. *J. Biol. Chem.* 268:21384–21388.

Nakamura, T., H. Furunaka, T. Miyata, F. Tokunaga, T. Muta, S. Iwanaga, M. Niwa, T. Takao, and Y. Shimonishi. 1988. Tachyplesin, a class of antimicrobial peptide from the hemocytes of the horseshoe crab *(Tachypleus tridentatus)*. Isolation and chemical structure. *J. Biol. Chem.* 263:16709–16713.

Nakamura, S., T. Morita, T. Harada-Suzuki, S. Iwanaga, K. Takahashi, and M. Niwa. 1982. A clotting enzyme associated with the hemolymph coagulation system of horseshoe crab *(Tachypleus tridentatus)*: its purification and characterization. *J. Biochem. (Tokyo)* 92:781–792.

Nguyen, N. Y., A. Suzuki, R. A. Boykins, and T. Y. Liu. 1986a. The amino acid sequence of Limulus C-reactive protein. Evidence of polymorphism. *J. Biol. Chem.* 261:10456–10465.

Nguyen, N. Y., A. Suzuki, S. M. Cheng, G. Zon, and T. Y. Liu. 1986b. Isolation and characterization of Limulus C-

reactive protein genes. *J. Biol. Chem.* 261:10450–10455.

Okino, N., S. Kawabata, T. Saito, M. Hirata, T. Takagi, and S. Iwanaga. 1995. Purification, characterization, and cDNA cloning of a 27-kDa lectin (L10) from horseshoe crab hemocytes. *J. Biol. Chem.* 270:31008–31015.

Pepys, M. B., and M. L. Baltz. 1983. Acute phase proteins with special reference to C-reactive protein and related proteins (pentaxins) and serum amyloid A protein. *Adv. Immunol.* 34:141–212.

Price, P. W. 1980. *Evolutionary Biology of Parasites.* Princeton: Princeton University Press.

Quigley, J. P., and P. B. Armstrong. 1983. An endopeptidase inhibitor, similar to mammalian α_2-macroglobulin, detected in the hemolymph of an invertebrate, *Limulus polyphemus. J. Biol. Chem.* 258:7903–7906.

———. 1985. A homologue of α_2-macroglobulin purified from the hemolymph of the horseshoe crab *Limulus polyphemus. J. Biol. Chem.* 260:12715–12719.

Ratcliffe, N. A., K. N. White, A. F. Rowley, and J. B. Walters. 1982. Cellular defense systems of the arthorpoda. In *The Reticuloendothelial System, A Comprehensive Treatise,* eds. N. Cohen and M. M. Segel. New York: Plenum Press. Pp. 167–202.

Ried, C., C. Wahl, T. Miethke, G. Wellnhofer, C. Landgraf, J. Schneider-Mergener, and A. Hoess. 1996. High-affinity endotoxin-binding and neutralizing peptides based on the crystal structure of recombinant *Limulus* anti- lipopolysaccharide factor. *J. Biol. Chem.* 271:28120–28127.

Rotstein, O. D. 1992. Role of fibrin deposition in the pathogenesis of intraabdominal infection. *Eur. J. Clin. MicroBiol. Infect. Dis.* 11:1064–1068.

Saito, T., M. Hatada, S. Iwanaga, and S. Kawabata. 1997. A newly identified horseshoe crab lectin with binding specificity to O-antigen of bacterial lipopolysaccharides. *J. Biol. Chem.* 272:30703–30708.

Saito, T., S. Kawabata, M. Hirata, and S. Iwanaga. 1995. A novel type of limulus lectin-L6. Purification, primary structure, and antibacterial activity. *J. Biol. Chem.* 270:14493–14499.

Salt, G. *The Cellular Defense Reactions of Insects.* 1970. Cambridge, England: Cambridge University Press.

Sastry, K., and R. A. Ezekowitz. 1993. Collectins: pattern recognition molecules involved in first line host defense. *Curr. Opin. Immunol.* 5:59–66.

Seki, N., T. Muta, T. Oda, D. Iwaki, K. Kuma, T. Miyata, and S. Iwanaga. 1994. Horseshoe crab (1,3)-β-D-glucan-sensitive coagulation factor G. A serine protease zymogen heterodimer with similarities to β-glucan-binding proteins. *J. Biol. Chem.* 269:1370–1374.

Shai, Y. 1999. Mechanism of the binding, insertion and destabilizastion of phospholipid bilayer membranes by α-helical antimicrobial and cell nonselective membrane-lytic peptides. *Biochim. Biophy. Acta* 1462:55–70.

Shrive, A. K., A. M. Metcalfe, J. R. Cartwright, and T. J. Greenhough. 1999. C-reactive protein and SAP-like pentraxin are both present in *Limulus polyphemus* haemolymph: crystal structure of *Limulus* SAP. *J. Biol. Chem.* 290:997–1008.

Shuster, C. N. 1978. The circulatory system and blood of the horseshoe crab *Limulus polyphemus* L.: a review. Washington, D.C.: U.S. Department of Energy, Federal Regulatory Commission. Pp. 1–63.

Sottrup-Jensen, L. 1987. α_2-Macroglobulin and related thiol ester plasma proteins. In *The Plasma Proteins: Structure, Function and Genetic Control*, ed. F. W. Putnam. Orlando, Fl.: Academic Press. Pp. 191–291.

———— 1989. α-Macroglobulins: structure, shape, and mechanism of

proteinase complex formation. *J. Biol. Chem.* 264:11539–11542.

Sottrup-Jensen, L., O. Sand, L. Kristensen, and G. H. Fey. 1989. The α-macroglobulin bait region. Sequence diversity and localization of cleavage sites for proteinases in five mammalian α-macroglobulins. *J. Biol. Chem.* 264:15781–15789.

Srimal, S., T. Miyata, S. Kawabata, and S. Iwanaga. 1985. The complete amino acid sequence of coagulogen isolated from Southeast Asian horseshoe crab, *Carcinoscorpius rotundicauda.* *J. Biochem. (Tokyo)* 98:305–318.

Starkey, P. M., and A. J. Barrett. 1982. Evolution of α_2-macroglobulin. The demonstration in a variety of vertebrate species of a protein resembling human α_2-macroglobulin. *Biochem. J.* 205:91–95.

Strickland, D. K., J. D. Ashcom, S. Williams, W. H. Burgess, M. Migliorini, and W. S. Argraves. 1990. Sequence identity between the α_2-macroglobulin receptor and low-density lipoprotein receptor-related protein suggests that this molecule is a multifunctional receptor. *J. Biol. Chem.* 265:17401–17404.

Swarnakar, S., R. Asokan, J. P. Quigley, and P. B. Armstrong. 2000. Binding of α_2-macroglobulin and limulin: regulation of the plasma hemolytic system of the American horseshoe crab, *Limulus. Biochem. J.* 347:679–685.

Tack, B. F. 1983. The β-Cys-γ-Glu thiolester bond in human C3, C4, and α_2-macroglobulin. *Springer Sem. Immunopathol.* 6:259–282.

Tai, J. Y., R. C. J. Seid, R. D. Huhn, and T. Y. Liu. 1977. Studies on Limulus amoebocyte lysate II. Purification of the coagulogen and the mechanism of clotting. *J. Biol. Chem.* 252:4773–4776.

Takagi, T., Y. Hokama, T. Morita, S. Iwanaga, S. Nakamura, and M. Niwa. 1979. Amino acid sequence studies on horseshoe crab *(Tachypleus tridentatus)* coagulogen and the mechanism

of gel formation. *Prog. Clin. Biol. Res.* 29:169–184.

Tan, N. S., B. Ho, and J. L. Ding. 2000. High-affinity LPS binding domain(s) in recombinant factor C of a horseshoe crab neutralizes LPS-induced lethality. *FASEB J.* 14:859–870.

Tan, N. S., M. L. Ng, Y. H. Yau, P. K. Chong, B. Ho, and J. L. Ding. 2000. Definition of endotoxin binding sites in horseshoe crab factor C recombinant sushi proteins and neutralization of endotoxin by sushi peptides. *FASEB J.* 14:1801–1813.

Tennent, G. A., P. J. Butler, T. Hutton, A. R. Woolfitt, D. J. Harvey, T. W. Rademacher, and M. B. Pepys. 1993. Molecular characterization of *Limulus polyphemus* C-reactive protein. I: Subunit composition. *Eur. J. Biochem.* 214:91–97.

Thogersen, I. B., G. Salvesen, F. H. Brucato, S. V. Pizzo, and J. J. Enghild. 1992. Purification and characterization of an α-macroglobulin proteinase inhibitor from the mollusc *Octopus vulgaris. Biochem. J.* 285:521–527.

Tokunaga, F., T. Muta, S. Iwanaga, A. Ichinose, E. W. Davie, K. Kuma, and T. Miyata. 1993a. Limulus hemocyte transglutaminase. cDNA cloning, amino acid sequence, and tissue localization. *J. Biol. Chem.* 268:262–268.

Tokunaga, F., M. Yamada, T. Miyata, Y. L. Ding, M. Hiranaga-Kawabata, T. Muta, S. Iwanaga, A. Ichinose, and E. W. Davie. 1993b. Limulus hemocyte transglutaminase. Its purification and characterization, and identification of the intracellular substrates. *J. Biol. Chem.* 268:252–261.

Van Leuven, F. 1984. Human α_2 macroglobulin. *Mol. Cell. Biochem.* 58:121–128.

Warren, H. S., M. L. Glennon, N. Wainwright, S. F. Amato, K. M. Black, S. J. Kirsch, G. R. Riveau, R. I. Whyte, W. M. Zapol, and T. J. Novitsky. 1992. Binding and neutralization of endotoxin by *Limulus* anti-lipopolysaccharide factor. *Infect. Immun.* 60:2506–2513.

13 Clotting Cells and *Limulus* Amebocyte Lysate

Bang, F. B. 1953. The toxic effect of a marine bacterium on *Limulus* and the formation of blood clots. *Biol. Bull.* 105:447–448.

———. 1956. A bacterial disease of *Limulus polyphemus. Bull. Johns Hopkins Hosp.* 98:325–351.

Bang, F. B., and B. Bang. 1982. Pathologic principles revealed by study of natural diseases of invertebrates. In *Progress in Clinical and Biological Research,* vol. 29: *Biomedical Applications of the Horseshoe Crab (Limulidae),* eds. E. Cohen, F. B. Bang, J. Levin, J. J. Marchalonis, T. G. Pistole, R. A. Prendergast, C. N. Shuster, Jr., and S. W. Watson. Pp. 289–296.

Bonaventura, J., C. Bonaventura, and A. Tesh. 1982. *Progress in Clinical and Biological Research,* vol. 81: *Physiology and Biology of Horseshoe Crabs: Studies on Normal and Environmentally Stressed Animals.* New York: Alan R. Liss.

Cohen, E., F. B. Bang, J. Levin, J. J. Marchalonis, T. G. Pistole, R. A. Prendergast, C. N. Shuster, Jr., and S. W. Watson, eds. 1979. *Progress in Clinical and Biological Research,* vol. 29: *Biomedical Applications of the Horseshoe Crab (Limulidae).* New York: Alan R. Liss.

Cooper, J. F., and J. C. Harbert. 1975. Endotoxin as a cause of aseptic meningitis after radionuclide cisternography. *J. Nucl. Med.* 16: 809–813.

Cooper, J. F., J. Levin, and H. N. Wagner, Jr. 1970. New rapid in vitro test for pyrogen in short-lived radiopharmaceuticals. *J. Nucl. Med.* 11:301.

———. 1971. Quantitative comparison of in vitro *(Limulus)* and in vivo (rabbit) methods for the detection of endotoxin. *J. Lab. Clin. Med.* 78: 138–148.

———. 1972. The limulus test for endotoxin (pyrogen) in radiopharma-

ceuticals and biologicals. *Bull. Parenter. Drug Assoc.* 26: 153–162.

Dabbah, R., E. Ferry, D. A. Gunther, R. Hahn, R. L. Sanford, P. Mazur, M. Neely, P. Nicholas, J. S. Pierce, J. Slade, S. Watson, and M. Weary. 1980. Pyrogenicity of *E. coli* 055:B5 endotoxin by the USP rabbit test—a HIMA collaborative study. *J. Parenter. Drug Assoc.* 34: 212–216.

Ding, J. L., C. Chai, A. W. M. Pui, and B. Ho. 1997. Expression of full length and deletion homologues of *Carcinoscopius rotundicauda* Factor C in *Saccharomyces cerevisiae:* immunoreactivity and endotoxin binding. *J. Endotoxin Res.* 4:33–43.

Federal Register Notice. November 4, 1977. Licensing of *Limulus* amebocyte lysate. Use as an alternative for rabbit pyrogen test. 42FR213.

Federal Register Notice. May 16, 1980. Additional standards for diagnostic substances for laboratory tests. 45FR97.

Flint, O. 1994. A timetable for replacing, reducing and refining animal use with the help of in vitro tests: the *Limulus* amebocyte lysate test (LAL) as an example. In *Alternatives to Animal Testing. New Ways in the Biomedical Sciences,* ed. C. A. Reinhardt. New York: John Wiley and Sons. Pp. 27–41.

Highsmith, A., R. L. Anderson, and J. R. Allen. 1982. Application of the *Limulus* amebocyte lysate assay in outbreaks of pyrogenic reactions associated with parenteral fluids and medical devices. In *Endotoxins and Their Detection with the Limulus Amoebocyte Lysate Test,* eds. S. W. Watson, J. Levin, and T. J. Novitsky. New York: Alan R. Liss. Pp. 287–299.

Hochstein, H. D., E. A. Fitzgerald, F. G. McMahon, and R. Vargas. 1994. Properties of U.S. standard endotoxin (EC-5) in human male volunteers. *J. Endotoxin Res.* 1:52–56.

Hochstein, H. D., D. F. Mills, A. D. Outschoorn, and S. C. Rastogi. 1983. The processing and collaborative assay of a reference endotoxin. *J. Biol. Stand.* 11:251.

Hochstein, H. D., E. B. Seligmann, R. E. Marquina, and R. Rivera. 1979. *Limulus* amebocyte lysate testing of normal serum albumin (human) released in the United States since 1975. *Devel. Biol. Stand.* 44:35.

Howell, W. H. 1885. Observations upon the chemical composition and the coagulation of the blood of *Limulus polyphemus, Callinectes hastatus,* and *Cucumaria* sp. *Johns Hopkins Circulation* 5:4.

Hurley, J. C. 1994. Concordance of endotoxemia with gram-negative bacteremia in patients with gram-negative sepsis: a meta-analysis. *J. Clin. Microbiol.* 32:2120–2127.

———. 1995a. Endotoxemia: methods of detection and clinical correlates. *Clin. Microbiol. Rev.* 8:268–292.

———. 1995b. Reappraisal with meta-analysis of bacteremia, endotoxemia, and mortality in gram-negative sepsis. *J. Clin. Microbiol.* 33: 1278–1282.

Hurley, J., and J. Levin. 1999. In *Endotoxin in Health and Disease,* eds. H. Brade, S. M. Opal, S. N. Vogel, and D. C. Morrison. New York: Marcel Dekker. Pp. 841–854.

Iwanaga, S., S. Kawabata, and T. Muta. 1998. New types of clotting factors and defense molecules found in horseshoe crab hemolymph: their structures and functions. *J. Biochem.* 123:1–15.

Ketchum, P. A., J. Parsonnet, L. S. Stotts, T. J. Novitsky, B. Schlain, and D. W. Bates. AMCC SEPSIS Project. 1997. Utilization of a chromogenic *Limulus* blood assay in a multi-center study of sepsis. *J. Endotoxin Res.* 4:9–16.

Levin, J. 1982. The Limulus test and bacterial endotoxins: some perspectives. In *Endotoxins and Their Detection with the Limulus Amebocyte Lysate Test,* eds. S. W. Watson, J. Levin, and T. J. Novitsky. New York: Alan R. Liss. Pp. 7–24.

———. 1985. The history of the development of the *Limulus* amebocyte test. In *Bacterial Endotoxins. Structure, Biomedical Significance, and Detection* with *Limulus Amebocyte Lysate Test,* eds. J. W. ten Cate, H. R. Büller, A. Sturk, and J. Levin. New York: Alan R. Liss. Pp. 3–28.

———. 1987. The *Limulus* amebocyte test: perspectives and problems. In *Detection of Bacterial Endotoxins with the Limulus Amebocyte Lysate Test,* eds. S. W. Watson, J. Levin, and T. J. Novitsky. New York: Alan R. Liss. Pp. 1–23.

———. 1988. The horseshoe crab: a model for gram-negative sepsis in marine organisms and humans. In *Bacterial Endotoxin. Pathophysiological Effects, Clinical Significance, and Pharmacological Control,* eds. J. Levin, H. R. Büller, J. W. ten Cate, J. H. van Deventer, and A. Sturk. New York: Alan R. Liss. Pp. 3–15.

Levin, J., and F. B. Bang. 1964a. The role of endotoxin in the extracellular coagulation of *Limulus* blood. *Bull. Johns Hopkins Hosp.* 115:265–274.

———. 1964b. A description of cellular coagulation in the *Limulus. Bull. Johns Hopkins Hosp.* 115:337–345.

———. 1968. Clottable protein in *Limulus:* its localization and kinetics of its coagulation by endotoxin. *Thromb. Diath. Haemorrh.* 19:186–197.

Levin, J., H. R. Buller, J. W. ten Cate, S. J. H. van Deventer, and A. Sturk, eds. 1988. *Bacterial Endotoxins. Pathological Effects, Clinical Significance and Pharmacological Control.* New York: Alan R. Liss.

Levin, J., T. E. Poore, N. S. Young, S. Margolis, N. P. Zauber, A. S. Townes, and W. R. Bell. 1972. Gram-negative sepsis: detection of endotoxemia with the *Limulus* test. *Ann. Intern. Med.* 76:1–7.

Levin, J., T. E. Poore, N. P. Zauber, and R. S. Oser. 1970a. Detection of endotoxin in the blood of patients with sepsis due to gram-negative bacteria. *N. Eng. J. Med.* 283:1313–1316.

Levin, J., P. A. Tomasulo, and R. S. Oser. 1970b. Detection of endotoxin in human blood and demonstration of an inhibitor. *J. Lab. Clin. Med.* 75:903–911.

Loeb, L. 1902. The blood cells and inflammatory processes of *Limulus. J. Med. Res.* 2:145–158.

Mascoli, C., and M. Weary. 1979. Limulus amebocyte lysate (LAL) test for detecting pyrogens in parenteral injectable products and medical devices: advantages to manufacturers and regulatory officials. *J. Parenter. Drug Assoc.* 33:81–95.

McClosky, W. T., C. W. Price, W. J. Van Winkle, H. Welch, and H. O. Calvery. 1943. Results of the first USP collaborative study of pyrogens. *J. Am. Pharm. Assoc.* 32:69–73.

Miyazaki, Y., S. Kohno, H. Koga, M. Kaku, K. Mitsutake, S. Maesaki, A. Yasuoka, K. Hara, S. Tanaka, and H. Tamura. 1992. G Test, a new direct method for diagnosis of Candida infection: comparison with assays for beta-glucan and mannan antigen in a rabbit model of systemic candidiasis. *J. Clin. Lab. Anal.* 6:315–318.

Morita, T., S. Tanaka, T. Nakamura, and S. Iwanaga. 1981. A new $(1–3)$-β-D-glucan-mediated coagulation pathway found in *Limulus* amebocytes. *FEBS Letters* 129:318–321.

Noguchi, H. 1903. On the multiplicity of serum haemagglutins of cold blooded animals. *Zentral. Bact. Abt.* 1 Orig. 34:286–288.

Novitsky, T. J. 1994. *Limulus* amebocyte lysate (LAL) detection of endotoxin in human blood. *J. Endotoxin Res.* 1:253–263.

——— 1999. Endotoxin detection in body fluids: chemical versus bioassay methodology. In *Endotoxin in Health and Disease,* eds. H. Brade, S. M. Opal, S. N. Vogel, and D. C. Morrison. New York and Basel: Marcel Dekker. Pp. 831–839.

Obayashi, T., M. Yoshida, T. Mori, H. Goto, A. Yasuoka, H. Iwasaki, H. Teshima, S. Kohno, A. Horiuchi, A. Ito, H. Yamaguchi, K. Shimada, and T. Kawai. 1995. Plasma $(1–3)$-β-D-glucan measurement in diagnosis of invasive deep mycosis and fungal febrile episodes. *The Lancet* 345:17–20.

Pearson, F. C. 1990. Detection of endotoxemia. In *Clinical Applications of the Limulus Amoebocyte Lysate Test,* ed. R. B. Prior. Boca Raton: CRC Press. Pp. 51–66.

Poole S., P. Dawson, and R. E. Gaines Das. 1997. Second international standard for endotoxin: calibration in an international collaborative study. *J. Endotoxin Res.* 4:221–231.

Prior, R. B. 1990. *Clinical Applications of the Limulus Amoebocyte Lysate Test.* CRC Press. 177pp.

Public Health Service. 1977. Food and Drug regulations for biological products. Publ. No. 21CFR610.90(b).

Rastogi, S. C., E. B. Seligmann, H. D. Hochstein, J. H. Dawson, L. G. Farag, and R. E. Marquina. 1979. Statistical procedures for evaluating the sensitivity of *Limulus* amoebocyte lysate. *Applications of Environmental Microbiology* 38: 911.

Rudbach, J. A., F. Aktya, R. J. Elin, H. D. Hochstein, M. K. Luoma, E. C. B. Milner, and K. R. Thomas. 1976. Preparation and properties of a national reference endotoxin. *J. Chem. Microbiol.* 3:21–25.

Rudloe, A. 1983. The effect of heavy bleeding on mortality of the horseshoe crab, *Limulus polyphemus,* in the natural environment. *J. Invert. Pathol.* 42:167–176.

Shirodkar, M. V., A. Warwick, and F. B. Bang. 1960. The *in vitro* reaction of *Limulus* amoebocytes to bacteria. *Biol. Bull.* 118:324–337.

Shuster, C. N., Jr. 1962. Serological correspondence among horseshoe "crabs" (Limulidae). *Zoologica* 47 (pt. 1):1–9.

———. 1979. *The Circulatory System and Blood of the Horseshoe Crab.* U.S. Dept. Energy/Federal Energy Regulatory Commission (DOE/FERC/0014): 63pp.

Siebert, F. B. 1923. Fever producing substance in some distilled waters. *Am. J. Physiol.* 67:90–104.

Steere, A. C., M. K. Rifaat, E. B. Seligmann, L. F. Barker, H. D. Hochstein, G. Friedland, P. Dasse, K. O. Wustrack, and K. J. Axnick. 1978. Pyrogen reactions associated with the infusion of normal serum albumin (human). *Transfusion* 18:102–107.

Sturk, A., S. J. H. van Deventer, J. W. Cate, L. G. Thijs, and J. Levin, eds. 1991. *Bacterial Endotoxins. Cytokine Mediators and New Therapies for Sepsis.* New York: Wiley-Liss.

ten Cate, J. W., H. R. Büller, A. Sturk, and J. Levin, eds. 1985. *Bacterial Endotoxin. Structure, Biomedical Significance, and Detection with Limulus Amebocyte Lysate Test.* New York: Alan R. Liss.

U.S. Dept. Health and Human Services. December, 1987. Guideline on the validation of the *Limulus* amebocyte lysate test as an end-product endotoxin test for human and animal parenteral drugs, biological products, and medical devices.

U.S. Pharmacopoeia. 1995. Bacterial Endotoxin Test 23, NF18:1696.

Wainwright, N. R., R. J. Miller, E. Paus, T. J. Novitsky, M. A. Fletcher, T. M. McKenna, and T. Williams. 1990. Endotoxin binding and neutralizing activity by a protein from *Limulus polyphemus.* In *Cellular and Molecular Aspects of Endotoxin Reactions,* eds. A. Notwotny, J. J. Spitzer, and E. J. Ziegler. Amsterdam: Elsevier Science Publishers. Pp. 315–325.

Watson, S. W., J. Levin, and T. J. Novitsky, eds. 1982. *Endotoxins and Their Detection with the Limulus Amebocyte Lysate Test.* New York: Alan R. Liss.

———. 1987. *Detection of Bacterial Endotoxins with the Limulus Amebocyte Lysate Test.* New York: Alan R. Liss.

Young, N. S., J. Levin, and R. A. Prendergast. 1972. An invertebrate coagulation system activated by endotoxin: evidence for enzymatic mediation. *J. Clin. Invest.* 51:1790–1797.

14 King Crab Fertilizer

Atlantic States Marine Fisheries Commission. 1998. Stock Assessment Report No. 98–01:11.

Beers, F. W. 1872. Topographical map of Cape May Co., New Jersey. Beers, Comstock & Cline (publishers, New York).

Berkson, J., and C. N. Shuster, Jr. 1999. The horseshoe crab: the battle for a true multiple-use resource. *Fisheries* 24:6–10.

Botton, M. L., R. E. Loveland, and T. R. Jacobsen. 1988. Beach erosion and geochemical factors: influence on spawning success of horseshoe crabs *(Limulus polyphemus)* in Delaware Bay. *Mar. Biol.* 99:325–332.

Brewer, T. M. 1840. *Wilson's American Ornithology.* Boston: Otis, Broaders & Co.

Byles, A. D. 1859. Map of Kent County, Delaware. Published in Philadelphia, Pa.

Cook, G. H. 1857. Geology of the County of Cape May, State of New Jersey. Trenton, N.J.: Office of the True American. Pp. 105–112.

Dickinson, K. A. 1958. Response to a request from Joseph L. Parkhurst, Jr. about the horseshoe crab fishery. Cape May County Historical Museum.

Earll, R. E. 1887. New Jersey and its fisheries. In *The Fisheries and Fishing Industries of the United States* (U.S. Commission of Fish and Fisheries, Washington, D.C.), ed. G. B. Goode. Section II: 397.

Federal Writers' Project, WPA. 1938. *Delaware. A Guide to the First State.* Hastings House. Pp. 400–401.

Fowler, H. W. 1908. The king crab fisheries in Delaware Bay. In *Annual Report, NJ State Museum, 1907*:111–119 + pls. 59–65.

Goode, G. B. 1887. *The Fisheries and Fishing Industries of the United States.* Washington, D.C.: U.S. Commission of Fish and Fisheries. Section II.

Milne, L. J., and M. J. Milne. 1947. Horseshoe crab—is its luck running out? *Fauna* (Zoological Society, Philadelphia) 9:66–72.

Montgomery, G. S. 1974. *Re King Crabs at Bowers Beach.* Delaware Discovered Series, Vol. 1:5.

National Ocean Survey, NOAA/USDOC. 1977. Chart 12304 (Delaware Bay).

Rathbun, R. 1887. The horseshoe crab fishery. In *The Fisheries and Fishery Industries of the United States,* ed. G. B. Goode. Washington, D.C.: U.S. Commission of Fish and Fisheries. Section V, Vol. II. Pp. 652–657.

Reed, A. 1986. Dias Creek. Cape May County, N.J., *Mag. Hist. Geneal.* 8: 433–441.

Shuster, C. N., Jr. 1997. Abundance of adult horseshoe crabs, *Limulus polyphemus,* in Delaware Bay, 1850–1990. In *Proceedings of the Horseshoe Crab Forum: Status of the Resource,* eds. J. Farrell and C. Martin. University of Delaware Sea Grant College Program (DEL-SG-05–97). Pp. 5–14.

———. 2001. Two perspectives: horseshoe crabs during 420 million years, worldwide, and the past 150 years in the Delaware Bay area. In *Limulus in the Limelight. A Species 350 Million Years in the Making and in Peril?,* ed. J. T. Tanacredi. New York: Kluwer Academic/Plenum Publishers. Pp. 17–40.

Shuster, C. N., Jr., and M. L. Botton. 1985. A contribution to the population biology of horseshoe crabs, *Limulus polyphemus,* in Delaware Bay. *Estuaries* 8:363–372.

Smith, H. M. 1891. *Notes on the King-Crab Fishery of Delaware Bay.* Bulletin, U.S. Fisheries Commission 9(19): 363–370.

The Sunday Bulletin, Philadelphia: June 21, 1953.

Works Progress Administration (WPA). 1935–1936. Riparian & Stream Survey, Cape May County Coast Line Survey (Sheets 17–21, 24–29, and 32 of 33). On file in the Aerial Photo and Map Library, Tidewater Management Bureau, State of New Jersey Department of Environmental Protection and Energy, CN 401, Trenton, N.J. 08625–0401.

Works Progress Administration (WPA). 1938. Cape May–Cumberland Counties: West Creek (Sheet 2 of 5). On file in the Aerial Photo and Map Library, Tidewater Management Bureau, State of New Jersey Department of Environmental Protection and Energy, CN 401, Trenton, N.J. 08625–0401.

15 Horseshoe Crab Conservation

American Fisheries Society. 2002. Program: 132nd Annual Meeting, Baltimore, Maryland (horseshoe crab poster session).

Atlantic States Marine Fisheries Commission. 1998. Interstate Fishery Management Plan for Horseshoe Crab. Fishery Management Report No. 32.

Atlantic States Marine Fisheries Commission. 1999. Horseshoe Crab Stock Assessment for Peer Review. Stock Assessment Report No. 98–01 (Supplement).

Atlantic States Marine Fisheries Commission. 2001a. 2000 State Compliance Reports and 2001 State Management Proposals. Interstate Horseshoe Crab Management Plan.

Atlantic States Marine Fisheries Commission. 2001b. Addendum I to the Interstate Fishery Management Plan for Horseshoe Crab. Atlantic States Marine Fisheries Commission. ASMFC approves Addendum II to the Horseshoe Crab FMP: Addendum provides for State-to-State quota transfers. News Release (24 April).

Attenborough, David. 1979. *Life on Earth: A Natural History.* London: William Collins Sons & Co. and the British Broadcasting Corporation.

———. 1986. *Life on Earth: A Natural History.* A Warner (WB) Home Video (color/233 minutes; based on the television series of 1979). London: British Broadcasting Corporation and BBC Enterprises.

Baptist, J. P., O. R. Smith, and J. W. Ropes. 1957. Migration of the horseshoe crab, *Limulus polyphemus,* in Plum Island Sound, Massachusetts. U.S. Department of the Interior, Fish and Wildlife Service, Special Fisheries Report. *Fisheries* 220:1–11.

Barlow, R. B., M. K. Powers, H. Howard, and L. Kass. 1986. Migration of *Limulus* for mating: relation to lunar phase, tide height, and sunlight. *Biol. Bull.* 170:310–329.

Berkson, J., and C. N. Shuster, Jr. 1999. The horseshoe crab: the battle for a true multiple-use resource. *Fisheries* 24:6–10.

Botton, M. L. 2001. The conservation of horseshoe crabs: what can we learn from the Japanese experience? In *Limulus in the Limelight. A Species 350 Million Years in the Making and in Peril?* ed. J. T. Tanacredi. New York: Kluwer Academic/Plenum Publishers. Pp. 41–51.

Botton, M. L., and H. H. Haskin. 1984. Distribution and feeding of the horseshoe crab, *Limulus polyphemus,* on the continental shelf off New Jersey. *Fish. Bull.* 82:383–389.

Botton, M. L., M. Hodge, and T. I. Gonzalez. 1998a. High tolerance to tributylin in embryos and larvae of the horseshoe crab, *Limulus polyphemus. Estuaries* 21:340–346.

Botton, M. L., K. Johnson, and L. Helleby. 1998b. Effects of copper and zinc on embryos and larvae of the horseshoe crab, *Limulus polyphemus. Arch. Environ. Contam. Toxicol.* 35:25–32.

Botton, M. L., and R. E. Loveland. 2003. Abundance and dispersal potential of horseshoe crab *(Limulus polyphemus)* larvae in the Delaware estuary. *Estuaries* (in press).

Botton, M. L., R. E. Loveland, and T. R. Jacobsen. 1988. Beach erosion and geochemical factors: influence on spawning success of horseshoe crabs *(Limulus polyphemus)* in Delaware Bay. *Mar. Biol.* 99:325–332.

———. 1994. Site selection by migratory shorebirds in Delaware Bay, and

its relationship to beach characteristics and abundance of horseshoe crab *(Limulus polyphemus)* eggs. *Auk* 111:605–616.

Botton, M. L., and J. W. Ropes. 1987a. The horseshoe crab, *Limulus polyphemus,* fishery and resource in the United States. *Mar. Fish. Rev.* 49:57–61.

———. 1987b. Populations of horseshoe crabs on the northwestern Atlantic continental shelf. *Fish. Bull.* 85:805–812.

———. 1988. An indirect method for estimating longevity of the horseshoe crab *(Limulus polyphemus)* based on epifaunal slipper shells *(Crepidula fornicata). J. Shellfish Res.* 7:407–412.

Botton. M. L., C. N. Shuster, Jr., K. Sekiguchi, and H. Sugita. 1996. Amplexus and mating behavior in the Japanese horseshoe crab, *Tachypleus tridentatus. Zool. Sci.* 13: 151–159.

Brockmann, H. J. 2002. An experimental approach to altering mating tactics in male horseshoe crabs *(Limulus polyphemus). Behav. Ecol.* 13:232–238.

Carmichael. R. H., D. Rutecki, and I. Valiela. 2003. Abundance and population structure of the Atlantic horseshoe crab *Limulus polyphemus* in Pleasant Bay, Cape Cod. *Mar. Ecol. Prog. Ser.* 246:225–239.

Corson, T. 2002. Stalking the American lobster. *Atlantic* 289:61–81.

Cronin, L. C. 1998. Reaction to the Blue Crab Symposium. *J. Shellfish. Res.* 17:587.

Ding, J. L., and B. Ho. 2001. A new era in pyrogen testing. *Trends Biotechnol.* 19:277–281.

Eagles, E. 2001. Issues and approaches in regulations of the horseshoe crab fishery. In *Limulus in the Limelight. A Species 350 Million Years in the Making and in Peril?* ed. J. T. Tanacredi. New York: Kluwer Academic/Plenum Publishers. Pp. 85–92.

Farrell, J., and C. Martin. 1997. *Proceedings of the Horseshoe Crab Forum: Status of the Resource.* University of Delaware Sea Grant College Program (DEL-SG-05-97).

Ferrari, K. M., and N. M. Targett. 2002. Chemical attractants in horseshoe crab, *Limulus polyphemus,* eggs: the potential for artificial bait. *J. Chem. Ecol.* 29:477–496.

Finn, J. J., C. N. Shuster, Jr., and B. J. Swan. 1991. *Limulus* spawning activity on Delaware Bay Shores 1990. Finn-Tech Industries brochure. P. 8 (reprinted in *Soc. Nat. Hist. Delaware* 1991:6).

Goode, G. B. 1887. *The Fisheries and Fishing Industries of the United States,* Section II. Washington, D.C.: U.S. Commission of Fish and Fisheries.

Himchak, P. J. 2001. Horseshoe crab management and resource monitoring in New Jersey, 1993–1998. In *Limulus in the Limelight. A Species 350 Million Years in the Making and in Peril?* ed. J. T. Tanacredi. New York: Kluwer Academic/Plenum Publishers. Pp. 103–115.

Hull, C. H. J., and J. G. Titus. 1986. Greenhouse effect, sea level rise, and salinity in the Delaware Estuary. Washington, D.C.: U.S. Environmental Protection Agency. EPA 230-05-86-010.

Itow, T., T. Igarashi, M. L. Botton, and R. E. Loveland. 1998a. Heavy metals inhibit limb regeneration in horseshoe crab larvae. *Arch. Environ. Contam. Toxicol.* 35:457–463.

Itow, T., T. Igarashi, R. E. Loveland, and M. L. Botton. 1998b. Developmental abnormalities in horseshoe crab embryos caused by exposure to heavy metals. *Arch. Environ. Contam. Toxicol.* 35:22–40.

Itow, T., H. Sugita, and K. Sekiguchi. 1991. The decrease of horseshoe crabs in the Seto Inland Sea and the cause. *Bull. Manage. Inform. Jobu Univ.* 4:29–46. [In Japanese]

Kerlinger, P., and D. Wiedner. 1991. The economics of birding at Cape May, New Jersey. In *Proceedings of the 2nd Ecotourism Symposium, Miami Beach, FL.,* ed. J. Kassler. New York: Holt, Reinhart and Winston. Pp. 324–334.

Kreamer, G. 2000. *Horseshoe Crab Workshop.* Division of Fisheries and Wildlife, Delaware Department of Natural Resources and Environmental Control.

Kungsuwan, A., Y. Nagashima, T. Noguchi, Y. Shida, S. Suvapeepan, P. Suwansakornkul, and H. Hashimoto. 1987. Tetrodotoxin in the horseshoe crab *Carcinoscorpius rotundicauda* inhabiting Thailand. *Nippon Suisan Gakkaishi* 53:261–266.

Kurtzke, C. 2001. Horseshoe crab surveys using underwater videography. In *Limulus in the Limelight. A Species 350 Million Years in the Making and in Peril?* ed. J. T. Tanacredi. New York: Kluwer Academic/Plenum Publishers. pp. 119–129.

Laughlin, R. B., Jr., and J. M. Neff. 1977. Interactive effects of temperature, salinity shock, and chronic exposure to no. 2 fuel oil on survival, development rate and respiration of the horseshoe crab, *Limulus polyphemus.* In *Fate and Effects of Petroleum Hydrocarbons in Marine Organisms and Ecosystems,* ed. D. A. Wolff. Oxford: Pergammon Press. Pp. 182–191.

Loveland, R. E. 2001. The life history of horseshoe crabs. In *Limulus in the Limelight. A Species 350 Million Years in the Making and in Peril?* ed. J. T. Tanacredi. New York: Kluwer Academic/Plenum Publishers. Pp. 93–101.

Loveland, R. E., M. L. Botton, and C. N. Shuster, Jr. 1997. Life history of the American horseshoe crab (*Limulus polyphemus* L.) in Delaware Bay and its importance as a commercial resource. In *Proceedings of the Horseshoe Crab Forum: State of the Resource,* eds. J. Farrell and C. Martin. University of Delaware Sea Grant College Program (DEL-SG-05-97). Pp. 15–22.

Manion, M. M., R. A. West, and R. E. Unsworth. 2000. *Economic Assessment of the Atlantic Coast Horseshoe Crab Fishery.* A report by Industrial Economics, Inc. (Cambridge, Mass.), prepared for the Division of Economics, U.S. Fish & Wildlife Service.

Michels, S. F. 1997. Summary of trends in horseshoe crab abundance in Delaware. In *Proceedings of the Horseshoe Crab Forum: State of the Resource,* eds. J. Farrell and C. Martin. University of Delaware Sea Grant College Program (DEL-SG-05-97). Pp. 26–29.

Morton, B. 1999. On turtles, dolphins and, now, Asia's horseshoe crabs. *Mar. Pollut. Bull.* 38:845–846.

Munson, R. E. 1998. Bait needs of the eel and conch fisheries in New Jersey and Delaware as determined from questionnaires filled out by commercial fishermen in those fisheries. Report to the Technical Committee, ASMFC Horseshoe Crab Management Plan.

National Marine Fisheries Service. 1998. Harvest data. National Oceanic and Atmospheric Administration, National Marine Fisheries Service, Northeast Fisheries Center Database. Gloucester, Massachusetts.

Neff, J. M., and C. S. Giam. 1977. Effects of Arochlor 1016 and Halowax 1099 on juvenile horseshoe crabs *(Limulus polyphemus).* In *Physiological Responses of Marine Biota to Pollutants,* eds. F. J. Vernberg, A. Calabrese, F. P. Thurberg, and W. B. Vernberg. New York: Academic Press. Pp. 21–35.

Oates, M. F. 1999a. Dollars on the Beach: The Horseshoe Crab Harvesting Controversy. ANEW, Inc. video time 27 mins.

———. 1999b. *Green Eggs and Sand: Horseshoe Crabs, Shorebirds, and Man.* ANEW. Video time 31 mins.

Oerlemans, J. 1989. A projection of future sea level. *Clim. Change* 15:151–174.

Pepe, M., and L. Chen. 2002. Endotoxin testing using recombinant Factor C. *LAL Review.* Winter issue (Bio-Whittaker, Inc.): 1–3.

Pierce, J. C., G. Tan, and P. M. Gaffney. 2000. Delaware Bay and Chesapeake Bay populations of the horseshoe crab *Limulus polyphemus* are genetically distinct. *Estuaries* 23:690–698.

Riepe, D. 2001. Horseshoe crabs: an ancient wonder of New York and a great topic for environmental education. In *Limulus in the Limelight. A*

Species 350 Million Years in the Making and in Peril? ed. J. T. Tanacredi. New York: Kluwer Academic/Plenum Publishers. Pp. 131–134.

Riska, B. 1981. Morphological variation in the horseshoe crab *Limulus polyphemus. Evolution* 35:647–658.

Ropes, J. W. 1961. Longevity of the horseshoe crab, *Limulus polyphemus* (L.). *Trans. Am. Fish. Soc.* 90:79–80.

Ropes, J. W., C. N. Shuster, Jr., L. O'Brien, and R. Mayo. 1982. *Data on the Occurrence of Horseshoe Crabs, Limulus polyphemus (L.), in NMFS-NEFC Survey Samples.* Woods Hole Laboratory Reference Document No. 82–23: 40pp.

Rudloe, A. 1971. *The Erotic Ocean.* World Publishing. Pp. 263–257.

———. 1983. The effect of heavy bleeding on mortality of the horseshoe crab, *Limulus polyphemus,* in the natural environment. *J. Invert. Pathol.* 42:167–176.

Saunders, N. C., L. G. Kessler, and J. C. Avise. 1986. Genetic variation and geographic differentiation in mitochondrial DNA of the horseshoe crab, *Limulus polyphemus. Genetics:* 613–627.

Schaller, S. 2002. *Survey of Maine Horseshoe Crab (Limulus polyphemus) Spawning Populations, 2001.* Report to Maine Department of Marine Resources (West Boothbay Harbor).

Selander, R. K., S. Y. Yang, R. C. Lewontin, and W. E. Johnson. 1976. Genetic variation in the horseshoe crab *(Limulus polyphemus),* a phylogenetic relic. *Evolution* 24:402–414.

Shuster, C. N., Jr. 1950. Observations on the natural history of the American horseshoe crab, *Limulus polyphemus,* Woods Hole Oceanographic Institution Contribution No. 564:18–23.

———. 1955. On morphometric and serological relationships within the Limulidae, with particular reference to *Limulus polyphemus* (L.). Ph.D. thesis, New York University (1958 Diss. Abstr. 18:371–372).

———. 1982. A pictorial review of the natural history and ecology of the horseshoe crab *Limulus polyphemus,* with reference to other Limulidae. In *Progress in Clinical and Biological Research,* vol. 81: *Physiology and Biology of Horseshoe Crabs: Studies on Normal and Environmentally Stressed Animals,* eds. J. Bonaventura, C. Bonaventura, and S. Tesh. New York: Alan R. Liss. Pp. 1–52.

Shuster, C. N., Jr., and M. L. Botton. 1985. A contribution to the population biology of horseshoe crabs, *Limulus polyphemus,* in Delaware Bay. *Estuaries* 8:363–372.

Smith, D. R., P. S. Pooler, R. E. Loveland, M. L. Botton, S. F. Michels, R. G. Weber, and D. B. Carter. 2002a. Horseshoe crab *(Limulus polyphemus)* reproductive activity on Delaware Bay beaches: interaction with beach characteristics. *J. Coastal Res.* 18:730–740.

Smith, D. R., P. S. Pooler, B. L. Swan, S. Michels, W. R. Hall, P. Himchak, and M. Millard. 2002b. Spatial and temporal distribution of horseshoe crab *(Limulus polyphemus)* spawning in Delaware Bay: implications for monitoring. *Estuaries* 25:115–125.

Sokoloff, A. 1978. Observations on populations of the horseshoe crab, *Limulus* (Xiphosura) *polyphemus. Res. Popul. Ecol.* 19:222–236.

State of Delaware. 1959. *Intrastate Water Resources Survey,* Section XIX: Water quality in the lower Delaware River with special emphasis upon pollution aspects. Pp. 19–68.

Strobel, C. J., and A. H. Brenowitz. 1981. Effects of Bunker C oil on juvenile horseshoe crabs *(Limulus polyphemus). Estuaries* 4:157–159.

Swan, B. L., and C. Giordano. 2001. Tagging data on the spatial and temporal distribution of horseshoe crabs *(Limulus polyphemus)* in the Middle Atlantic Bight. Unpublished manuscript.

Swan, B. L., W. R. Hall, and C. N. Shuster, Jr. 1991–2001. Reports on the annual survey of horseshoe crabs spawning on the shores of Delaware Bay. Limited distribution.

Titus, J. G. 1990. Greenhouse effect, sea level rise, and barrier islands: case study of Long Beach Island, New Jersey. *Coastal Mgmt.* 18:65–90.

Titus, J. G., R. A. Park, S. P. Leatherman, J. R. Weggel, M. S. Greene, P. W. Mausel, S. Brown, C. Gaunt, M. Trehan, and G. Yohe. 1991. Greenhouse effect and sea level rise: the cost of holding back the sea. *Coastal Mgmt.* 19:171–204.

U.S. Army Corps of Engineers. 1997. Villas & Vicinity, NJ Interim Feasibility Study, Draft Feasibility Report and Environmental Assessment. U.S. Army Corps of Engineers, Philadelphia District.

U.S. Department of Commerce, National Oceanic and Atmospheric Administration, National Marine Fisheries Service. 2001. 50 CFR Part 697. Atlantic Coastal Fisheries Cooperative Management Act Provisions; Horseshoe Crab Fishery; Closed Area. Federal Register 66(23): 8906–8911.

Walls, E. A., J. Berkson, and S. A. Smith. 2002. The horseshoe crab, *Limulus polyphemus:* 200 million years of existence, 100 years of study. *Rev. Fisheries Sci.* 10:39–73.

Weis, J. S., and A. Ma. 1987. Effects of the pesticide diflubenzuron on larval horseshoe crabs, *Limulus polyphemus. Bull. Environ. Contam. Toxicol.* 39:224–228.

Widener, J. W., and R. B. Barlow. 1999. Decline of a horseshoe crab population on Cape Cod. *Biol. Bull.* 197:300–301.

ACKNOWLEDGMENTS

2 Nesting Behavior; and
3 Male Competition and Satellite Behavior

I would like to thank our many hard-working field assistants: K. Abplanalp, S. Wineriter, C. Solare, E. Botsford, M. Meador Penn, A. Alonso, B. Biederman, M. Marquez, R. Shammami, L. Niko, D. Weaver, B. Ploger, J. Hardcastle, M. Sack, M. Moore, L. Eberhardt, P. Martin, L. Wysocki, M. Stowe, M. Groom, R. Podolski, T., P., and L. Rider, and the students in the animal behavior and marine biology courses for their field assistance. D. Stauch, K. Achey, S. Luis, A. McNeil, M. Stens, K. Sundar, J. Infante, A. Milton, and C. Kosarek helped with the paternity analysis. We are grateful to F. Sone, Drs. F. Nordlie, C. Lanciani, L. McEdward, E. Maurmeyer, C. Oyen, H. Mehta, and C. Shuster for their comments, advice, and assistance on various aspects of this research. The research was supported by NSF OCE90-06392, Sigma Xi Grants-In-Aid of Research, the University of Delaware College of Marine Sciences, the University of Florida Foundation, the Department of Zoology, and the Seahorse Key Marine Laboratory. The research in Florida was conducted under special use permits from the Cedar Keys National Wildlife Refuge. The Delaware research was conducted with permission from the Cape Henlopen State Park.

5 Growing Up Takes about Ten Years and Eighteen Stages

The valuable assistance of Dr. Hiroaki Sugita of the University of Tsukuba, Japan, enabled us to coordinate the preparation of this chapter. Over the years the staff of the National Marine Fisheries Service aquarium at Woods Hole, Mass., has kept us informed of any unusual specimens or special studies. We are especially grateful to Mr. Thomas L. Morris, Jr., for this courtesy.

7 A History of Skeletal Structure

Fossils, mainly of *Mesolimulus walchi*, were examined by Shuster in the following collections: in Germany, through the courtesy of Dr. Gunter Viohl, Jura Museum, Eichstätt; Mr. Helmut Leich, private collection, Bochum; Mr. B. A. Jurgen Schmitt, private collection, Frankfurt-am-Main; Dr. Theo Kress, Museum beim Solnhofen Aktien-verein; Dr. Heinz Malz (1964), Senckenberg Museum, Frankfurt-am-Main; Mr. Georg Berger, Museum Berger, Eichstätt; Mr. Eduard Schopfel, private collection, JUMA, Gungolding-Altmühltal. Specimens of *Eupoops danae* were examined at the Museum of Natural History, The Smithsonian Institution, Washington, D.C., courtesy of Jann Thompson, Collections Manager.

Bill Baird (formerly) National Museums of Scotland, Chambers Street, Edinburgh; Dr. R. A. Fortey, Natural History Museum, London; and Dr. Paul A. Selden, University of Manchester, were valuable counselors to Anderson.

8 Throughout Geologic Time

This chapter also benefited from the contributions of persons and institutions acknowledged in Chapter 7. Additionally, we appreciated the assistance of librarians at the Smithsonian Museum of Natural History, Washington, D.C., U.S. Geological Survey, Reston, Va., and Virginia Institute/School of Marine Science, The College of William and Mary, Gloucester Point, Va.—especially, at the latter, the friendly help of Charles McFadden and staff and use of the Horseshoe Crab Document Center.

9 Coping with Environmental Changes

This review is dedicated to the memory of Dr. Charlotte Preston Mangum, mentor, colleague, and long-time friend, whose imagination and energy inspired much of the work described here. Support for writing this chapter came from National Science Foundation grants (IBN-9407261 and IBN-9807539 to DWT and IBN-9727835 to RPH) and from the Foster G. McGaw Foundation (DWT).

10 Diseases and Symbionts

The authors wish to thank Dr. Marion Georgi, Department of Parasitology, College of Veterinary Medicine, Cornell University, Ithaca, N.Y.; Dr. Libero Ajello, Director, Division of Mycotic Diseases, Center for Disease Control, Atlanta, Ga.; Dr. John Waterberry, Woods Hole Oceanographic Institute, Woods Hole, Mass.; Dr. Arthur Humes, Boston University Marine Program, Woods Hole, Mass.; Dr. Craig A. Harms; and Michelle McCafferty for their assistance in this work. Vicki Grantham, Dr. James Clark, and Dr. Paul Orndorff assisted with the editing and word processing of the manuscript.

This study was supported in part by a grant from the Division of Research Resources National Institutes of Health (P40-RR1333–06) and a special grant to Dr. Gregory A. Lewbart from the Frederick B. Bang Scholarship, Marine Biological Laboratory, Woods Hole, Mass.

12 Internal Defense against Pathogenic Invasion

Original research reported in this article was supported by Grant No. MCB-97-26771 from the National Science Foundation.

14 King Crab Fertilizer

We thank Lee E. Widjeskog (New Jersey Department of Environmental Protection), who located WPA maps of Cape May and Cumberland Counties and aided in consulting these maps. Many of the details concerning the setting of the pound nets have not been previously published. Important insights into the construction and operation of the fishery utilizing pounds were obtained during oral history interviews with Kenneth D. Schellinger (Green Creek, N.J.), Myron McGuigan (Dennisville, N.J.), Russell A. Conover (Dias Creek, N.J.), and Howard E. Walizer (Dennisville, N.J.).

Thanks also to the many other persons who helped us locate information for this chapter: Dr. Ingrid Ratsep (Public Service Electric & Gas, Estuary Enhancement Program); Michael Ryan (Aerial Photo and Map Library, Tidelands Management Bureau, N.J. Dept. of Environmental Protection); Myrna M. Adamczyk (Cape May County Surrogate Court); Jane H. Boone and John Moyer (Bowers Beach Maritime Museum, Del.); Ione Williams (Cape May County Historical Society); and Mark L. Botton for locating the slide on the Camp fertilizer plant. A special thanks goes to Walter Canzonier (N.J. Aquaculture Association, Port Norris, N.J.) for comments and suggestions on an earlier version of this chapter.

15 Horseshoe Crab Conservation

We are indebted to those who have provided information and critiqued earlier drafts; alphabetically, they are David B. Carter (Delaware Coastal Management Program), Glenn Gaurvy (Ecological Research and Development Group), Pete Himchak (New Jersey Department of Environmental Protection), Jon Hulburt (Wilmington, Del.), Gary Kreamer (Delaware Department of Natural Resources and Environmental Control and participants in the *Green Eggs and Sand* initiative of teachers, scientists, watermen, fisheries managers, natural resource agencies, education specialists, and nonprofit organizations from Delaware, Maryland, and New Jersey), Tom O'Connell (Maryland Department of Natural Resources), Stewart Michels (Delaware Department of Natural Resources and Environmental Control), Thomas J. Novitsky (Associates of Cape Cod), and Benjie Lynn Swan (Limuli Laboratories).

Michael F. Oates (Anew) has been notable for his tireless and energetic videotaping of reactions of the public and stakeholders to the problems associated with the horseshoe crab fishery and the Horseshoe Crab Management Plan. Concurrently, he developed documentary and educational materials and a video for a curriculum on *Green Eggs and Sand.*

INDEX